'By the theory of natural selection all living species have been connected with the parent-species of each genus… and these parent-species… have in their turn been similarly connected with more ancient species; and so on backwards… But assuredly, if this theory be true, such have lived upon this earth.'

Charles Darwin, *On the Origin of Species by Means of Natural Selection* (1859), pp. 281–82

The Complete World of Human Evolution

Chris Stringer • Peter Andrews

with 432 illustrations, 180 in color

Half-title: *Charles Darwin caricature from* 'The Hornet', *1871.*

Title-page: *The reconstructed figure of* Homo heidelbergensis *from Boxgrove, England, with the modern quarry in the background.*

This page: *Palaeontologists search for fossils in the Fayum Depression, Egypt.*

First published in 2005 in hardcover in the United States of America by Thames & Hudson Inc., 500 Fifth Avenue, New York, New York 10110

thamesandhudsonusa.com

Library of Congress Catalog Card Number 2004110563

ISBN-13: 978-0-500-05132-0
ISBN-10: 0-500-05132-1

Printed and bound in China by Everbest Printing Co Ltd

Contents

III Interpreting the Evidence

Introduction

When the first fossil human finds were discovered and recognized over 150 years ago, evolutionary ideas, palaeontology and archaeology were still in their infancy. Now, as well as a growing fossil record, there is a panoply of different approaches to reconstructing human prehistory. We have a large body of comparative data from our primate relatives and from human societies the world over that allow us to flesh out prehistory by building models of the past. Our anatomy and that of our fossil relatives can be examined in every aspect; forensic-style studies can be applied to fossil sites to show how past populations lived and died; and past events can be reconstructed and dated in unprecedented detail. In this book, we hope to show the way that scientists approach the reconstruction of human evolution, and highlight the main areas of growth in both the fossil record and in our means of interpreting that record.

The authors have been active researchers in palaeoanthropology for over 30 years, and we have

In 1991 the discovery of a fossil jawbone beneath a medieval village at Dmanisi in Georgia confirmed the presence of ancient humans there, and this was followed by the discovery of two skulls in 1999. In 2001, this skull (right), resembling Homo habilis *from Africa, was found. Dmanisi is now one of the richest sites of human fossils anywhere in the world. (Below) Georgian palaeontologist David Lordkipanidze and colleagues examine a new find.*

(Right) Evidence is sometimes buried deep. Here, 11 m (36 ft) below the surface, archaeologists are excavating evidence of early humans in the Liang Bua cave on the Indonesian island of Flores.

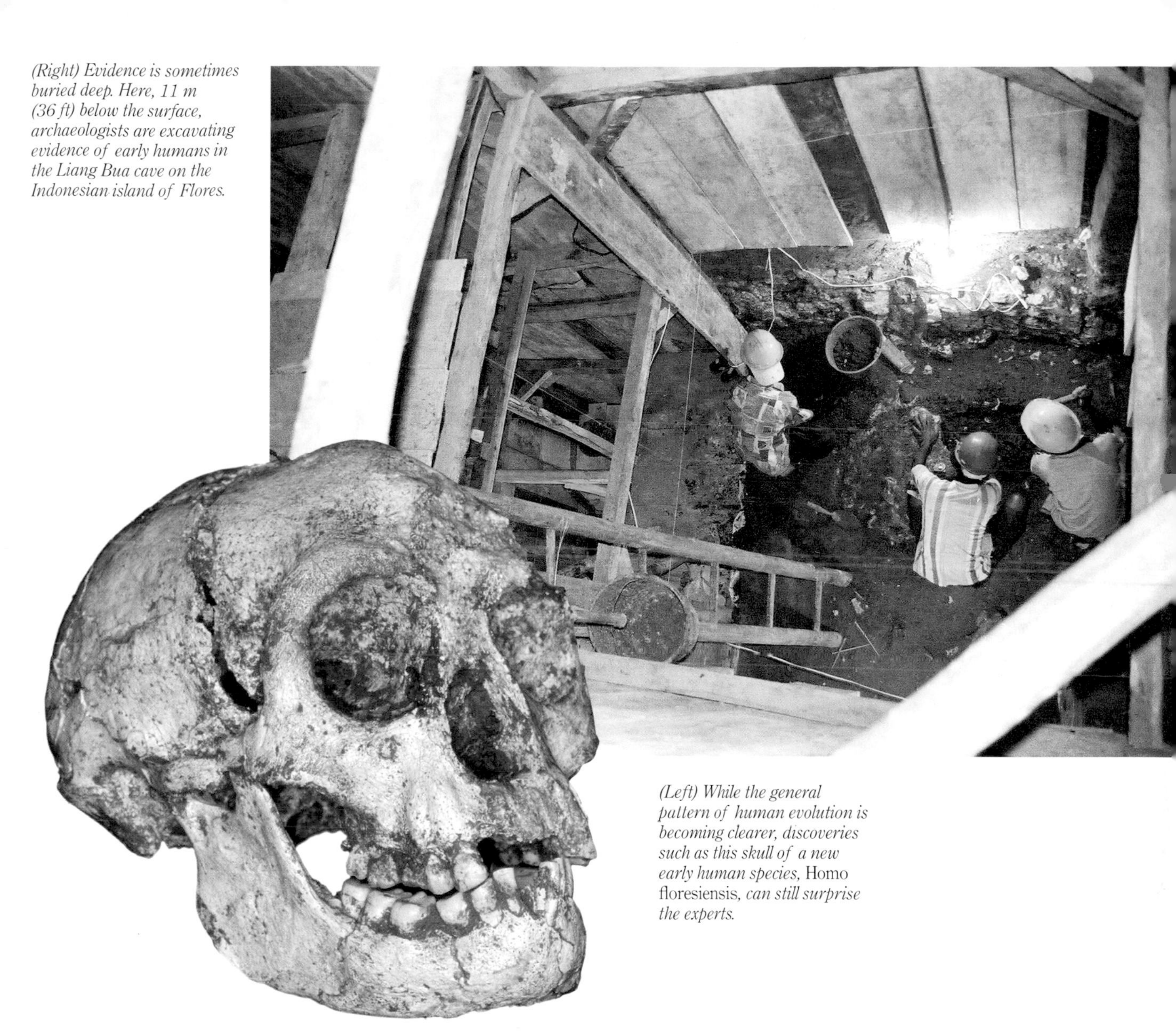

(Left) While the general pattern of human evolution is becoming clearer, discoveries such as this skull of a new early human species, Homo floresiensis, *can still surprise the experts.*

been privileged to witness some of the most significant discoveries and breakthroughs in the reconstruction of our evolution. To begin with, there have been remarkable fossil finds illuminating the early evolution of the apes, in Africa and beyond – these illustrate the great diversity of early apes, and show how successful and wide-ranging these were, compared with their much more limited present representatives. New finds from various regions of Africa are fleshing out the possible beginnings of the human evolutionary line, with several candidate species vying for the title of our earliest ancestor. The surprisingly early spread of humans from Africa is well illustrated by the unexpectedly rich site of Dmanisi in Georgia, dating close to 2 million years ago. Here, skulls, jaws and parts of skeletons of several very primitive human fossils have been found beneath the remains of a medieval village. The small brain size and very basic stone tool kit of these early humans are surprising since it had been assumed that significant advances in intelligence and technology would have been necessary for early humans to make the first moves out of their ancestral African homelands.

The origin and spread of our own species, again from Africa, is being revealed by fossil discoveries ranging from South Africa to Australia. But how little we yet know about later human evolution in regions such as Southeast Asia has been highlighted by the recent discovery of a remarkable skeleton of a primitive human form on Flores, one of the Indonesian Sunda Islands. The existence of this creature, named *Homo floresiensis* ('Man from Flores'), was completely unsuspected, and the fact that it apparently survived until less than 20,000 years ago means that modern humans dispersing through the region to Australia would have encountered these strange relatives.

Our ability to calibrate the growing fossil record has fortunately also developed in tandem, with increasing precision in the dating of volcanic rocks, so important for the earliest African phases of human evolution, and in radiocarbon dating, so important in the most recent stages after 40,000 years ago. In between, developments in techniques such as uranium series and electron spin resonance dating are helping to date periods beyond the range of radiocarbon, and where suitable volcanic rocks are not available.

Our capabilities of gleaning the maximum possible information from the fossil evidence have also increased in leaps and bounds over the last few years. CT scanning, a three-dimensional X-ray technique that was originally developed for medical

From the left, front and side views of skulls of Homo erectus *(Sangiran 17, Java);* Homo heidelbergensis *(Broken Hill, Zambia);* Homo neanderthalensis *(La Ferrassie, France); and* Homo sapiens *(recent, Indonesia).*

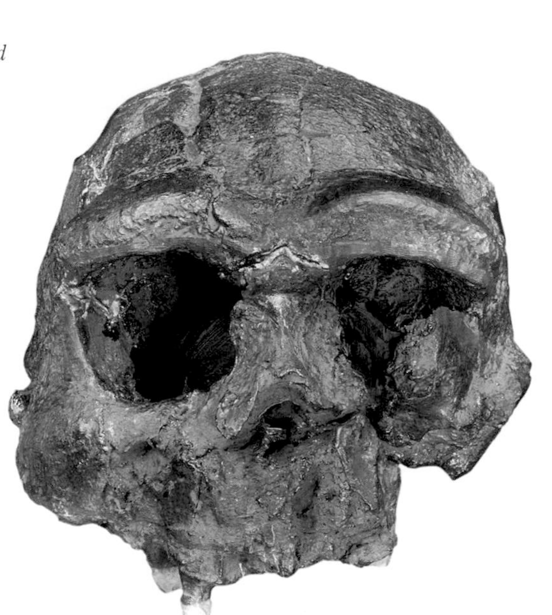

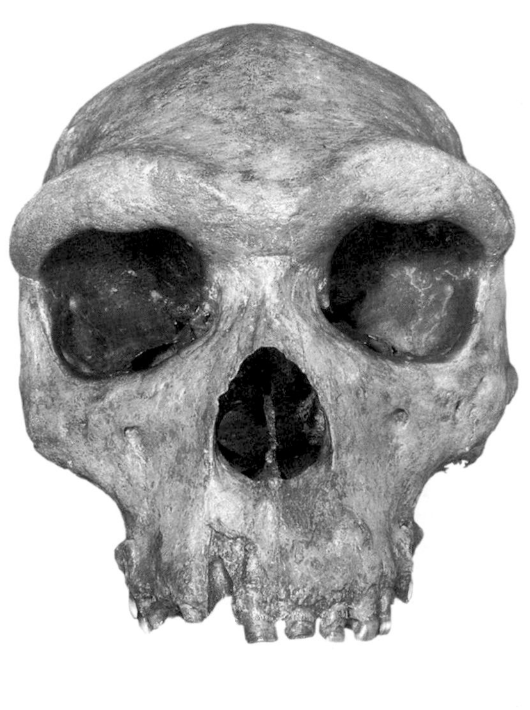

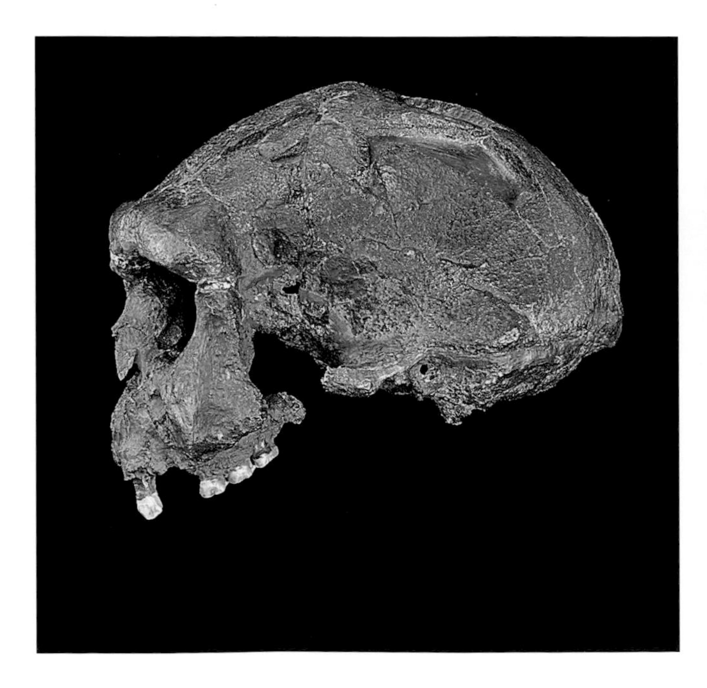

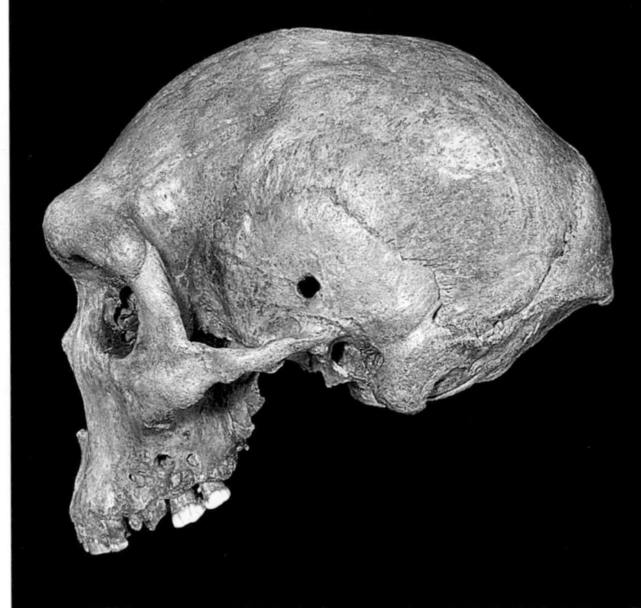

purposes, can now provide very accurate external and internal images of fossils, even where they are distorted or still partly encased in rock. Microscopic techniques allow us to study the formation and growth patterns of bones and teeth, while isotopic analyses permit unprecedented insights into ancient diets. And new methods for capturing and analyzing information about the size and shape of fossils are also helping us reconstruct our evolutionary history.

One other approach has also exerted a growing influence over the last 20 years – that of genetic analysis, which can now even be extended to 50,000-year-old Neanderthal fossils. However, most of the genetic data concerned is gathered from recent primates, especially humans, helping to confirm the closeness of our evolutionary relationship to the African apes, especially the chimpanzees, and the recent African ancestry of all people alive today, whatever their size, shape or colour.

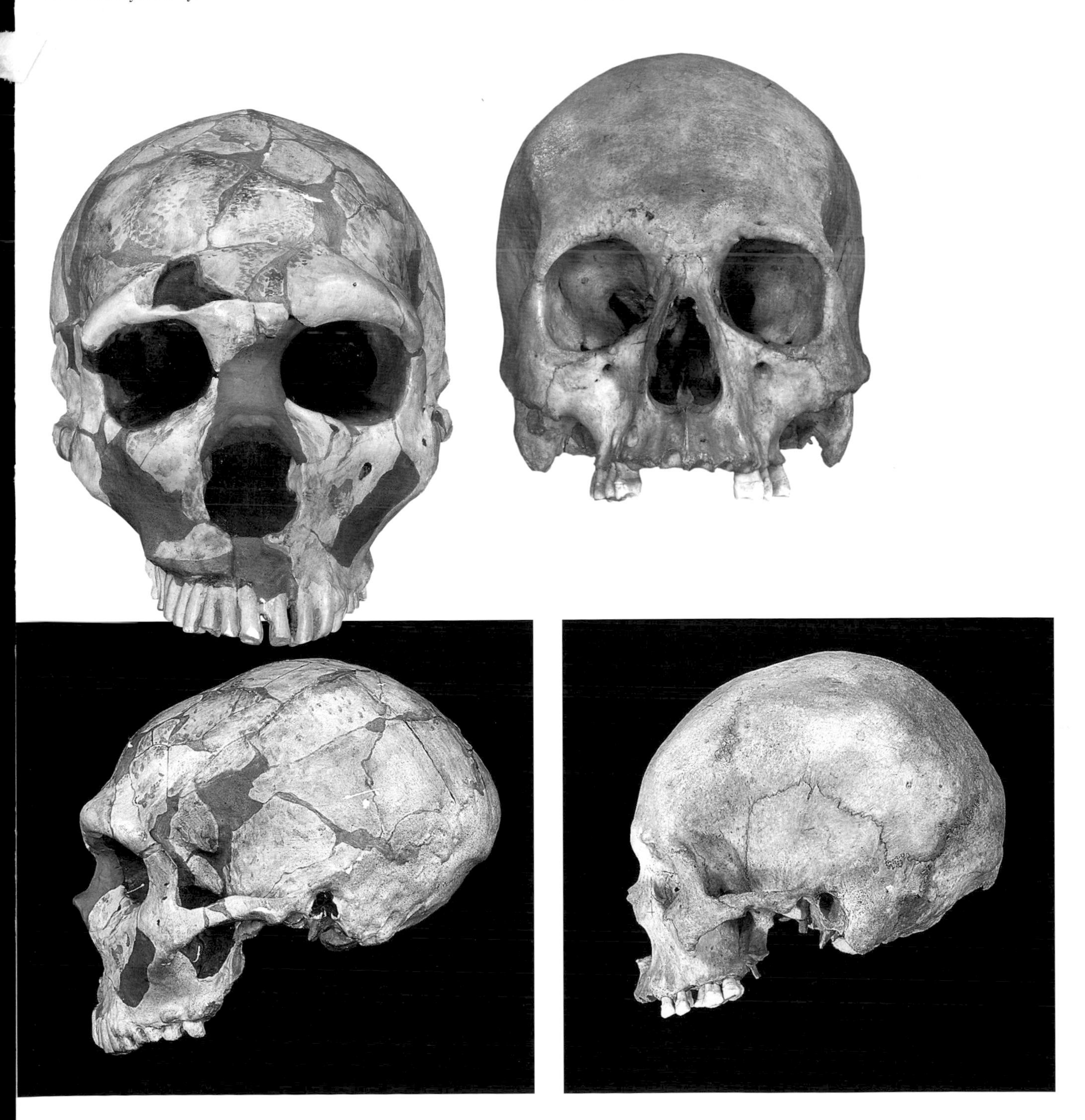

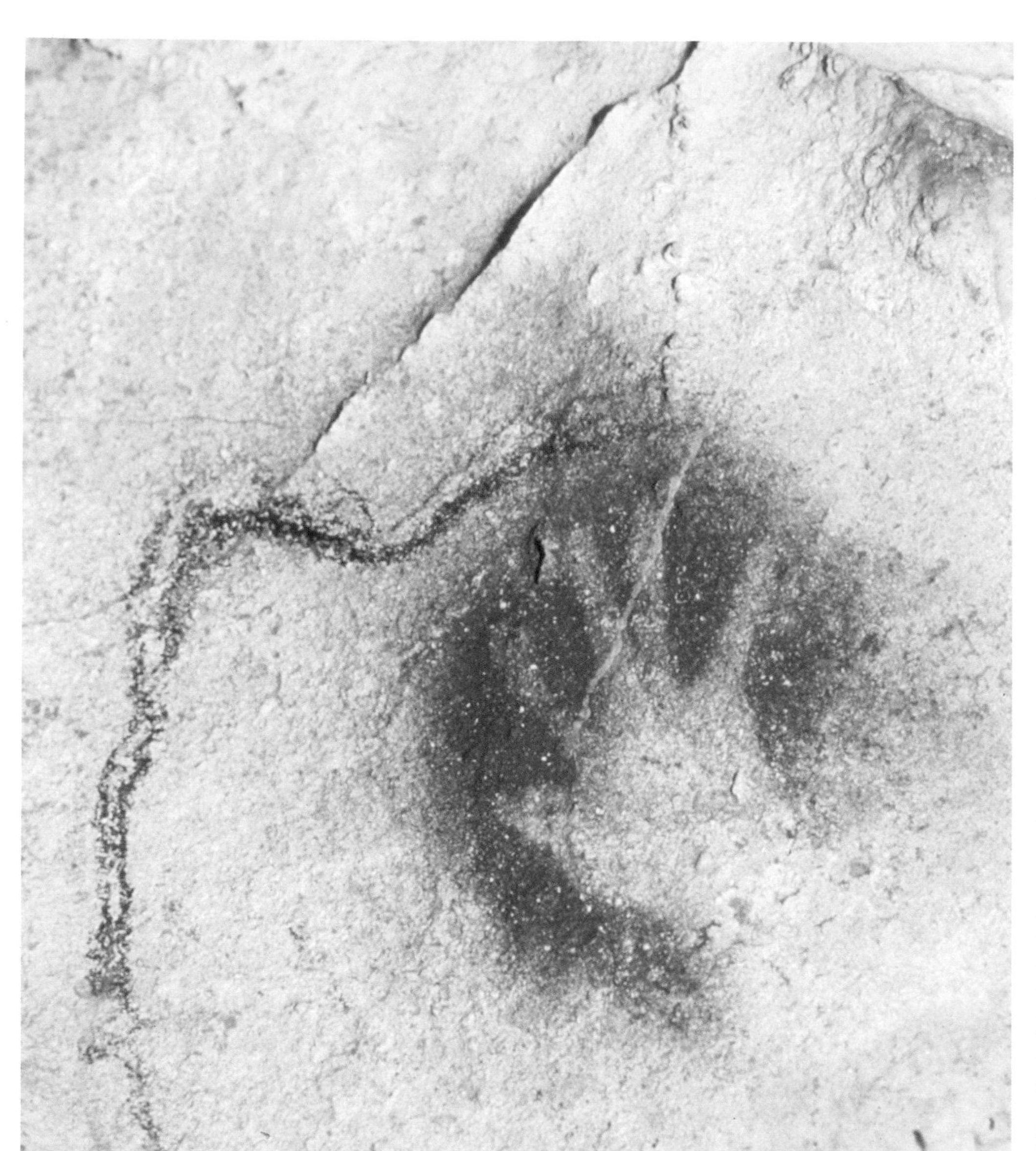

(Left) Hand stencils such as this one found deep inside Chauvet Cave in southeast France, were created by spraying pigments (by blowing or spitting) around a hand placed against the wall. The black profile represents part of a mammoth – it was drawn before the hand.

(Opposite) At Chauvet, a huge range of different fauna are represented. This cave bear (above) and rhinoceros (below) are joined by many other species, including lions, horses, ibex, bison, aurochs, mammoths, panthers, reindeer and even an owl. Additionally, as well as human hand prints, there are also insect-like geometric signs and some sexual images to be found. The purpose and meaning of all these paintings remains, in most respects, enigmatic.

As well as the growing fossil record, there has been a parallel expansion of archaeological data – thus the oldest stone tools and butchered bones are now dated at about 2.5 million years ago in East Africa, while symbolism and art was being produced in Africa at more than 75,000 years ago, with cave painting flourishing in the Upper Palaeolithic of western Europe after 30,000 years ago. Increasingly detailed studies of our living primate relatives show that they share many behavioural features with us including, in the case of chimpanzees, varied traditions of simple tool making and use, and the co-operative hunting of monkeys for their meat.

But despite all the advances summarized above, there are still many fascinating puzzles. The exact nature of the last common ancestor we shared with chimpanzees remains uncertain, as is the date at which it lived, and the environment from which it diversified. There are many different ideas about why our ancestors began the fundamental human adaptation of walking upright on two legs, but we are still far from knowing whether any of them are correct. And although we can be pretty sure it happened in Africa before 2 million years ago, we do not really know when, where and why the first members of the genus *Homo* evolved. Equally, for *Homo sapiens*, what processes were involved in the origin of our species, and how we eventually replaced other surviving humans, such as the Neanderthals, are still unknown. However, highlighting these deficiencies in our knowledge serves to focus research efforts on such fundamental unknowns and guarantees that more complete stories of human evolution than this one will be told in the future.

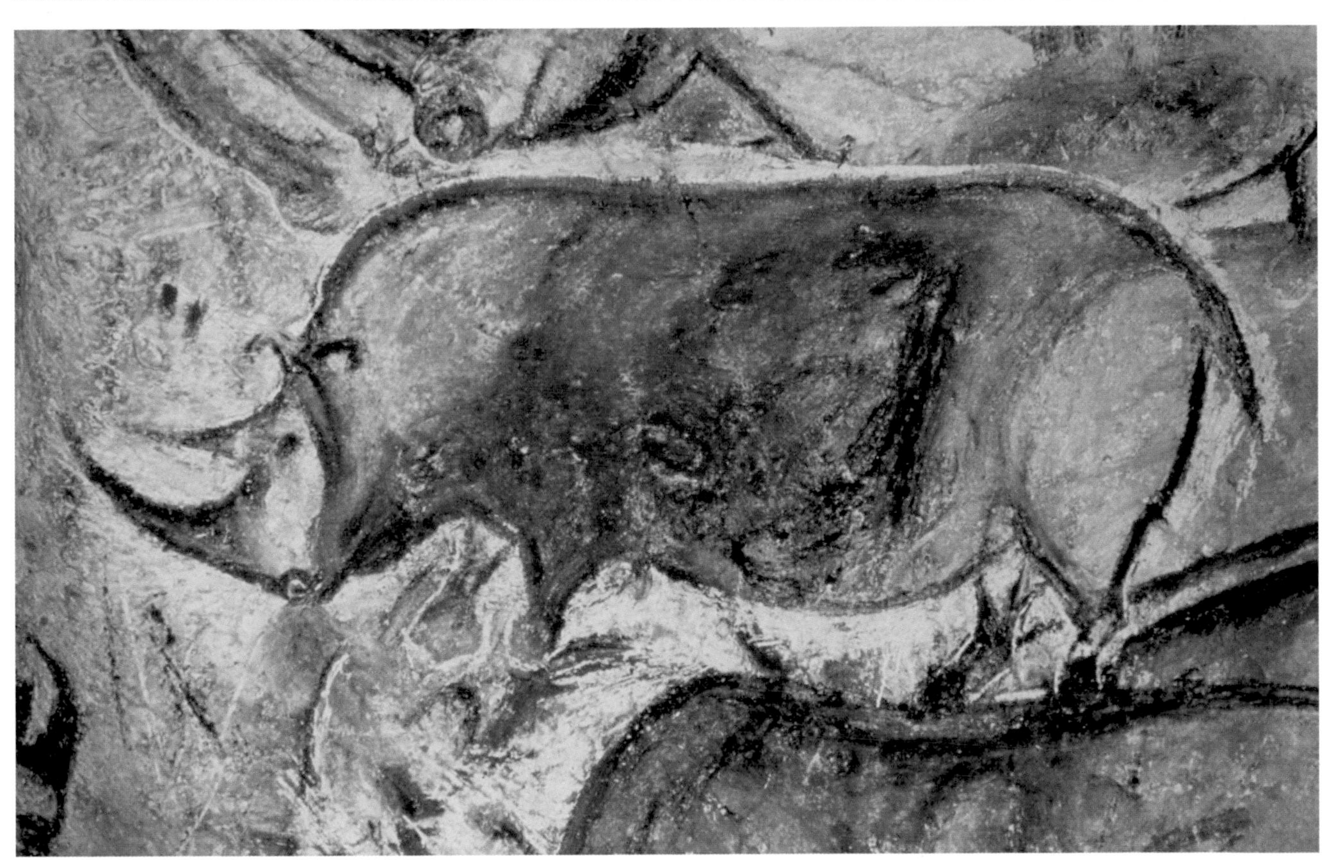

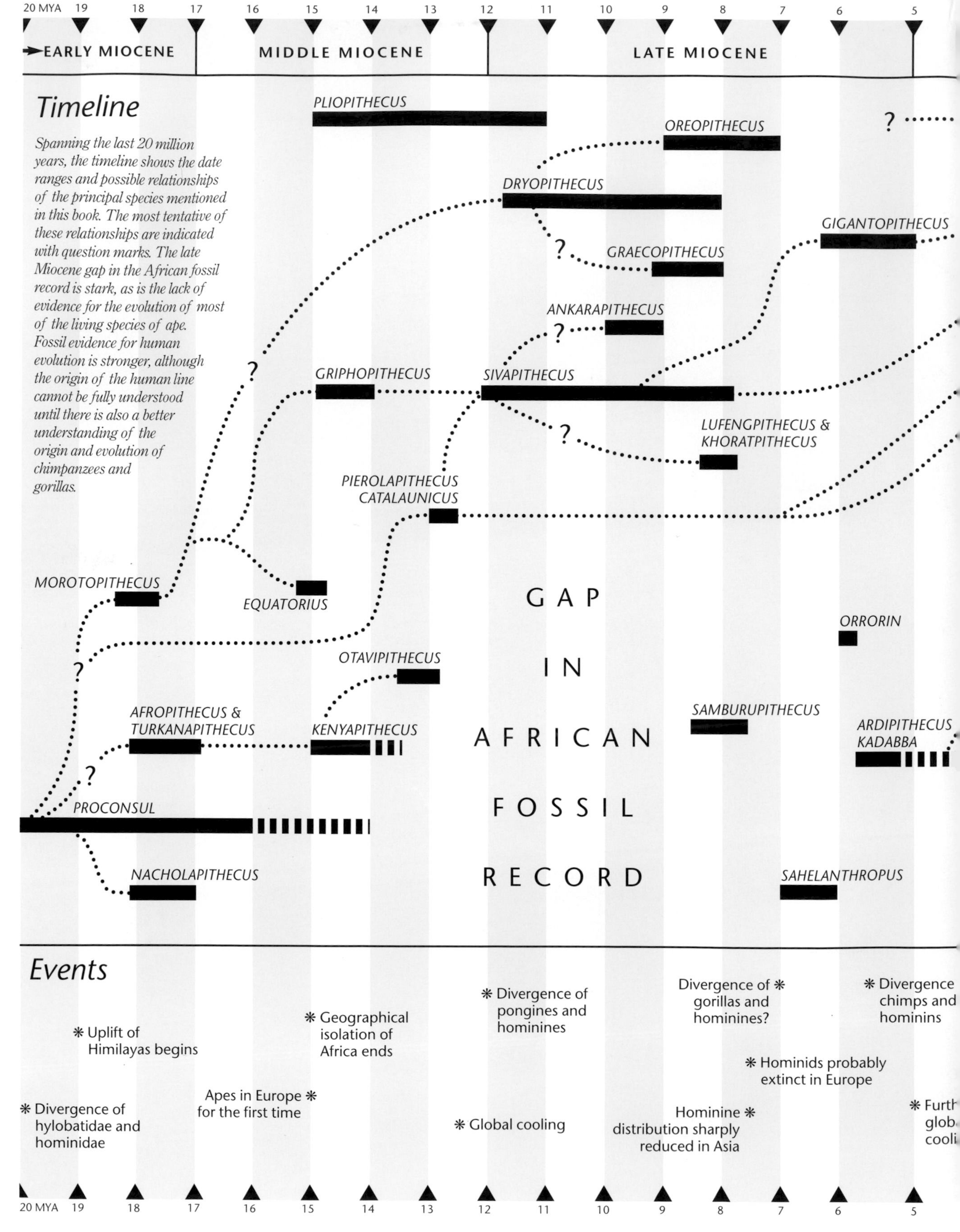
20 MYA
19
18
17
16
15
14
13
12
11
10
9
8
7
6
5
EARLY MIOCENE
MIDDLE MIOCENE
LATE MIOCENE
Timeline
Spanning the last 20 million years, the timeline shows the date ranges and possible relationships of the principal species mentioned in this book. The most tentative of these relationships are indicated with question marks. The late Miocene gap in the African fossil record is stark, as is the lack of evidence for the evolution of most of the living species of ape. Fossil evidence for human evolution is stronger, although the origin of the human line cannot be fully understood until there is also a better understanding of the origin and evolution of chimpanzees and gorillas.
PLIOPITHECUS
OREOPITHECUS
DRYOPITHECUS
GIGANTOPITHECUS
GRAECOPITHECUS
ANKARAPITHECUS
GRIPHOPITHECUS
SIVAPITHECUS
LUFENGPITHECUS & KHORATPITHECUS
PIEROLAPITHECUS CATALAUNICUS
MOROTOPITHECUS
EQUATORIUS
GAP IN AFRICAN FOSSIL RECORD
ORRORIN
OTAVIPITHECUS
SAMBURUPITHECUS
AFROPITHECUS & TURKANAPITHECUS
KENYAPITHECUS
ARDIPITHECUS KADABBA
PROCONSUL
NACHOLAPITHECUS
SAHELANTHROPUS
Events
Uplift of Himilayas begins
Divergence of hylobatidae and hominidae
Apes in Europe for the first time
Geographical isolation of Africa ends
Divergence of pongines and hominines
Global cooling
Divergence of gorillas and hominines?
Hominine distribution sharply reduced in Asia
Hominids probably extinct in Europe
Divergence chimps and hominins
Furth glob cooli

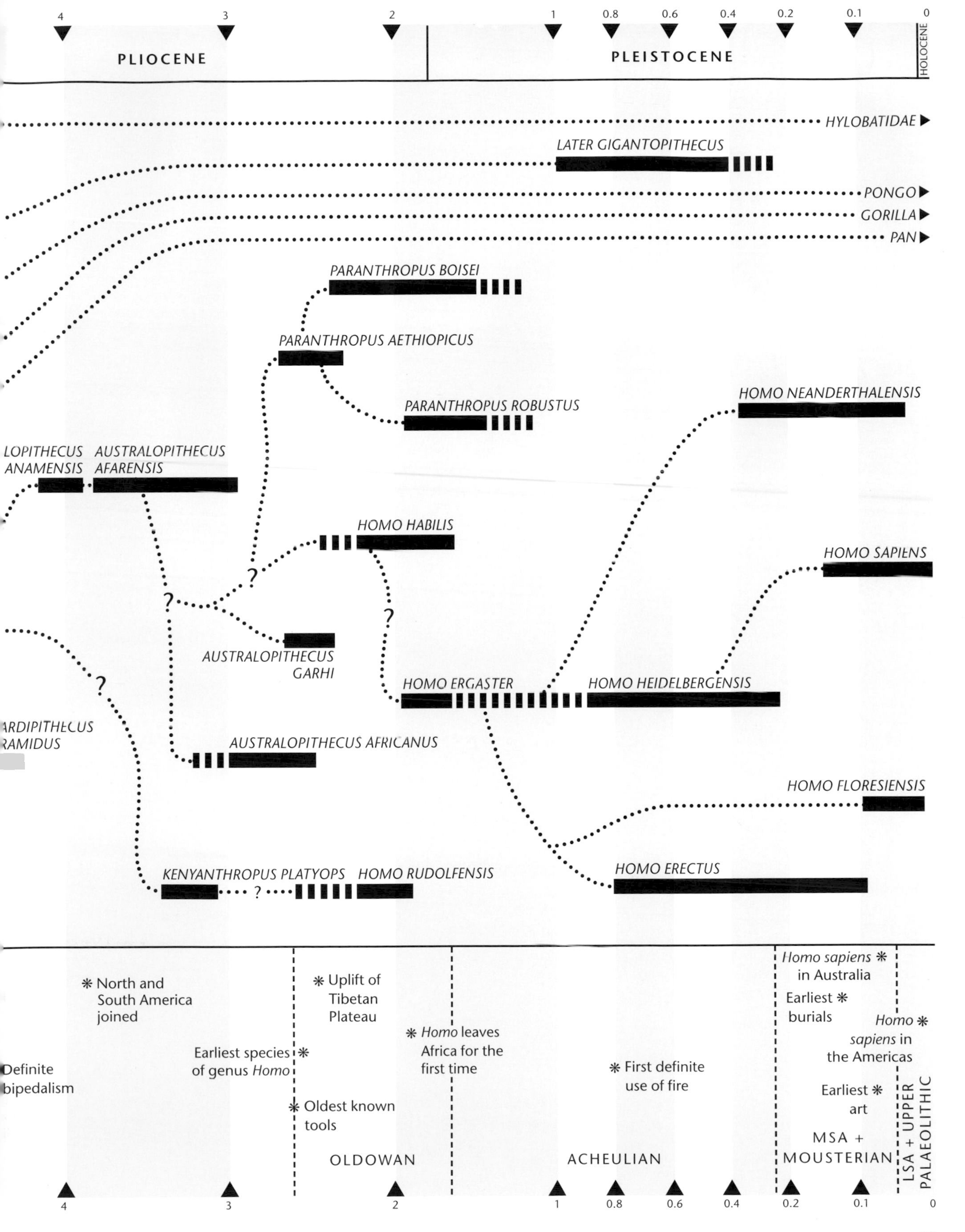

4
3
2
1
0.8
0.6
0.4
0.2
0.1
0
PLIOCENE
PLEISTOCENE
HOLOCENE
HYLOBATIDAE
LATER GIGANTOPITHECUS
PONGO
GORILLA
PAN
PARANTHROPUS BOISEI
PARANTHROPUS AETHIOPICUS
PARANTHROPUS ROBUSTUS
HOMO NEANDERTHALENSIS
LOPITHECUS ANAMENSIS
AUSTRALOPITHECUS AFARENSIS
HOMO HABILIS
HOMO SAPIENS
AUSTRALOPITHECUS GARHI
HOMO ERGASTER
HOMO HEIDELBERGENSIS
ARDIPITHECUS RAMIDUS
AUSTRALOPITHECUS AFRICANUS
HOMO FLORESIENSIS
HOMO ERECTUS
KENYANTHROPUS PLATYOPS
HOMO RUDOLFENSIS
North and South America joined
Uplift of Tibetan Plateau
Homo leaves Africa for the first time
Earliest species of genus Homo
Definite bipedalism
Oldest known tools
First definite use of fire
Homo sapiens in Australia
Earliest burials
Homo sapiens in the Americas
Earliest art
MSA + MOUSTERIAN
LSA + UPPER PALAEOLITHIC
OLDOWAN
ACHEULIAN

Palaeoanthropology is the study of the evidence for human evolution. Its timespan extends from modern humans, *Homo sapiens*, back to our possible ape ancestors in the Miocene Period around 20 million years ago or even further. In this book we will be looking at this evidence, but first we will introduce some general topics that are important for the interpretation of all stages of human evolution. We will look at the context in which fossils and artifacts are found: how sites are excavated, including discussion of some famous case histories; the environment associated with fossils, using some of the same examples; and geological time and the dating of fossil sites. We will also consider the importance of variation in fossil species, for it is this variation that provides the differences within populations on which natural selection operates.

We all know that living human populations vary, both within themselves and between people from different regions. This type of natural variation exists in any living species, and different environmental conditions may favour different aspects of it. For example people with dark skin are better able to tolerate strong sunlight – in the future they might be favoured if solar radiation were to become stronger or the ozone layer were to be destroyed. This is natural selection in operation, as it has operated through all of evolutionary history, and in order to understand how human beings evolved we need to examine how natural selection operated through the different stages of human evolution. For this we need both the fossil remains of our ancient relatives, which provide the evidence of past adaptations and variation, but we also need the context in which they are found, so that we can estimate how old they are and attempt to reconstruct the environment in which they were living. Only with all this information can we discover not just what our possible human ancestors looked like, but how they came to acquire their various adaptations and how some of them eventually became us.

The site of Boxgrove, in southern England, during excavations in 1995. A rich collection of stone tools and fossil bones, including human teeth and a shinbone dating from around 500,000 years ago, were discovered here.

1 In Search of Our Ancestors

Living Apes and Their Environment

For a human audience, watching chimpanzees can be both an amusing and disturbing experience, more so than with any other animal. Why should this be? The short answer is that chimpanzees are apes, the family to which human beings also belong, and not just any ape but the species that is most closely related to us. They share over 98 per cent of their genes with us, and in many aspects of their behaviour and biology they are very similar to ourselves. Probably no other species of monkey or ape could be trained to mimic human behaviour so closely, although the gorilla could come a close second.

Chimpanzees and gorillas live only in Africa, and as such they have long been considered to be the closest relatives of humans. In the last century, Charles Darwin placed human origins in Africa, sharing a common ancestor in Africa with chimpanzees and gorillas, although at that stage they were placed in different families, Hominidae and Pongidae. As it has come to be accepted that chimpanzees are more closely related to humans than are gorillas, putting them in the same family does not make sense. In this book we will include all great apes and humans in the family Hominidae and divide this into three subfamilies for orang utans, gorillas, and humans and chimpanzees. Subfamily Homininae is the name given for humans and chimps, abbreviated as hominine, and humans are distinguished from chimpanzees by inclusion in a separate tribe Hominini, abbreviated as hominin.

The final group of apes to be mentioned is the gibbon family, Hylobatidae, living today in eastern Asia. So, what is it that links these forms together? The most obvious characteristic is their lack of a tail. Most mammals have tails of one size or another, and the primates are no exception. The primates are the larger group to which apes and monkeys belong, and some of these, like the monkeys of South America, have long and prehensile (gripping) tails. In fact, all have tails except for the apes. There are other characters as well. For example, apes alone of the primates have an appendix, a mixed blessing since apes are prone to appendicitis just as we are. Another character that will be mentioned later in this book is the structure of the elbow region which is adapted for both stability and for mobility by a combination of characters of the joint.

The Asian apes

The gibbons are the smallest of the living apes, ranging in size from 4 to 13 kg (9 to 29 lb). They inhabit forests throughout tropical Asia, living in small family groups with just one male, one female and their offspring. They live almost exclusively in trees, eating mainly fruit and moving through the trees by a unique form of locomotion called brachiation, swinging underneath branches with their long and powerful arms. Males and females

Classification of Apes Used in this Book

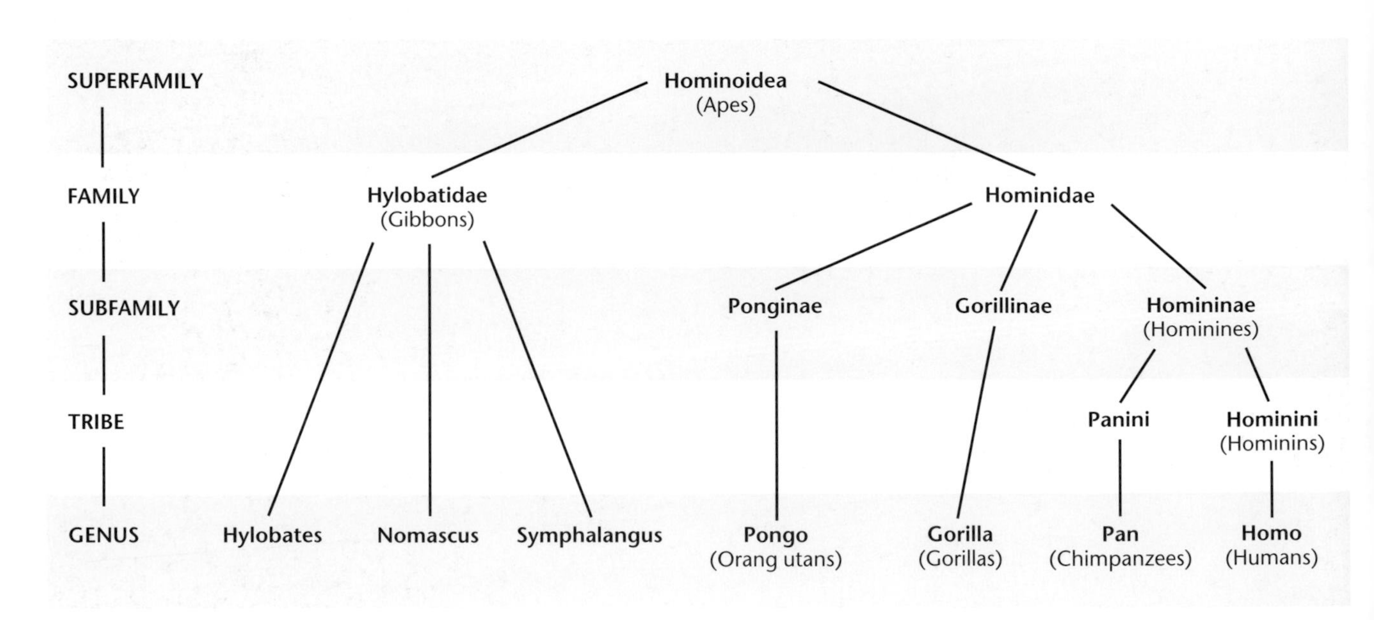

Living apes: chimpanzees (top left) and gorillas (bottom left) are found only in tropical Africa, gibbons (right) in the forests of Southeast Asia, and orang utans (centre) in Borneo and Sumatra.

share most activities, including territorial defence and use a complex system of vocalizations with which to maintain their social interactions. Up to seventeen species are currently recognized.

The orang utan is represented today by just one species, albeit with two rather distinct subspecies living in Sumatra and Borneo (although some recognize two separate species). It ranges in size from 40 to 140 kg (88 to 310 lb), much larger than gibbons, but it also is mainly arboreal, i.e. living in trees. They mostly eat fruit, but, unlike gibbons, orangs move in trees by slow and cautious four-handed climbing, for their legs and feet are almost as mobile as their arms and hands. Solitary animals, orang utan offspring stay with females while males live apart, the sexes only coming together to mate. They are generally silent animals, befitting their low level of social interaction.

(Below left) A gibbon showing the characteristic posture in brachiation, supporting itself by its arms.

(Below right) A Bornean orang utan showing how it uses all four limbs to climb in trees and the great flexibility of both its arms and legs.

The African apes

The gorilla is also represented today by a single species, with three subspecies not as distinct as those of the orang. They live in the tropical forests of west and central Africa, with one form restricted to the mountain forests on the borders between Rwanda and Zaire. Gorillas are the largest of the apes, weighing between 75 and 180 kg (165 and 400 lb), and they are also the most vegetarian, especially the mountain gorilla. They also tend to come to the ground most frequently, probably partly due to their great size, and they move on the ground by supporting their weight on the knuckles of their hands. They practise a form of polygamy, with one mature male having a number of females and their offspring making up the core group.

Comparison of adult chimpanzee and human skeletons. Humans have a broader pelvis and curved lower spine so that their centre of gravity is set back, enabling them to stand upright. Chimps by contrast lean forwards when they stand upright and cannot maintain that position for long. Humans also stand with their knees together so that their weight passes down through the centre of their body, and they have their toes lined up, lacking the opposable toes of chimps.

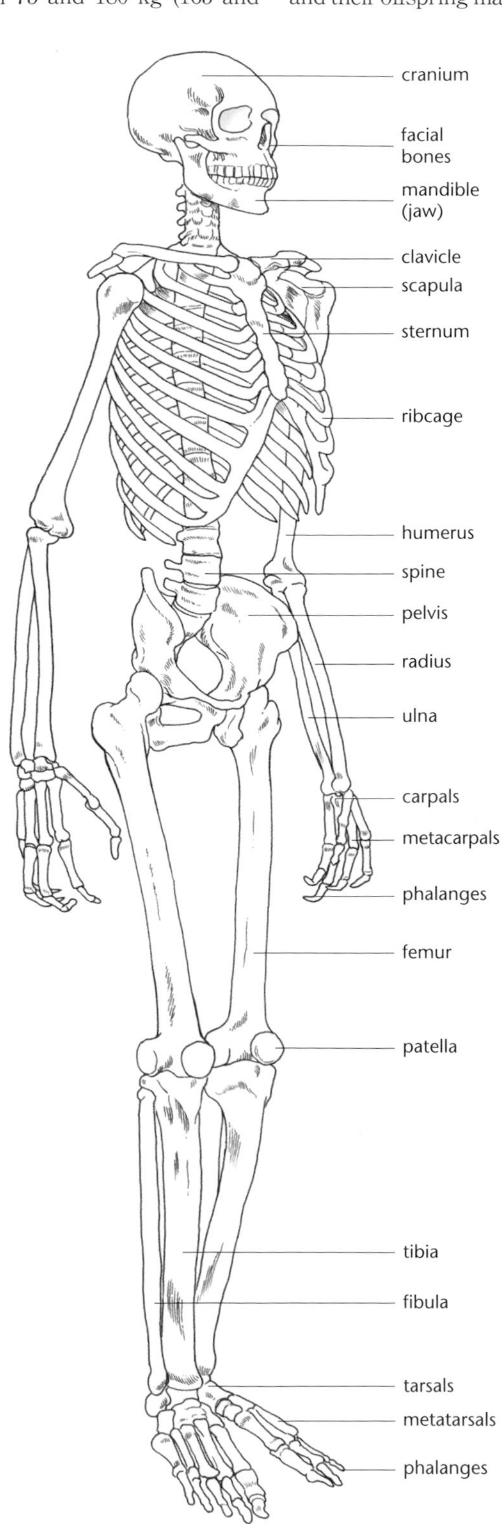

(Below) Chimpanzees live in noisy social groups, foraging for food partly on the ground and partly in trees.

Chimpanzees are divided into the bonobo, which used to be called the pygmy chimpanzee, and the common chimpanzee. They cover much the same range as gorillas but extend further into less richly forested tropical habitats. Like the gorilla they are knuckle-walkers, and this shared unique character has indicated to some scientists that they should be grouped together as distinct from other hominins. They are smaller than gorillas, eat more fruit, and live in much larger and more flexible groups centred on multi-male groups. Their social complexity is marked with advanced forms of communication by facial expression and vocalizations.

Modern humans

Humans today are represented by just one species, but as we shall see later in this book this was not always the case. We have yet another unique form of locomotion, walking upright on two feet, and at present this is taken as the earliest adaptation by which we can recognize human ancestors in the fossil record. We are probably African in origin, like chimpanzees and gorillas, although today we have of course spread across the world. We were probably originally adapted for a diet of fruit, again like the other apes. The racial differences between peoples of different continents were probably of very recent origin as the populations adapted to conditions quite unlike those in their African homeland.

(Above) Gorillas are gentle, sociable animals with a strictly vegetarian diet.

Human Variations

(Right) The results of computer analyses of skull measurements (top) are compared with analyses of genetic data. While skull shape may group African and Australian samples together, genetic data suggest sub-Saharan Africans are the most distinctive (and diverse) compared with the rest of humanity.

All humans alive today belong to one species. We are interfertile across the world, and share common bodily features such as a lightly built skeleton, a large brain housed in a high and relatively rounded skull, a small face tucked under the braincase, and a lower jaw with a chin. Our species was named *Homo sapiens* ('Wise man') by the great Swedish classifier of living things, Linnaeus. He recognized several varieties of our species, based on the continents in which they lived, and these have commonly been regarded as representing 'races'. Europeans and related peoples are said to belong to the 'Caucasoid' race (because the people of the Caucasus were supposedly perfect examples), orientals to the 'Mongoloid' race, black Africans to the 'Negroid' race, etc. There are obvious external differences between human populations in features such as skin colour, hair form, shape of the nose, eyes and lips, and physique, and these have formed the basis of racial classifications.

Cranial Cluster (28 groups)

Africa (*plus* Andaman Islands)
Australoids
Caucasoids
America (*plus* Europe and Buriats)
Japan
Asia (Northeast and Southeast)
Ainu, Guam, Eskimo
Polynesia

Genetic Cluster (42 groups)

Africa
Caucasoid
Northeast Asia
America
Southeast Asia
Pacific Islands
New Guinea, Australia

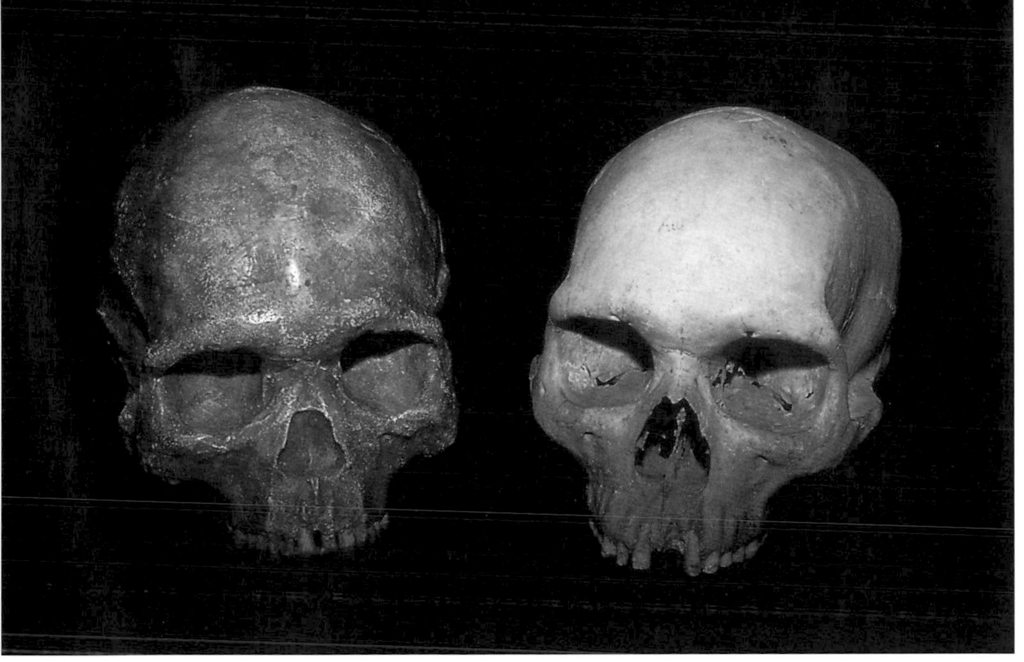

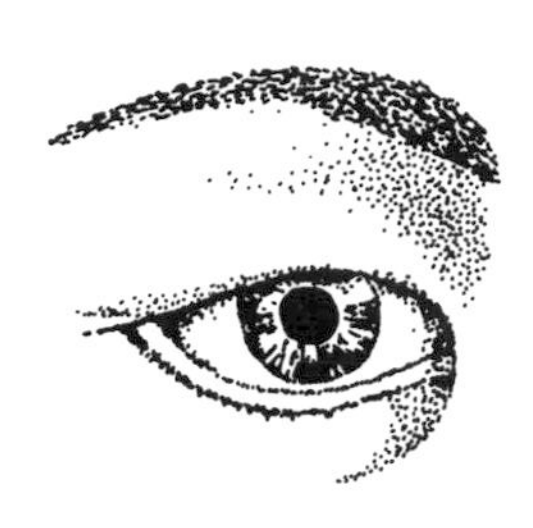

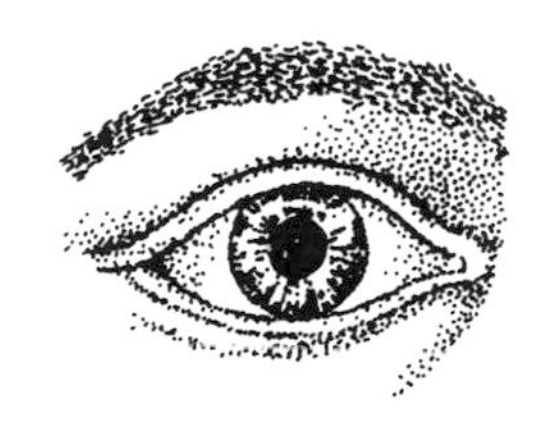

Regional differences are most apparent in the head and face: (above) the form of the eyelid common in oriental and Native American populations (top) compared with the most common shape found elsewhere. Shared characteristics are those that typify our species: in this comparison (above left) of a fossil skull from Fish Hoek (South Africa), left, and a recent Australian, distinctive features such as a large and rounded braincase, high forehead, small browridge and small retracted face are clearly visible in both.

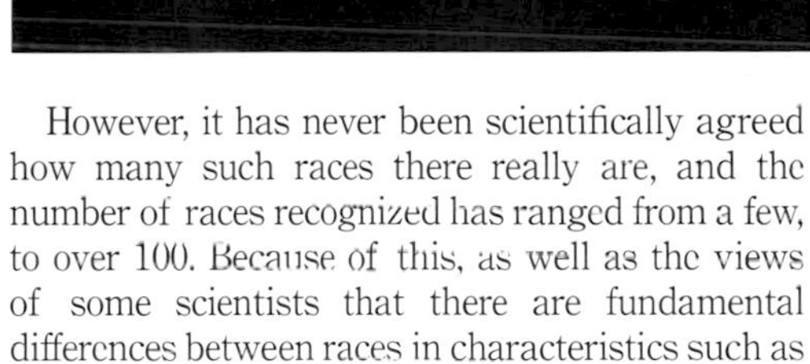

However, it has never been scientifically agreed how many such races there really are, and the number of races recognized has ranged from a few, to over 100. Because of this, as well as the views of some scientists that there are fundamental differences between races in characteristics such as behaviour and intelligence, the whole concept of race has become highly controversial. Even those who believe in rather rigid racial characteristics accept that the populations concerned can interbreed, and it is therefore acknowledged that the boundaries between such groups in areas where they overlap are hardly likely to be sharply defined. So, many anthropologists prefer to talk about 'regional', rather than 'racial' differences, and while acknowledging that there are clear physical differences between major regional groups, they prefer not to view the categories as rigid or absolute. In addition, major population movements, both historic and prehistoric, have overlain ancient patterns with new ones. Populations of European origin are now firmly established in the Americas,

(Left) Human beings come in many shapes, sizes and colours. However, under the skin, our skeletons and our genes show that we are all closely related. The variations we see today probably all developed within the last 200,000 years as our species diversified from an ancestral African population.

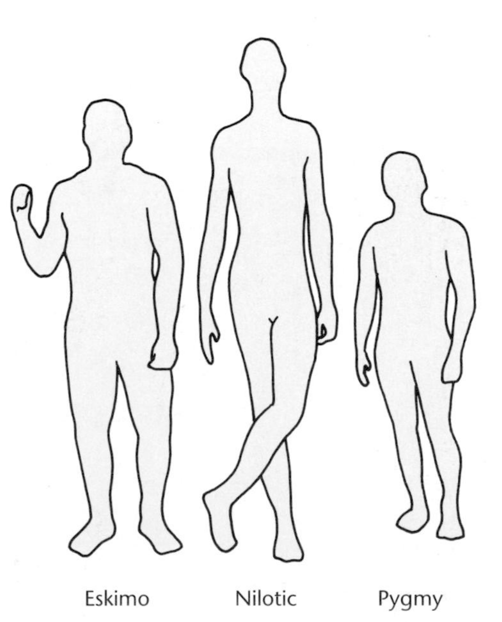

(Right) Different human body shapes are certainly partly related to the climate in which they have evolved: people with a shorter and rounder body physique (left) will retain heat better than those with a taller, lankier physique (centre). However, a distinctively smaller body may be favoured in closed tropical forest environments (right).

As well as the most remarked feature of skin colour, human populations vary in the shape of the nose, lips and cheekbones, as well as in the form and extent of head and facial hair. These illustrations contrast the facial appearance of an Ainu man, left, and a Khoisan ('Bushman') woman from southern Africa, right. The Ainu, aboriginal inhabitants of Japan, are particularly noted for the hirsuteness of the men.

southern Africa, Australia and New Zealand by colonization, while African-origin peoples have been transported to the Americas as slaves. In the more distant past, there is evidence that 'Caucasoid' peoples lived in western China, while peoples related to small and geographically restricted living populations, such as the Khoisan ('Bushman') of southern Africa and the Ainu of northern Japan, were probably much more widespread.

Genetic studies

Now we are able to look at population differences and population relationships in much greater detail through genetic studies, which look far beyond the external physical differences to the inherited coding for many characteristics of the body, such as proteins, enzymes and blood groups. Such studies also look at differences in long stretches of DNA (deoxyribonucleic acid) which do not code for anything at all. Features such as skin colour are controlled through the interaction of a few genes, whereas geneticists can potentially examine thousands. This means that concepts of 'race' and regionality are now being completely re-evaluated. Much research is also being carried out on the origin of regional characteristics. It now seems that these have evolved recently in human history, probably mostly within the last 50,000 years.

Several factors are believed to be behind the evolution of regional variations. Natural selection, Darwin's favoured mechanism of evolutionary change, is certainly one, for example accounting for at least some of the variation in skin colour across the world. Highly pigmented skin affords protection against the potentially harmful ultraviolet rays of the sun, which may cause skin damage, cancers and interference with folic acid production, and this advantageous characteristic will be favoured through natural selection. However, these rays are beneficial in moderation, since cells beneath our skin can use them to make essential vitamin D. So the degree of skin pigmentation can be seen as a balancing act between allowing sufficient sunlight to penetrate the skin, and preventing too much in conditions of high ultraviolet light. However, the relationship between skin colour and ultraviolet

(Below) This diagram shows how skin colour must play a part in balancing the amount of ultraviolet radiation received from the sun against the body's needs for the production of both vitamin D (facilitated by sunlight) and folic acid (damaged by ultraviolet radiation).

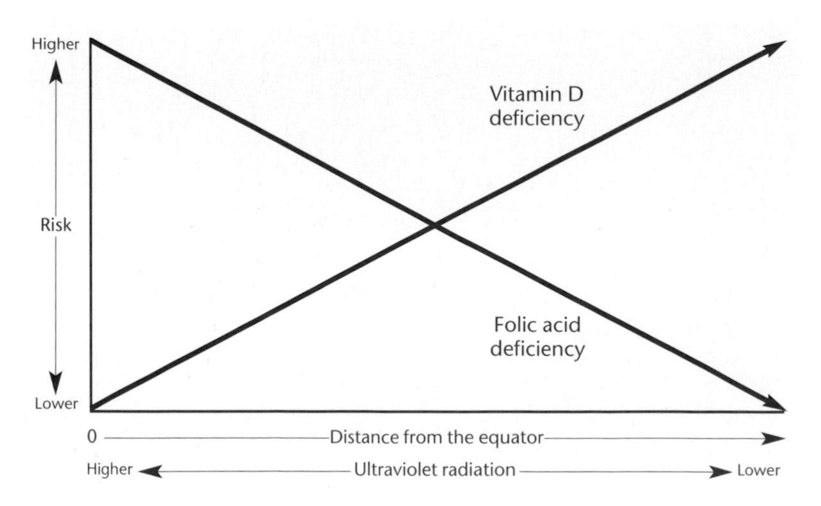

light in populations in their native regions is not a perfect one, and this may be because of past population movements confusing the original patterns, past differences in ultraviolet radiation, or because other factors than natural selection are at work.

One of these additional factors could be sexual selection, and this mechanism was also investigated by Darwin, and was one which he thought could account for the evolution of some 'racial' characteristics. In such cases, mate selection influenced by cultural preferences can gradually steer a population in a particular direction – for example, a preference for, say, males with larger or smaller noses, or women with paler or darker skin tones. Repeated over a number of generations, such individual choices can accumulate in a population, changing its typical characteristics. Further factors behind population differences are more random processes, called genetic drift and founder effect. In the former process, once two populations are separated, they may diverge simply through chance – once a particular change occurs, it may become accentuated in one population compared with another. In the second case, a very small and perhaps atypical sample of a large population may found a different population – for example, where a few individuals may have travelled on rafts from Southeast Asian islands to the completely uninhabited continent of Australia-New Guinea. The particular characteristics of those few founders were then multiplied many thousands of times in their descendants, who became the Australian Aborigines.

(Right) The magnificent tail of the male peacock is used in display and is believed to have evolved as a result of female mating preferences over countless generations – an example of sexual selection. Charles Darwin considered that many regional ('racial') differences in human appearance were probably also the result of sexual, rather than natural, selection. This view went out of favour for many years, but is now gaining increasing acceptance.

Palaeoanthropology

Palaeoanthropology is the study of all aspects of human evolution. It is a branch of anthropology that investigates the biological and geological background of human evolution, and as such it covers an enormously wide range of subjects. What they all have in common is the study of humans, their origin, evolution, adaptations, and behaviour. We will try and provide some idea of the scope of the subject here.

First of all, one cannot consider human evolution in isolation, and so the subject expands to consider evolution in general and primate evolution in particular. As we have seen, Primates is the name of the group to which apes and monkeys belong, and it has a history extending back at least 50 million years. Primate evolution can be studied in several different ways. Comparing the anatomy of all the living species tells us what their common ancestors were like. For example, living apes have been described as sharing three characters (among others): loss of the tail, presence of an appendix, and adaptations of the elbow joint. All living apes and humans have these characters, and it is a reasonable assumption that the common ancestor of the apes and humans also had them. The alternative is that the same characters developed independently in each lineage of ape, which is not very likely.

(Below left) Chimpanzees and humans share 98.8% of their DNA, so that the difference between them is only 1.2%. Chimpanzees are equally similar to gorillas, but gorillas are more different from humans and there are even greater differences between all three and the orang utan.

(Below right) The Proconsul *skull from East Africa found in 1948 by Louis and Mary Leakey.*

Distinguishing characters in this way, known as shared derived characters, is one of the core principles in identifying relationships between different species. The presence in apes and humans of these three features, compared with the presence of tails, lack of appendix and lack of ape-like characters of the elbow in all other primates, shows that these characters are unique to the apes and humans. Underlying this approach to character definition is the assumption that similarity in characters is based on common inheritance, but clearly in some cases it is not. The wings of birds and bats, for example, are similar in terms of function but structurally and genetically they are different and provide no evidence of relationship.

These principles can be applied to another new source of evidence that comes directly from genes. Changes in the genetic material DNA accumulate in the same way as, and in many cases are responsible for, anatomical changes, and by analysing these changes it is possible to base relationships directly on genes, the basic material involved in inheritance of characters.

Evidence from fossils

The evolutionary trees produced by both anatomical and molecular evidence provide the primary evidence for evolutionary relationships, and these can be tested by fossils. Fossil apes and fossil humans are described by palaeoanthropologists and compared with their living relatives, and if enough characters are present they can be fitted into the evolutionary trees. In these cases, the fossils show when and where changes occurred during evolution, because the fossils can be dated and we know where they are from. For example, we know

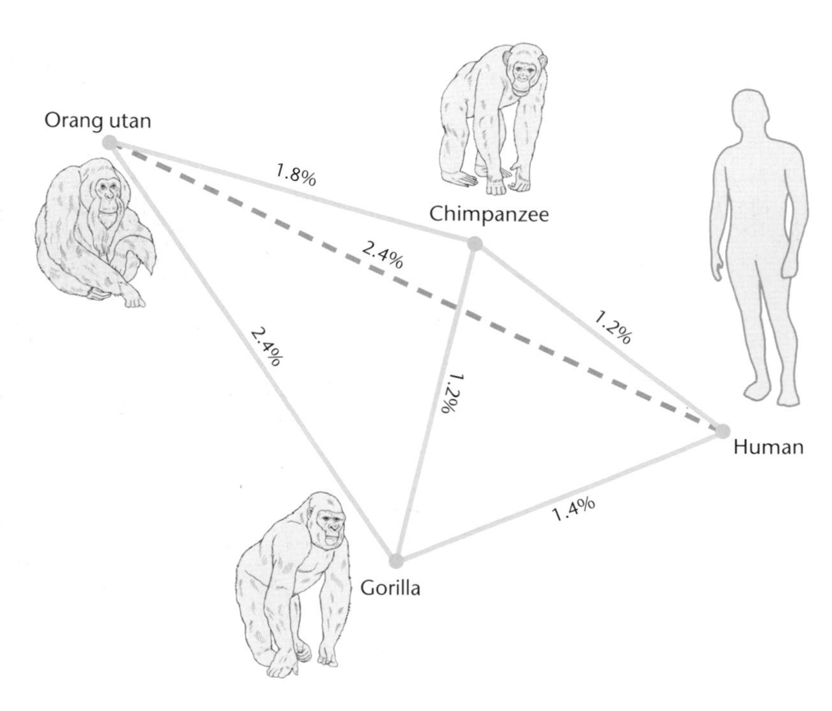

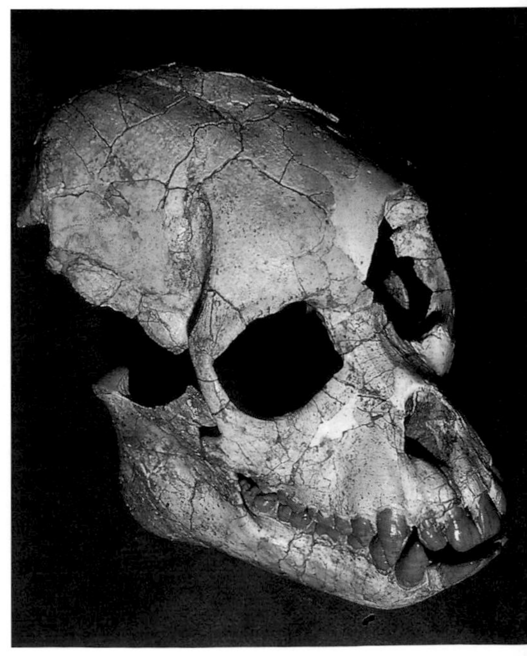

that humans originated in Africa, because the first 2 to 3 millions of years of fossil humans are entirely restricted to Africa, and the first appearance of fossil humans outside Africa dates from much later.

The fossils also give us information about the kinds of environment present in the past. The nature of the deposits in which fossils are found tell us about local conditions, whether, for instance, in a lake, in river deposits or in a soil. The fossils have to be related to the geology, however, and one of the ways this is done is by investigating the taphonomy of the site. Taphonomy is the study of how bones, or other parts of animals or plants, become preserved as fossils. For example, if the geological setting indicates fossils in lake sediments, the taphonomic analysis may show that the fossils arrived in the lake by being washed in by a river, carried in by a crocodile, or dropped in from an overhanging tree by a predator eating its prey up in the branches. Knowing what these processes are is important for understanding the nature of the fossil assemblages, because the kinds of animals washed in by a river may be different from the kinds of animals eaten by a crocodile and even more different from those carried up into a tree and eaten by predators.

After taking all these things into account, the fossil species present at a fossil site can be used to indicate the kind of habitat that was present. Many tree-living species, for example, would indicate a habitat with abundant trees; and many ground-living species would indicate a habitat with open spaces. The more that is known of the animal species in a fossil fauna, the more accurate is likely to be the environmental reconstruction of the fossil site.

(Right) Some early human ancestors were found in caves with animal bones, and it was thought at first that the animal bones were the remains of their food. Now it is understood that both were the prey of leopards, and this is a reconstruction where a leopard has carried its kill into a tree to protect it from the hyaenas below. The remains of the fossil hominin (human ancestor) will fall into the cave at the bottom of the picture to be preserved along with other animal remains.

(Below) Excavations in cave sites, such as this one in Gibraltar, can provide direct evidence of human prehistory in the form of bones and stones, but also contextual information about past climates and environments.

Behaviour and archaeology

Another quite distinct aspect of palaeoanthropology is the study of behaviour, both past and present. By studying the behaviour of living apes such as chimpanzees and gorillas, it is possible to infer the behaviour of fossil apes. Animals with similar body structures or similar body sizes tend to have similar social structures, so that comparing fossil animals with living ones tells us about past social structures.

The related subject of archaeology also fits in here, for archaeology is essentially the study of cultural artifacts that result from past human behaviour. Archaeology therefore complements, and is exclusive to, the study of human behaviour by these other means. Remains of animal and plant foods at archaeological sites provide information on past diets or the use of objects for purposes other than food. The development and history of diseases is also part of palaeoanthropology in that evidence of diseases may be identified by marks left on fossil bones, from anaemia, for example, or the shared presence of diseases may be indicated in related species, such as malaria in chimpanzees.

The Geological Timescale

Early geologists realized that thick deposits of sedimentary rocks must represent ancient periods in the history of the Earth, but they had no conception of how ancient they actually were, and no way of measuring the timescale involved. Such deposits were often assumed to be the result of the biblical deluge and flood. Now we have ways of measuring ancient time through radiometric clocks, which work through radioactive elements present in rocks. These have revealed that the Earth is at least 4,500 million years old.

The earliest stages of Earth history were lifeless. However, simple forms of life had appeared by 3,500 million years ago, while complex forms mainly evolved during the last 600 million years. As geologists built up a more complete record of the Earth's rocks, they divided its history into a number of Eras, with further subdivisions called Periods and Epochs. The common divisions, starting from the most ancient, are the Precambrian (by far the longest), Palaeozoic ('Ancient Life'), Mesozoic ('Middle Life'), Tertiary (so called because it was originally the third stage) and Quaternary (originally the fourth stage, which includes recent and present time). The Tertiary Period is in turn

Eras
^
Periods Epochs

(Below) *The Grand Canyon in Arizona, USA, is one of the most spectacular demonstrations of the depth of geological time. Although probably only carved out by the Colorado River during the last few million years, the rocks in the canyon walls range in age from about 270 million years old at the top to about 1.8 billion years at the bottom.*

Millions of years ago	Eon	Era	Period		Epoch
0–0.01	Phanerozoic	Cenozoic	Quaternary		Holocene
0.01–1.8	Phanerozoic	Cenozoic	Quaternary		Pleistocene
1.8–5	Phanerozoic	Cenozoic	Tertiary	Neogene	Pliocene
5–24	Phanerozoic	Cenozoic	Tertiary	Neogene	Miocene
24–37	Phanerozoic	Cenozoic	Tertiary	Palaeogene	Oligocene
37–58	Phanerozoic	Cenozoic	Tertiary	Palaeogene	Eocene
58–65	Phanerozoic	Cenozoic	Tertiary	Palaeogene	Palaeocene
65–142	Phanerozoic	Mesozoic	Cretaceous		
142–206	Phanerozoic	Mesozoic	Jurassic		
206–248	Phanerozoic	Mesozoic	Triassic		
248–290	Phanerozoic	Palaeozoic	Permian		
290–354	Phanerozoic	Palaeozoic	Carboniferous		
354–417	Phanerozoic	Palaeozoic	Devonian		
417–443	Phanerozoic	Palaeozoic	Silurian		
443–495	Phanerozoic	Palaeozoic	Ordovician		
495–545	Phanerozoic	Palaeozoic	Cambrian		
545–2500	Precambrian	Proterozoic			
2500–3800	Precambrian	Archean			
3800–4560	Precambrian	Hadean			

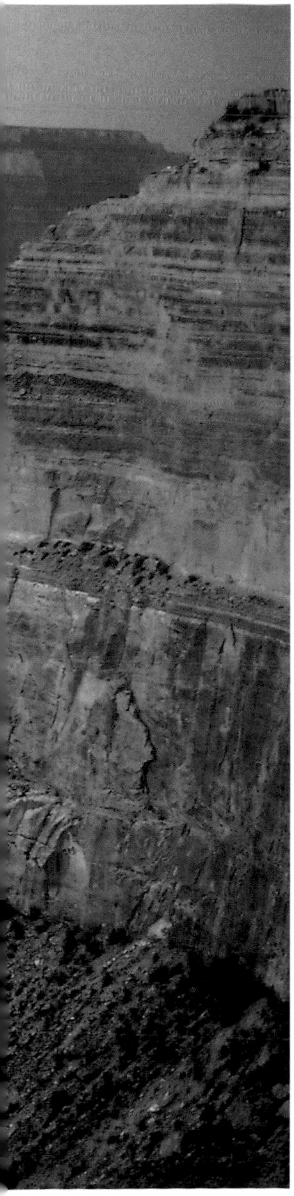

The Geological Column

There are various schemes for dividing up Earth history, which covers some 4.5 billion years. The system shown here is one of the most widely used, with periods and epochs designated by geological events and changes in the characteristic fossils preserved. The estimated ages for the various divisions of the geological column are shown in millions of years before the present.

divided into the Palaeocene, Eocene, Oligocene, Miocene and Pliocene Epochs, and the Quaternary into the Pleistocene and Holocene (Recent) Epochs.

Imagining the vastness of geological time

A common way to illustrate the vastness of deep geological time, and the extreme brevity of our own occupancy of this planet, is to consider Earth history as, say, a 24-hour clock, or a week of seven days, or as a whole year. If we use the last analogy, the Earth would have begun to form on 1 January. The first bacteria-like life would have appeared in February or March, but the first multicelled plants only in October, followed by complex animals such as trilobites in November. Towards the end of November, jawed fishes had evolved, and the first animals were colonizing land at the beginning of December. The dinosaurs were dominant in the middle of December, and became extinct around 26 December, when the primate group was beginning its evolution. Monkeys had appeared by 29 December, and apes during 30 December. Our line of evolution would have separated from that of our nearest living relatives, the chimpanzees, around midday on 31 December, with the first modern humans appearing within the last 20 minutes of that day. Our species colonized Australia and Europe during the last six minutes, and farming began just before the last minute. On this timescale, the Christian era would have begun

It is very difficult for us to grasp the immensity of time involved in Earth history and the evolution of life. A common device to illustrate this is to represent geological time as a clock or calendar. In this example, the whole 4.56 billion years of Earth history is equated with one year. Each day represents the passing of 12.5 million years, each hour around 500,000 years, each minute 8,500 years, and each second 150 years. On this scale, modern humans only appeared in the last 20 minutes of the last day.

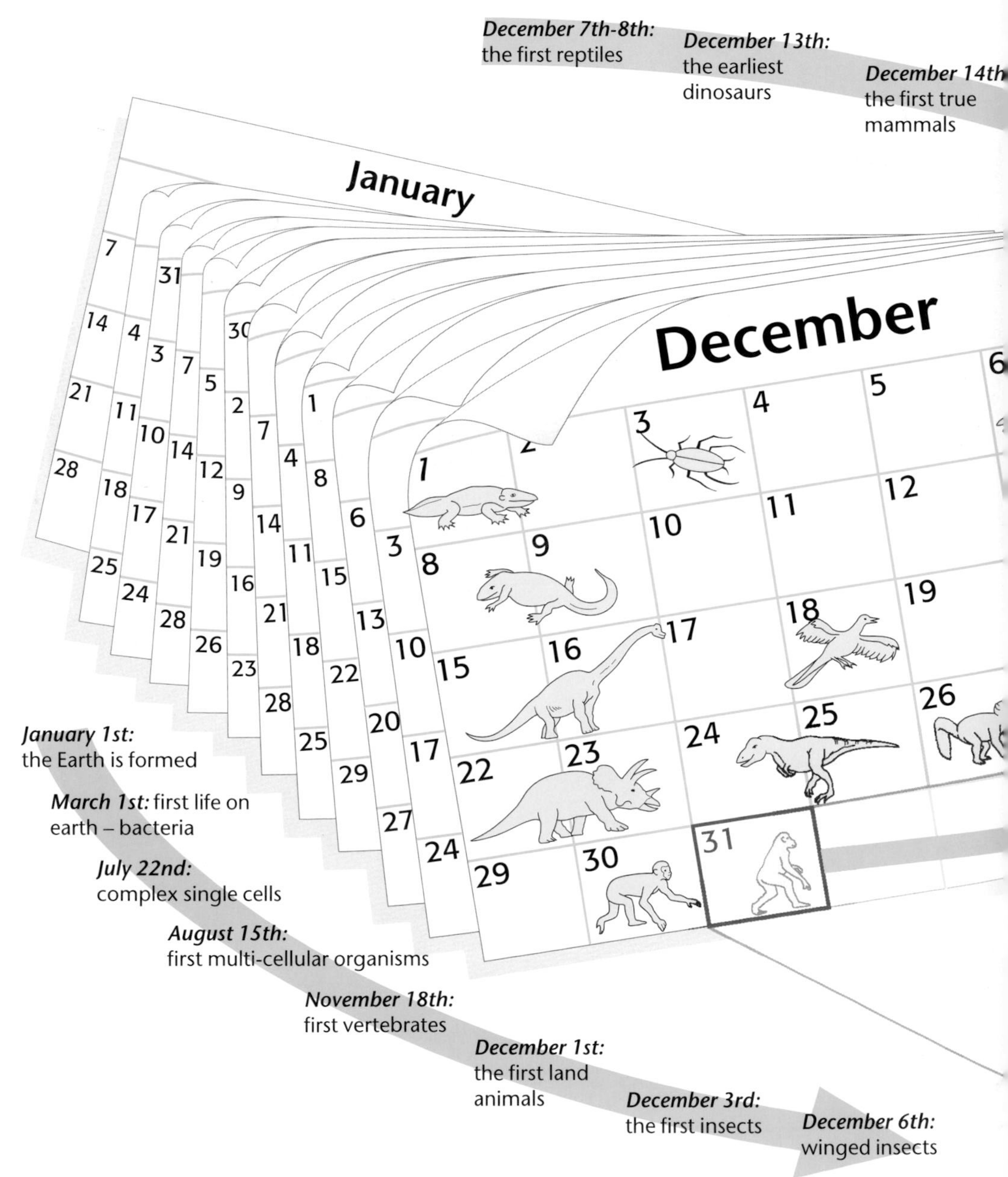

about 15 seconds before midnight on 31 December, while we have all lived during the last second! So we, and even our closest relatives, the primates, are very much newcomers in evolution.

It is easy to fall into the trap of believing that the aim of evolution was to produce humans, but examining the fossil record shows that, while there has certainly been an increase in complexity overall, there has been no coherent pattern or goal for evolution. Instead many groups have thrived for long periods and then been completely, or almost completely, swept away. Many scientists now believe that the largest waves of extinctions, such as at the end of the Permian period, about 250 million years ago, and at the end of the Cretaceous period, about 65 million years ago, were largely the result of random effects such as asteroid or comet impacts on the Earth. If large enough, these would have seriously disturbed the Earth's climates and ecosystems. However, other scientists believe that changes in the Earth itself, whether geological or climatic, were also important factors behind such massive extinctions.

Hominin evolution during the last few million years has often been viewed as a ladder of progress, inexorably leading to us, with our large brains, high intelligence and complex behaviour. Yet, the fossil record demonstrates that human evolution, like that of many other groups, was more like a bush than a ladder, and was characterized by successive radiations of different species, only a few of which persisted or evolved into new forms. Our recent 'success' (in our terms, but certainly not for many of the rest of the Earth's species, whose survival we threaten) has been very brief, and an impartial observer from another planet who looked at our ancestors of even 100,000 years ago, would probably have considered us very unpromising material for world domination.

cember 15th-25th:
osaurs flourish

December 18th:
the first birds

December 21st:
flowering plants

December 26th:
dinosaurs extinct;
the first primates

December 29th:
Himilayan uplift
begins; the first
monkeys

December 30th:
the first apes

7

14

21

28

12.00
humans and
chimps diverge

23.40
first modern
humans

23.54
humans to
Europe and
Australia

23.58.30
Ice Age ends
and first
Americans

23.59
animals
domesticated
and first farmers

23.59.46
Christian era
begins

23.59.57
Columbus to
America

23.59.59
our lifetimes

Throughout the Earth's history, our planet has been bombarded with debris from the rest of the solar system – it is believed that the moon was created by one such impact. Once life had evolved, it was vulnerable to major perturbations in the environment such as in atmospheric composition or sea or land temperatures, and some experts believe that massive impact events triggered the largest extinctions in life history such as at the end of the Permian period, about 250 million years ago, and at the end of the Cretaceous, about 65 million years ago.

Dating the Past

There are two main categories of dating – relative and radiometric (sometimes called absolute) dating. The first relates an object or layer to another object or layer in past deposits. Unless there has been major disturbance, a layer in a geological sequence is always younger than the layer below it. Thus Bed II at Olduvai Gorge in Tanzania (see pp. 68–71) overlies, and is therefore younger than, Bed I. Bed I contains fossils of the early hominins *Homo habilis* and *Paranthropus boisei*, and they are therefore assumed to be the same age as Bed I. However, they and Bed I are in turn younger than the layer of volcanic rock which lies at the base of the whole Olduvai sequence. The fossil animals in Olduvai Bed I are similar to some of those found in rocks which contain fossils of *Paranthropus boisei* at the site of Koobi Fora in northern Kenya, and these are, therefore, assumed to be of approximately the same age – in other words they can be directly related to one another – or correlated. But none of these relationships can tell us how old Bed I actually is – relative dating can only tell us whether it is older than, younger than, or about the same age as another layer or fossil.

The Serengeti Plain with Mount Lemagrut, an extinct volcano, in the background. Olduvai Gorge is in the middle distance on the left of the picture.

Radiometric dating: potassium-argon and radiocarbon

To go further than a relative date, we need a sort of geological clock which will tell us how far back

some rocks were laid down, or how long it is since an animal or plant died. Such clocks are called radiometric, because they measure time using natural radioactivity. An example of such a technique is potassium-argon dating, which can be used on volcanic rocks. Potassium partly consists of an unstable form called potassium-40, and this isotope gradually changes over many millions of years into the gas argon. When there is a volcanic eruption, the liquid lava or hot ash contains a small proportion of potassium-40, and when the lava or ash cools and solidifies, the unstable form of potassium begins to change into argon. Provided this argon gas is trapped in the volcanic layer, the amount produced can be used as a natural measure of time since the volcanic rock was deposited. Using this technique, the lava at the base of Olduvai Bed I has been dated at nearly 1.9 million years old. It is believed that Bed I was deposited soon after this time, so the *Homo habilis* and *Paranthropus* fossils it contains are, therefore, about 1.8 million years old. By correlation this should be the approximate age of the comparable deposits at Koobi Fora, and this has, in fact, been confirmed by radiometric dating.

The most famous radiometric dating method is radiocarbon dating. This relies on the fact that an unstable form of carbon called carbon-14 is constantly produced in the upper layers of the Earth's atmosphere by cosmic-radiation. This unstable form of carbon gets taken into the bodies of living things, along with the much more common, and stable carbon-12. However, when the plant or animal dies, no more carbon-14 is taken in, and the amount left begins to break down by radioactive decay, such that the amount present halves about every 5,700 years (this is called the half-life). So measuring the amount of carbon-14 left in, say, a piece of charcoal or a fossil bone,

The site of Olduvai Gorge contains one of the key records of recent African geological history, and this record has been calibrated by a range of dating techniques. Volcanic deposits can be dated using the potassium-argon technique, and they may also contain a snapshot of the particular state of the Earth's magnetic field at the time. Such data can be used not only to assemble the record at Olduvai but also correlate that record with sites elsewhere.

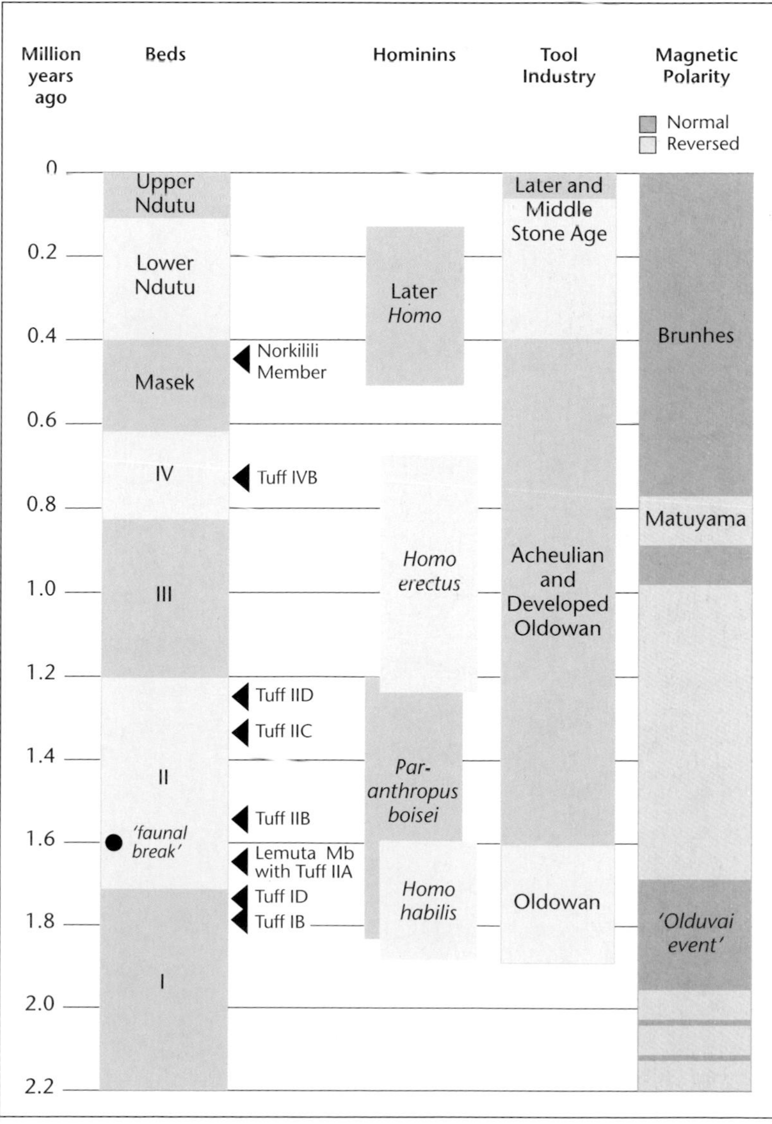

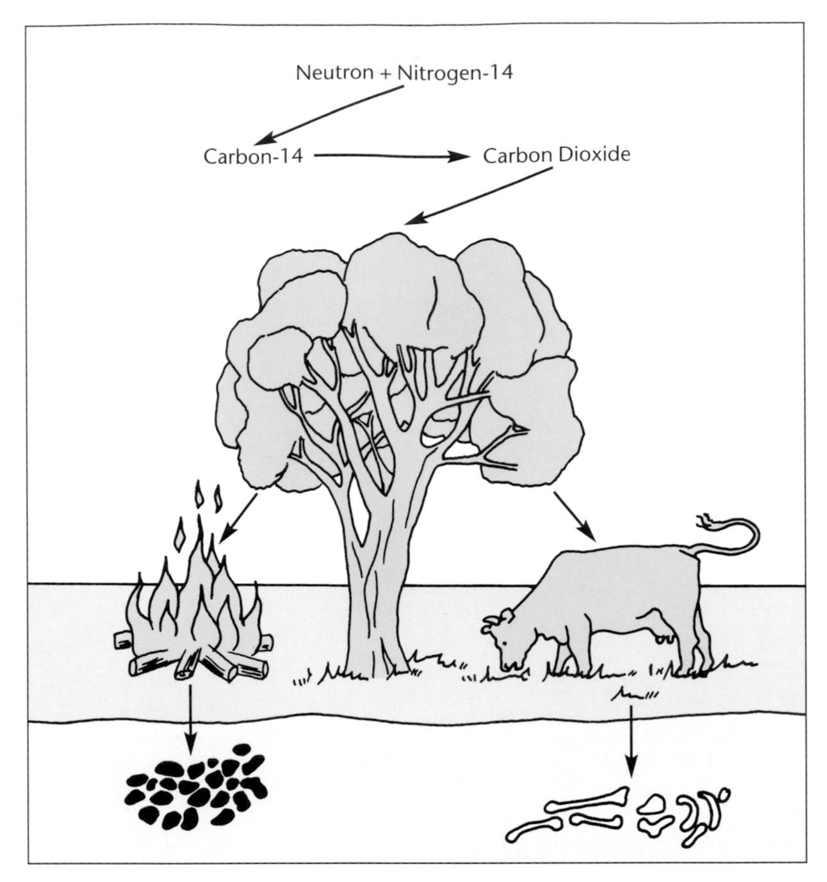

allows us to estimate how long it is since the plant or animal concerned was alive. The method cannot be used on very ancient materials because the amount of carbon-14 left behind is too small to measure accurately, and hence radiocarbon dating may become unreliable beyond about 30,000 years ago. Moreover, the assumption of constant production of carbon-14 in the past is not entirely true, so scientists talk of dates in 'radiocarbon years' rather than real years. Nevertheless, comparisons with other methods suggest that radiocarbon dating, while not exact over the whole of the last 40,000 years, is quite reliable.

Other radiometric dating methods

More recently, other radiometric methods have been developed to date both fossil and archaeological materials beyond the limits of radiocarbon dating. These include uranium-series dating, based on the decay of different forms of uranium. Accumulation and measurement of so-called daughter products is possible in substances like stalagmites and corals. The former has been very useful in cave sites, while the latter has been used to examine past changes in sea levels around tropical and sub-tropical coasts. A number of other methods depend on the fact that crystalline substances such as a burnt flint or the enamel of a tooth store up damage from the radiation they receive from their surroundings once they are buried. The accumulated amount can be

(Above) A radioactive and unstable form of carbon, carbon-14, is constantly being produced by the action of cosmic rays on the Earth's atmosphere. It is then taken into the bodies of living things in the form of carbon dioxide. Once there, in materials such as wood or bone, carbon-14 begins to decay, disappearing at the rate of about 50 per cent every 5,700 years. Its decay thus creates a natural 'clock' that can be used to estimate the age of fossils.

(Left) The geochronologist Rainer Grün stands in deposits at Qafzeh Cave, Israel. Working with colleagues, he applied the techniques of electron spin resonance and uranium series dating to fossil teeth from Qafzeh, estimating their age at about 100,000 years. Such work contributed to a major re-evaluation of the whole sequence of human evolution in the Middle East.

(Opposite) The Radiocarbon Accelerator at The Research Laboratory for Archaeology, Oxford University, can detect and measure very small quantities of radiocarbon. This has allowed direct dating of important relics, from fossil skulls to the Turin Shroud.

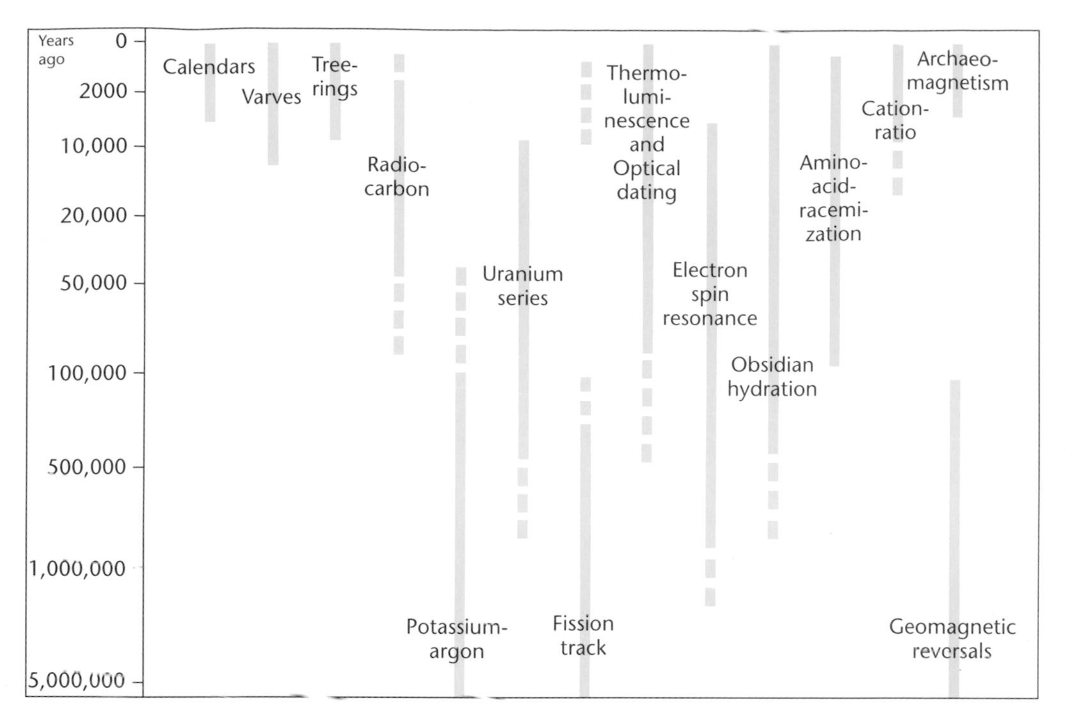

Dating the Past

Archaeologists and palaeontologists have a range of dating techniques at their disposal, each with their own range of applications, strengths and weaknesses. Beyond 5 million years ago, potassium-argon is the primary dating method, although fission track dating (also based on the operation of a radioactive clock) can also be used. Geomagnetic reversals can provide a useful check on dates found using other methods.

measured in the flint by a laser-beam (optically stimulated luminescence (OSL)) or by heating (thermoluminescence (TL)), or detected in the tooth enamel by microwave radiation (electron spin resonance (ESR)). Provided the rate of accumulation of radiation damage in the flint or tooth can be estimated from the site in which it was buried, the length of time it has been in the ground (for example in a Neanderthal fireplace or butchery site) can be estimated.

Studying Animal Function

When we look at function, we are looking at characters in terms of the purpose they fulfil in the life of an animal. Take the wings of bats and birds: the function of wings is to enable their owners to fly. It tells us that wings evolved to this end, but it tells us nothing about the actual evolution of the structures and nothing about the evolutionary relationships of flying animals. The study of function is, therefore, the antithesis of evolutionary relationship because there are only so many functions animals are able to perform, and there is a great deal of repetition, or convergence, as it is called, in evolution. Characters that are convergent tell us nothing about the evolutionary relationship of the animals that possess them.

When Charles Darwin formulated his ideas on natural selection as the principal means of change in evolution, he was implicitly referring only to functional characters. Characters having no function would be missed by natural selection because there would be nothing to select. This apparent truism is the basis of much of the molecular work aimed at understanding evolutionary relationships between animals. Human genes contain around 3 billion items of genetic information (nucleotide pairs), and only about 10 per cent of this contains information relating to human function of any kind. The other 90 per cent is essentially neutral, so that the changes that occur from random mutations accumulate at the level of the mutation rate. By counting the changes between related groups of animals (or taxa), for example between chimpanzees and humans, and knowing the mutation rate, it is then possible to estimate the time of their divergence from their common ancestor.

(Below) A gibbon at home in the trees using both its powerful muscular arms and its long legs to give support. Their distinct form of locomotion is called brachiation.

(Below right) Proboscis monkey leaping across a river, Borneo.

(Left) Chimpanzees and gorillas have adapted to ground living while still being quite mobile in trees by adopting an unusual behaviour, supporting themselves on the knuckles of their hands. This is called knuckle-walking.

(Right) Four distinctive locomotion types in primates. Along the top, prosimians (lower primates) leap from supports in a vertical position using powerful hind legs; second row, quadrupedal locomotion on the ground; third row, a gibbon brachiating; bottom row, chimpanzee knuckle-walking.

Function in primates

The main emphasis on function in primates has traditionally been on their different forms of locomotion and on the different foods they eat. Most primates have adaptations of their limb skeletons for climbing in trees. Characters such as gripping hands and feet, mobile shoulders, hip and elbow joints, and stereoscopic vision are of crucial importance to active tree-dwelling primates, and these are characters generally present in most primates. Some species that have adapted to ground living have lost some of these adaptations at the expense of developing longer limbs for running on the ground, for example the patas monkey, or the loss of gripping feet, as in humans. Adaptations for leaping are seen in some of the prosimian primates,

(Below) Yellow baboons in the Kwahi river, Botswana, show the quadrupedal form of locomotion.

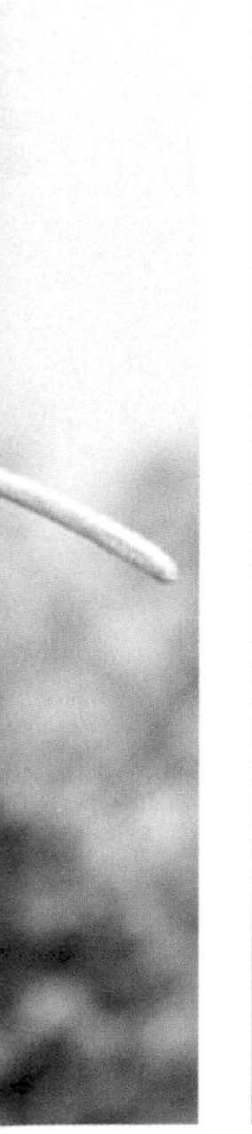

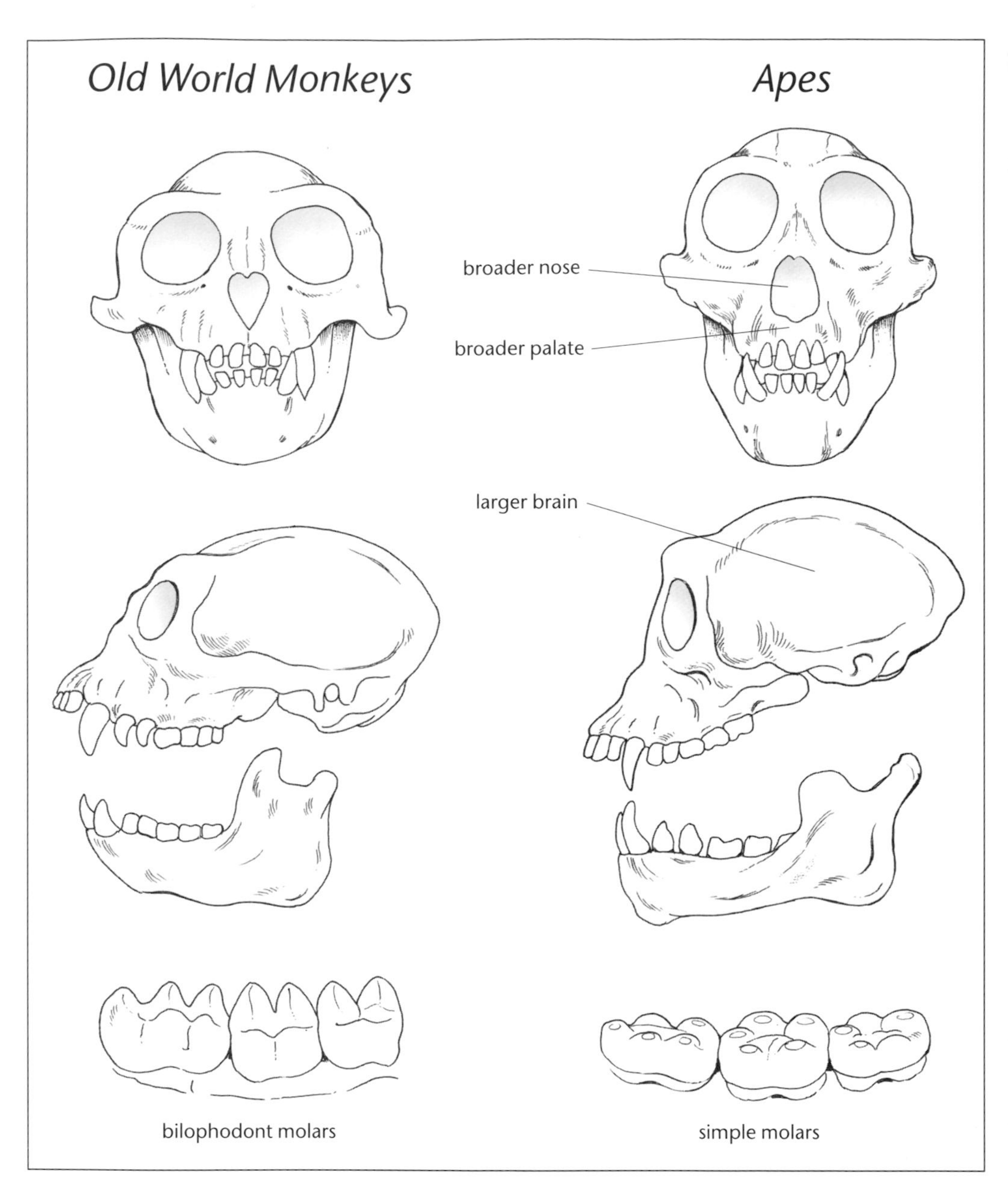

(Above) Apes and monkeys differ in many ways, and in terms of their skull form and teeth it is the monkeys that are most specialized and the apes that are closer to the ancestral catarrhine condition. Monkeys have developed bilophodont teeth, that is teeth with two transverse crests for cutting up coarse vegetation, whereas apes have retained simple low crowned teeth.

(Above right) Chimpanzees have mobile lips and large front teeth which are very effective at stripping and opening up fruits.

with their elongated and immensely strong legs, and adaptations for hanging suspended from branches are seen in some apes, which have elongated and powerful arms. Any of these characters that can be identified in fossil apes provides direct evidence about the probable form of locomotion of the fossil apes. By this means we are able to identify early fossil apes as four-footed, tree-living, above-branch primates. Similarly we can identify the points at which some apes became more terrestrial, others moved to suspension below branches as in living apes, and others still became bipedal on the way to modern humans.

Adaptations of the teeth and jaws reflect strongly the type of food eaten. Insect-eating primates have sharp pointed teeth, as in many of the lower primates like lemurs and bushbabies. Molar tooth areas become larger as primates eat more vegetation, and in the extreme cases the teeth become ridged and have long shearing crests for cutting up leaves and other vegetation, as in monkeys. A few species have moved towards eating tougher vegetation, and have high crowned teeth to resist wear, and complex crown structures to grind up the tough food. A similar dichotomy is seen in the fruit-eating apes, where species eating soft fruits have low crowned teeth with flat surfaces for crushing the fruit, while species eating harder fruits have higher crowns and thicker enamel to resist the greater stress put on the teeth. These forms have correspondingly more massive jaws to accommodate these stresses. Again, it is possible to

identify all these features in fossil apes, so that early forms can be identified as soft fruit eaters, some later ones becoming hard fruit eaters and a few adapting to a diet with more herbaceous vegetation.

Many of these features are linked with body size in predictable ways. Insectivorous species are generally small, herbivorous ones generally large, with frugivores (fruit eaters) in between. Arboreal forms tend to be small and terrestrial forms larger, although there are many exceptions to this. The relationships of these ecological attributes to the environment are also highly predictable and will be described later.

Certain difficulties arise from the use of functional characters to show evolutionary relationships. The problem of convergence has been mentioned above, but there are ways of identifying this, for example by looking at the way characters develop during the growth of individuals to see if they follow the same pathway. In the example of bats and birds, the way the wings of bats develop in juveniles is quite different from the way bird wings develop, and there is no question that they represent the same character. In any event, and whatever the outcome of this argument, it remains true that the uses of function for interpreting behaviour and past ecologies are valuable in their own right.

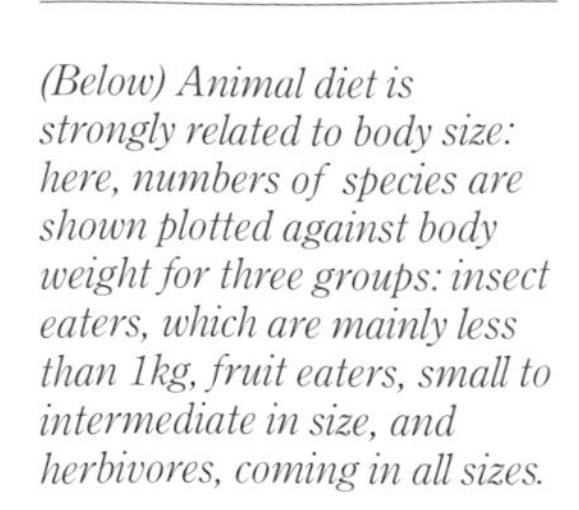

(Below) Animal diet is strongly related to body size: here, numbers of species are shown plotted against body weight for three groups: insect eaters, which are mainly less than 1kg, fruit eaters, small to intermediate in size, and herbivores, coming in all sizes.

Size Counts – Animal Diet v. *Body Size*

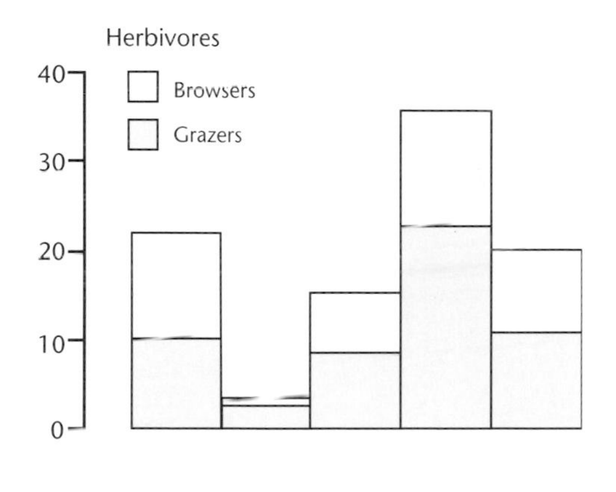

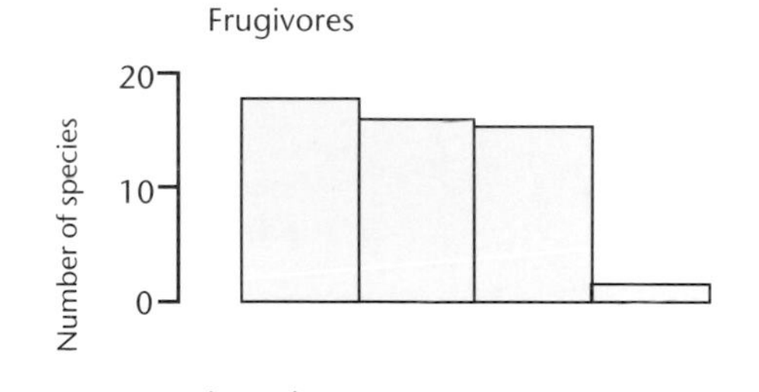

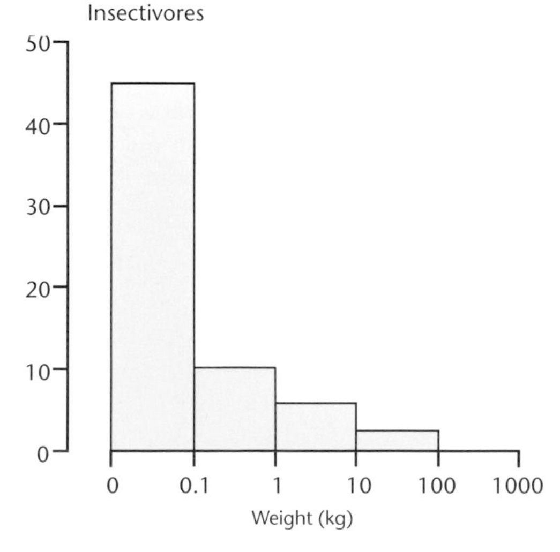

Excavation and Analytical Techniques

When Darwin wrote The Descent of Man, *there was virtually no fossil record to support his arguments. Thanks to excavations around the world during the last 120 years, there is now a relative wealth of material documenting human evolution.*

Scientists like us work with fossils, and to do this we first have to find them. This means finding fossil sites of the age we are interested in, excavating in a way that gets the kind of information we need to analyse the fossils, and then examining the fossils back in the Natural History Museum in London, where we both work, to obtain the information on functional morphology (the study of the structure of organisms) and phylogenetic analysis (study of their evolutionary history) that has been mentioned earlier. The methods we use for this multitude of purposes include both field and laboratory techniques.

In the field

The methods we use in the field are similar to those used in archaeology. Sometimes we know that a fossil site exists before we start work, and sometimes we just know that deposits of a certain age are present and we then have to explore the area to find out if there are fossils there. This can be a very speculative venture, for many areas that look really promising have no fossils at all. Most anthropologists have worked on an excavation at one time or another that produced nothing for all their efforts.

Having found an area or site that does look promising, the area is surveyed. In the old days this was done using old-fashioned surveying equipment but more recently satellite and laser technology are being employed. Maps and plans of the site are drawn and positions of potentially fossil-bearing areas marked on the plans. Then a decision has to be made about where to start digging. This can be quite difficult since one cannot see under the surface of the ground, but there are two ways of approaching the problem. One is to look for the

greatest concentration of fossils on the surface of the ground, as this may indicate that there are more where they came from under the surface. The other way, usually more reliable, is to dig a trench into the deposits, and this has the added benefit of providing additional information about the stratigraphy.

Excavations are usually marked out in metre squares. Specimens found during excavation are measured with respect to fixed datum points so that plans can be drawn showing where they came from. Excavation techniques vary, but usually a sharp pointed trowel is used for coarse work and dental picks for finer work, and where there is a danger that specimens may be damaged by excavation, delicate wooden implements may be used to clean around them. The key to finding specimens before they pop out of the ground (which is bad since by doing so their position and alignment in the ground is lost) is to keep the surface of the excavation clean by continual brushing with paint brushes.

As specimens are found, they may be drawn, photographed and measured before removal. This

(Right) After an area has been excavated, the sediment is collected and screened to find any fossils that may have been missed. Sediment may be dry-screened, but where water is available, as at this site in Turkey, it is more effective to wet-screen.

(Below) Excavation at a Miocene site in Turkey showing meticulous excavation by dental pick and paint brush. Three students are working in three separate metre squares, recording their finds as they go along.

Excavation is a slow and meticulous process involving many measurements and careful preparation and preservation of specimens, often while still in the ground. Here, Peter Andrews excavates at Maboko Island, Kenya.

is particularly important if the specimens are already broken in the ground so that when they are lifted they will fall into little pieces. Typical measurements record the size of the specimens, and their direction of alignment in the soil; the angle at which they are resting may also be measured, which is called their angle of dip. Also, of course, their position in the excavation is measured and marked on the site plan. When the specimens are removed, they are given a catalogue number and the measurements are recorded against that number in the site catalogue. The sediment that is removed in the course of excavation is screened – either dry or with water – to recover any specimens missed during the excavation. Usually that completes the field end of operations.

Fossils recovered from an excavation are measured, as here by the Finnish palaeontologist Mikael Fortelius, who is measuring the teeth of a rhino specimen from Paşalar in Turkey.

In the laboratory

After the fossils are shipped to the laboratory, they are cleaned and placed in boxes or tubes, depending on their sizes, ready for examination. There are various methods of examination of increasing detail and precision. The basic instruments are digital calipers for measuring the fossils and a binocular light microscope, which is a low-power (up to x40 to x80 magnification) instrument, for seeing most surface features of bones. For higher magnifications a scanning electron microscope (SEM) is used. Electrons have shorter wavelengths than light, and so high-precision images can be made at extremely high magnifications. The SEM is used to detect taphonomic modifications on the surface of bone, for example the effects of digestion or the marks left by trampling, and it is used also for identifying patterns of microwear on teeth, the fine scratches left by food as it was eaten by the animal during life. Some SEMs may be combined with a microprobe which can identify the mineral composition of the fossil and provide information about the diagenesis of the specimens, or in other words the physical and chemical changes that took place during and after deposition.

These techniques are all non-destructive, but more detailed analysis usually requires some measure of destruction of fossils. Trace element analysis provides information about the environment; isotope analysis of tooth enamel does likewise and also may indicate the kind of food eaten by the animal during life; and if there is any organic matter left in the fossils, extraction of DNA has the potential to add all kinds of information. A recent spectacular finding of DNA in a Neanderthal bone from Germany, revealing dramatic differences from living humans, shows how useful it can be (see also pp. 180–181). If enough DNA can be extracted, it has the potential to demonstrate relationships between individual animals, their sex and possibly even their diseases.

Even after this work there is still much to be done by way of preserving and conserving collections of fossils. We work in a museum where there are many millions of fossils found in countless excavations from previous years, and for these to be kept available for future generations of anthropologists to study, or more usually to compare with as yet unexcavated collections in the future, they have to be kept in secure and well-controlled storage.

Fossils are often examined using a scanning electron microscope (SEM). The SEM can be used to show details of the surface of the bone or tooth, such as the type of wear found on tooth surfaces, and it may also be used to see details of very small fossils such as rodent teeth.

New Techniques for Studying Fossils

Scientists investigating fossil remains have an ever-growing battery of techniques to help them study their finds in greater detail. They can now potentially directly date a fossil using a range of techniques (see pp. 30–33). These include radiocarbon dating, if the specimen is probably no older than about 40,000 years, uranium series dating by placing it in a gamma ray counter, or electron spin resonance (ESR) dating, by using a fragment of its tooth enamel. Computers have allowed the compilation and rapid analysis of large quantities of data obtained from fossils. The slow and methodical use of traditional metal measuring instruments, similar to engineering calipers, is now giving way to the rapid recording of data by electronic, sonic or laser light sensors, which can relate points on surfaces very precisely in three dimensions, and record them directly into computers for recall. The resultant network of points can essentially reconstruct the shape of the object, such as a fossil skull, and compare it on screen with others. Morphing techniques can be used to illustrate the amount of change in shape required to, say, evolve one specimen into another, or to grow a series of specimens through their life cycle.

Scanning fossils

Once the technique of X-raying or radiographing objects became widespread, hidden internal information about fossils could be studied for the first time – for example, the shape of sinus chambers within skulls, or the form of tooth roots within jawbones. Now, a powerful new X-ray technique has become available from medicine, called computerized tomography. The resultant images are called CT scans, and these can be presented on computer screens, printed, or even transformed into solid replicas by a technique called stereolithography. They are providing unprecedentedly detailed internal images of fossils, and the images can be manipulated to 'remove' rock that is still obscuring a fossil, or to reconstruct an incomplete fossil.

For example, in 1926, Neanderthal fossil remains were found in Gibraltar, at a site called Devil's Tower. They consisted of parts of the upper and lower jaws and braincase of a child. The teeth of the child matched those of a modern five-year-old in their state of development. However, the original assumption that they represented a single child of about five was challenged in 1982 by the suggestion that these bones might represent the remains of two

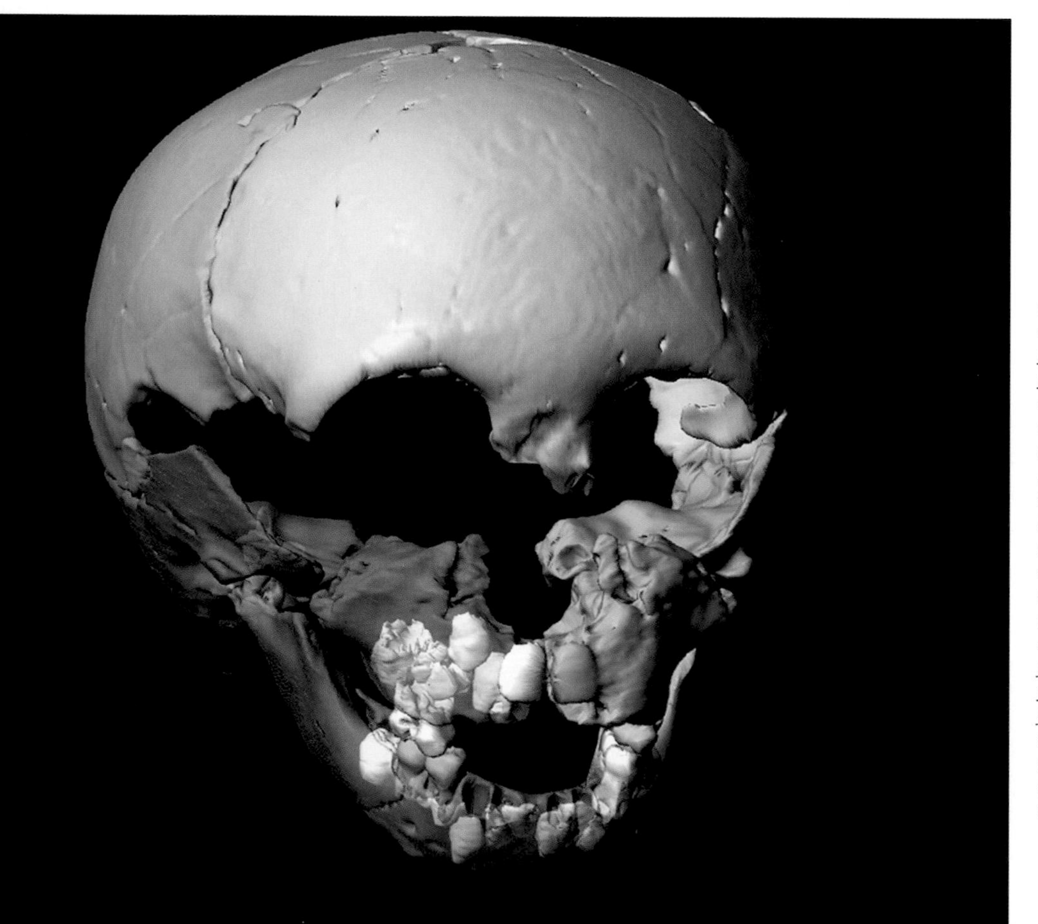

(Left) In 1926, five bones of a child's skull were excavated from deposits below the north face of the Rock of Gibraltar, near Devil's Tower. They were recognized as a rare fossil example of a young Neanderthal. Later research raised the possibility that the bones of two children of different maturity had been mixed up. However, detailed study of the bones and the teeth within the jaws suggested that they all came from a child aged about 4 years at death. This was further confirmed when computerized tomography (CT) was used to reconstruct the whole skull, showing that the bones did belong together.

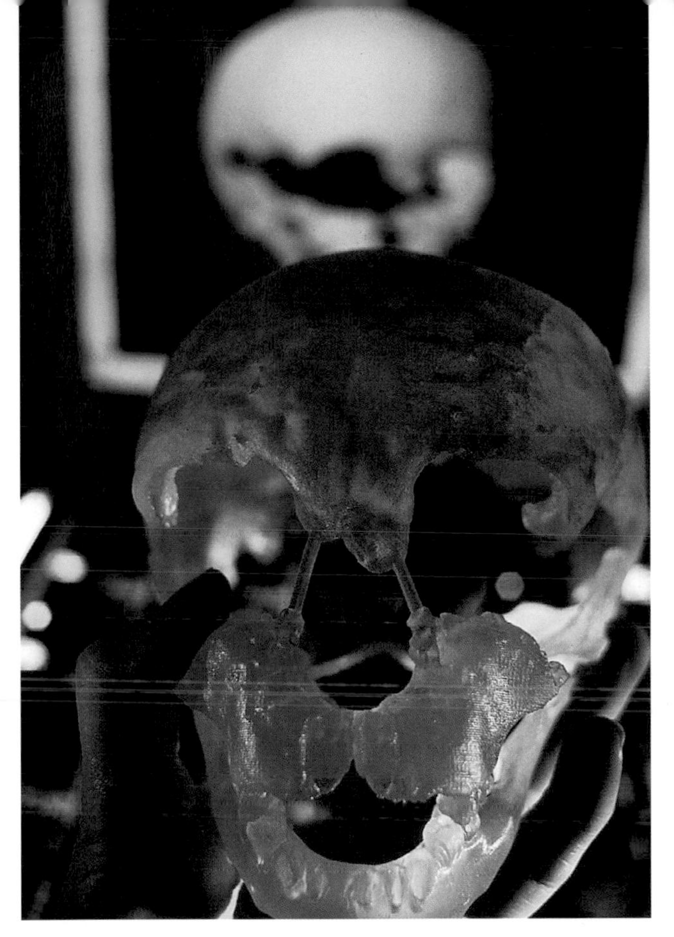

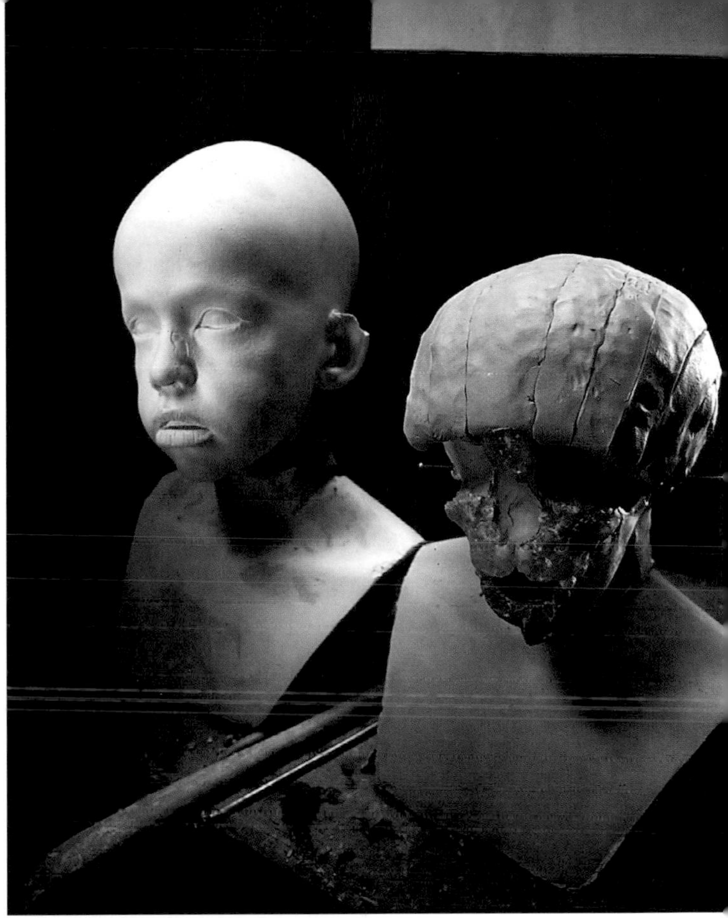

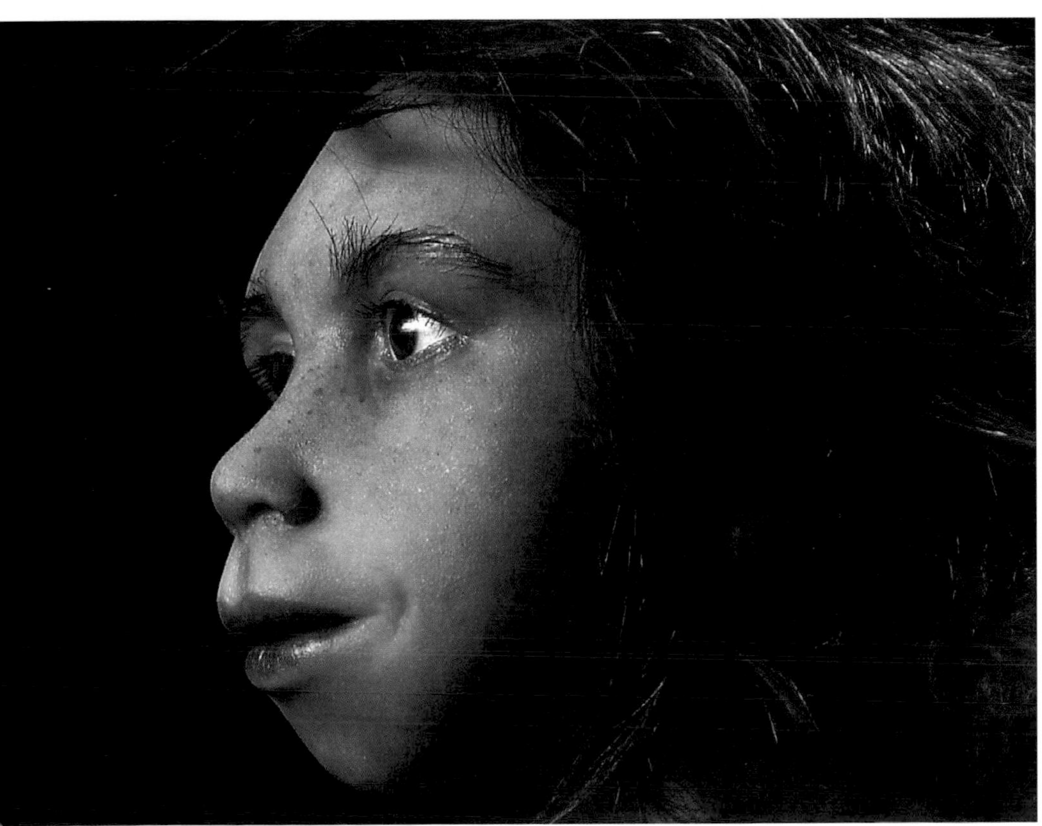

Computerized tomography can be used to create a plastic replica, called a stereolithograph, of the whole fossil or of parts buried within it, such as unerupted teeth or the bones of the inner ear. Here (above left), a replica has been created of the Devil's Tower Neanderthal skull, as a basis on which to reconstruct the child's features. Using modelling clay, muscle, fat and skin can then be built up over the skull (above right). For some features good data are available from comparative anatomy, but for others, such as the shape of the ears and lips, and the colour of the skin and eyes, educated guesses must be used. Unless a freak of preservation gives us a Neanderthal body in ice or a peat-bog, some of the details will remain conjectural. However, the addition (left) of hair, eyes and skin tones brings the finished reconstruction to life.

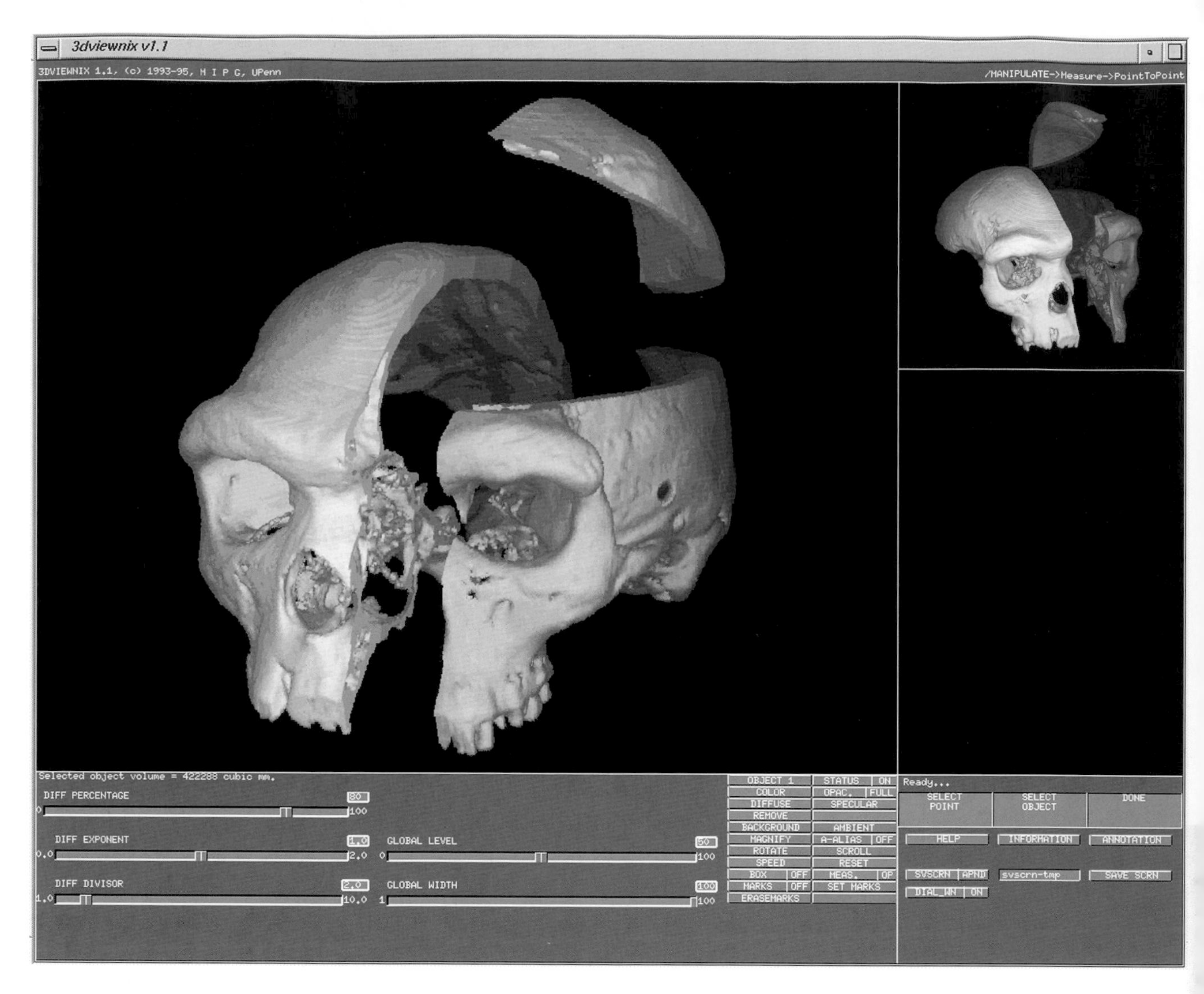

Traditionally, fossils have been measured with tapes and calipers, and brain size estimated by the use of seeds or ball bearings, or plaster replication. The advent of CT has meant that the size and shape of fossils can be recorded with unprecedented accuracy, inside and out. Here the internal anatomy of the Broken Hill skull from Zambia is revealed.

children – one aged about three years at death (represented by one bone) and the other about five (the rest of the bones). CT scanning was applied to the remains in 1995, and this revealed new anatomical data, producing a three-dimensional reconstruction of the whole skull for the first time.

The work confirmed that the bones do belong to a single child, since they all fitted well into the overall reconstruction. The child's brain cavity size and shape could be visualized and measured very accurately, and was an impressive 1,400 millilitres (ml), about the same as that of an average adult man of today. The scans revealed that the child had probably suffered a broken jaw which had healed, but which had caused a disruption of the normal tooth development. CT scans also showed that the still-hidden inner ear bones of this child – and of every other Neanderthal studied since – are distinct in shape from those of modern people, and from those imaged in the fossil skulls of our probable ancestors.

Some other special techniques have helped to show how old the child was at death. First, a precise resin replica was made of the front surface of the child's unerupted, but exposed, upper incisor. This was then examined using a scanning electron microscope (SEM), and the growth lines on the tooth were photographed and counted. Since these lines form at about eight-day intervals, it was possible to estimate how long the tooth had taken to form before its development was interrupted by the death of the child. This count showed that the child died at about four years of age.

For many other sites, SEM images have also allowed the detailed study of bone surfaces, showing cut marks made by stone tools; the estimation of age at death from sections of bones and teeth; the identification of ancient diseases from the traces left behind; and the diet of early hominins from wear patterns on their teeth, or plant particles lodged on their surfaces.

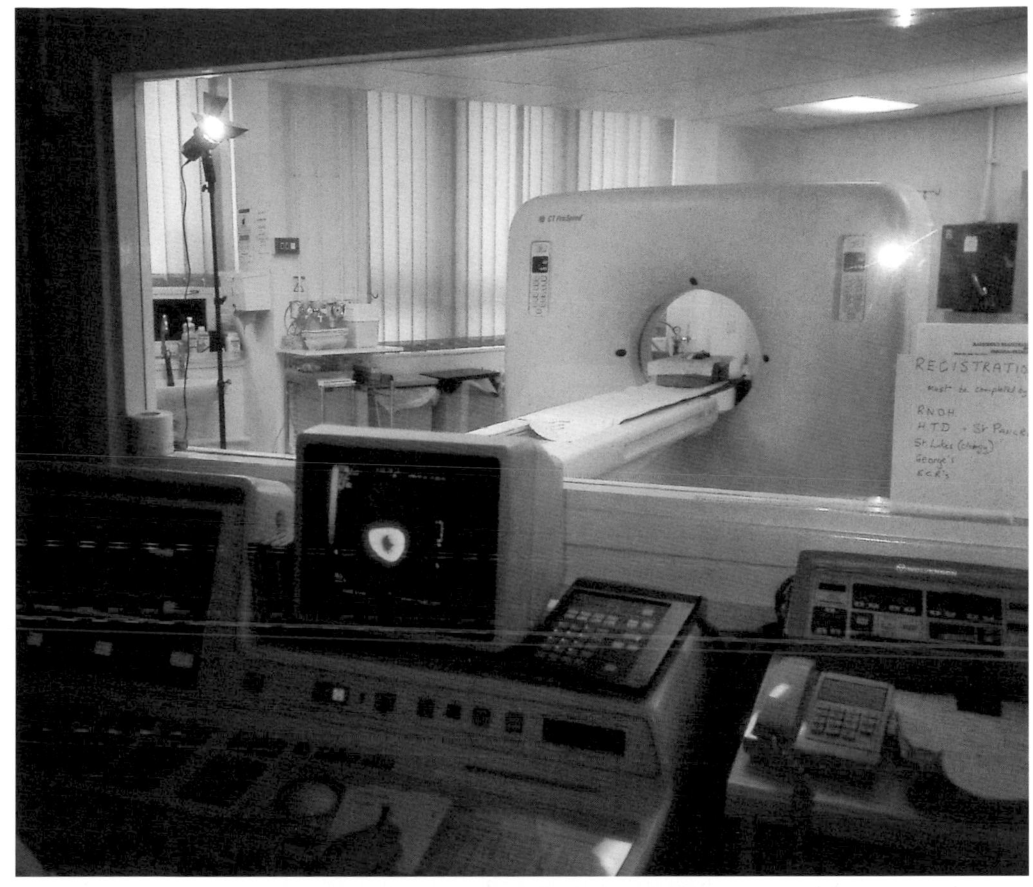

(Right) When new fossils are discovered, it is now routine to subject them to CT scanning as part of their anatomical study. Here the Boxgrove tibia, found in 1993, undergoes CT scanning at University College Hospital in London. The scanner can be seen through the glass partition, while a cross-sectional image of the bone, showing its great thickness, appears on the screen.

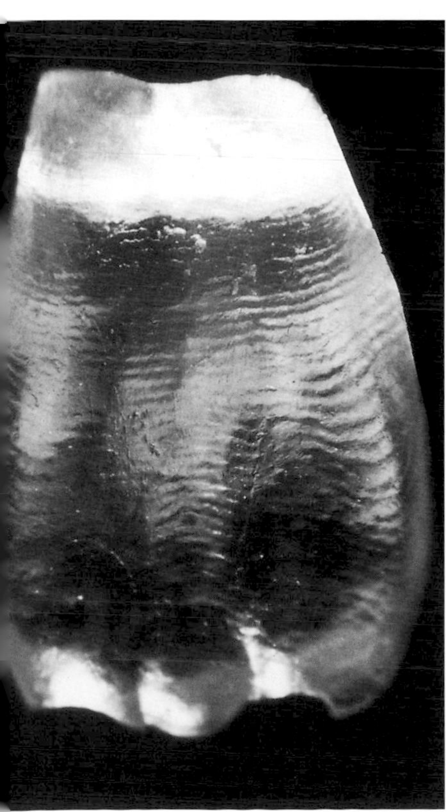

(Left) Microscopic techniques allow detailed study of the anatomy of fossils and help to reconstruct life histories and behavioural information in unprecedented detail. On the left is a replica of a fossil hominin incisor tooth, showing the surface lines known as perikymata. These correspond to underlying growth cycles lasting about eight days and so, like the rings of trees, counts of these can be used to estimate age at death and rates of growth. On the right is an image taken by a scanning electron microscope of the front surface of one of the incisors found at Boxgrove. The scratch marks revealed on the tooth are probably the result of stone tools being used to cut through plant or animal tissues held between the front teeth, and occasionally marking the tooth beneath. Studies of the directionality of cuts on fossil teeth indicate that prehistoric people, like us, were predominantly right-handed.

Taphonomy: How Fossils are Preserved

Taphonomy is the study of how bones or other animal or plant remains become preserved as fossils. The term comes from the Greek *taphos* meaning burial and *nomos* meaning law. There is a long path from living organisms to fossils. At each stage of the path there are many processes that can act either to destroy or to add information associated with the remains, and knowledge of these processes and the effects they produce is essential to the interpretation of fossil sites.

The sequence of taphonomic change illustrated runs from living animals at the top to fossil bones on a museum shelf at the bottom. It is important to remember that bones were once parts of living animals, and many of the properties of the bones stem from their location in the animals' bodies, their age at death, their state of health and even the way the animal died. Poor nutrition, disease and old age can all reduce the chances of bones surviving

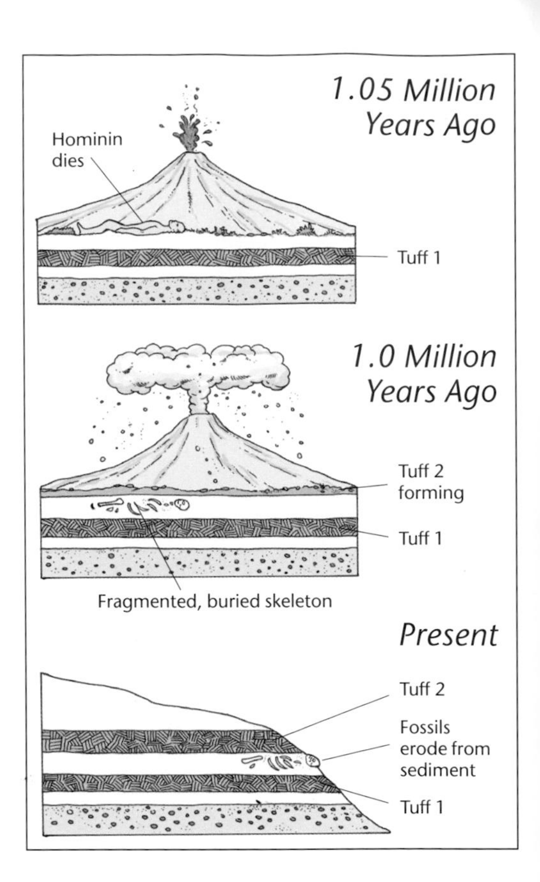

(Above) The process by which a hominin fossil came to be buried in sediment and preserved below an air fall tuff from an active volcano. The bottom diagram shows the remains being exposed by present-day erosion.

(Below) On the left is the sequence of events connecting a living community of animals and their recovery as fossils, ending up on museum shelves, while on the right are the various types of taphonomic modification that may cause changes in faunal composition. The sum of these modifications make up the taphonomic profile of a site.

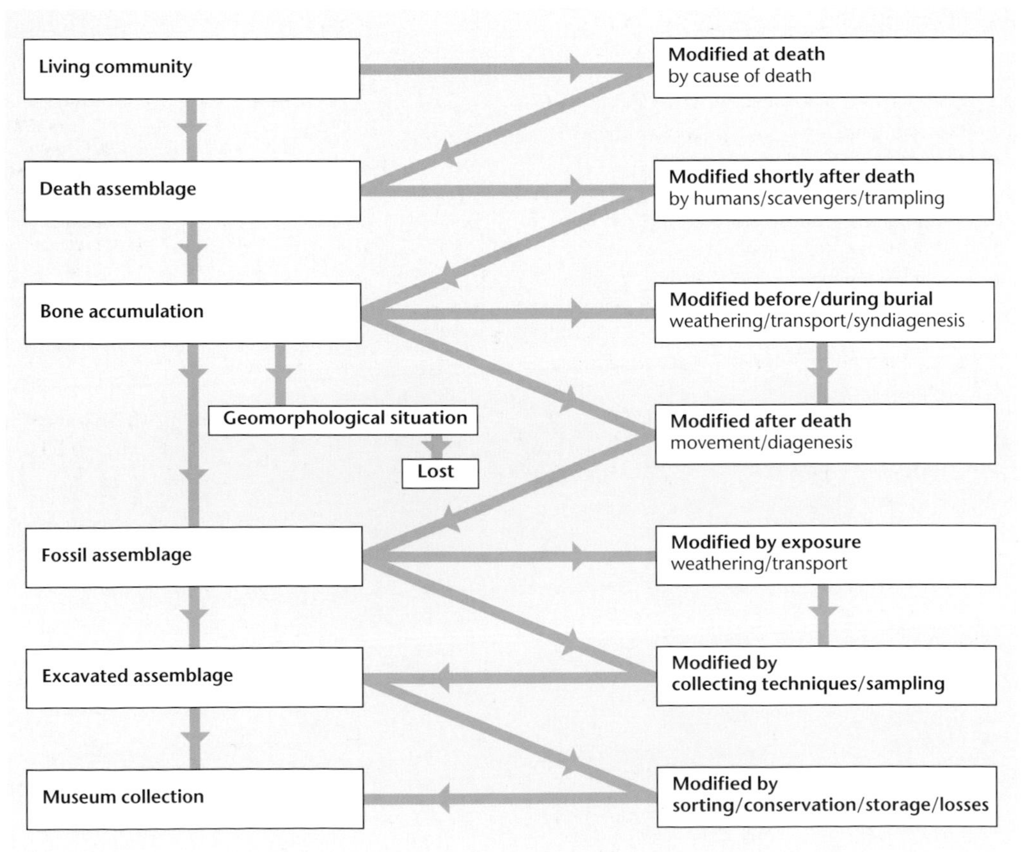

(Opposite above) Nine human postcranial bones that have been split and chewed by spotted hyaenas, producing damage typical of this scavenger. The bones are from the modern Kajiado cemetery in Kenya which the hyaenas habitually raided.

taphonomic pressures to become fossils. So also can predation, if an animal is killed and eaten: the bones may become broken by bone-crushing carnivores like hyenas, they may be eaten and digested by predators that swallow bones whole like crocodiles, snakes, or, in the case of small mammal predation, owls. Many bones are destroyed by these processes, but conversely those bones that survive with only partial damage carry the marks of the damage into the fossil record. Bone breakage by hyaenas is readily distinguishable from bone breakage by lions, for example, and so if a bone from a lion kill is preserved as a fossil it will carry with it not just the information as to the species of animal it came from, but how it died and what ate it.

One source of confusion is that predators rarely finish a meal completely, and they may leave some remains for scavengers to eat. It is difficult, often impossible, to distinguish the effects of primary predation from secondary scavenging, but this is an important distinction when the behaviour of carnivores is being considered, for scavengers simply take what is lying around on the ground whereas predators select their prey on the basis of a whole set of hunting criteria. Prey assemblages of predators are, therefore, predictable and testable, while scavenger assemblages are not. There is some indication that predators take the most nutritious parts of the bodies of their prey, leaving the left-overs for scavengers, so that the two assemblages should consist of different bone types, but in practice this does not always hold true.

Weathering of bones

Other processes start to work on bones even before they have had all the meat taken off them. They may be broken by trampling, transported by wind or water, and early stages of weathering by exposure to sun, rain and wind may further weaken them. These processes accelerate as bones are completely exposed to the elements from their protecting skin and cartilage, and in general the greater the exposure, the greater the damage, but

(Below) Cut mark mimics on a bovid limb bone. The bone comes from an animal that died at Draycott in Somerset and was resting on a rocky substrate that was a pathway for other animals. The effect of trampling caused the bone to rotate about its long axis, and where it rubbed against stones, numerous parallel scratch marks were produced perpendicular to the long axis of the bone.

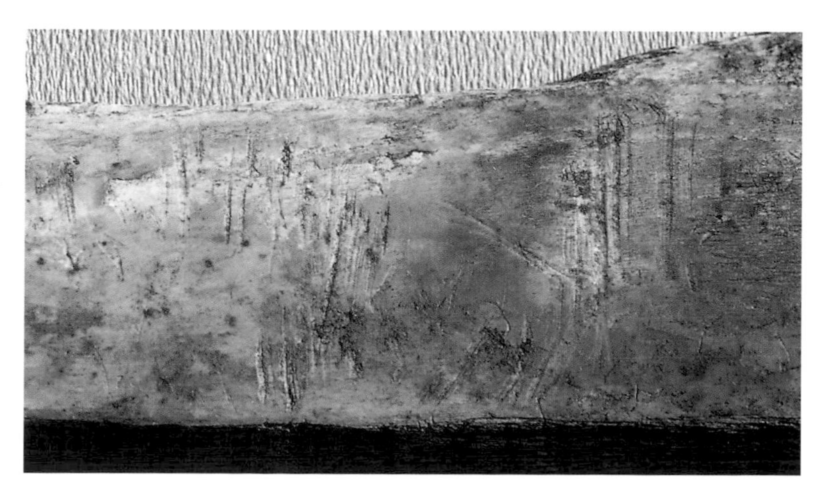

(Left) SEM micrograph of one of the cut mark mimics from Draycott showing multi-faceted striations with bone raised up on one shoulder, both typical of human-made cut marks.

(Below) Bone scatter commonly found on open sites in East Africa today. The site is near Lainyamok in the Kenya Rift Valley.

there are many limiting factors to be considered. Weathering depends on the type of climate, and it is more rapid in its effects in the tropics than in temperate climates. Also, any protection afforded by vegetation or physical barriers greatly affects the rate and degree of weathering. Bones may also be protected from trampling by physical barriers and vegetation, and the location of the bones with respect to normal routes of travel by potential tramplers is also of significance.

It is probable that damage from earlier processes increases the liability of bones being damaged by later processes, such as weathering. A bone digested or broken by predators is probably more susceptible to weathering than is fresh bone. It has been demonstrated that weathered bones become abraded much more quickly and to greater degrees than fresh bones, and it is likely that this multiple action is the general rule, although such sequential events have been little studied by taphonomists.

Weathering continues even after burial in soil, for soils are biologically active environments. The effects of subsurface weathering, however, are quite different from surface weathering and are readily identifiable, so that it is often possible to tell how quickly a fossil bone was buried prior to fossilization. Later on, a whole new set of physical and

chemical processes comes into play, depending on the nature of the sediments in which the bones are buried, for example their acidity, mineral composition and degree of protection from the atmosphere. It is at this stage that fossilization actually occurs, with the replacement of the organic content of the bone (only about 10 per cent by volume) with whatever minerals are present in the water flowing through the sediment. The mineral content may also be leached out, and the hydroxyapatite that makes up a proportion of the bone replaced by the ground-water minerals, most commonly calcium carbonate. This process is usually referred to as diagenesis.

Some fossil sites have been investigated in some detail and the processes leading to the accumulation of fossils identified. One example is shown here from an excavation we undertook from 1976 to 1984 in a cave at Westbury-sub-Mendip in Somerset.

After fossilization, the fossils are still subject to taphonomic processes, particularly weathering, as the fossil-bearing sediments erode and expose the fossils at the surface of the ground again. Further damage often ensues during collection, such as excavation or screening damage, and it even continues on to museum shelves if temperature and humidity are not strictly controlled. All in all it is sometimes surprising that any fossils survive at all!

The taphonomic profile of the cave sequence at Westbury-sub-Mendip, Somerset, by which the history of the site could be reconstructed.

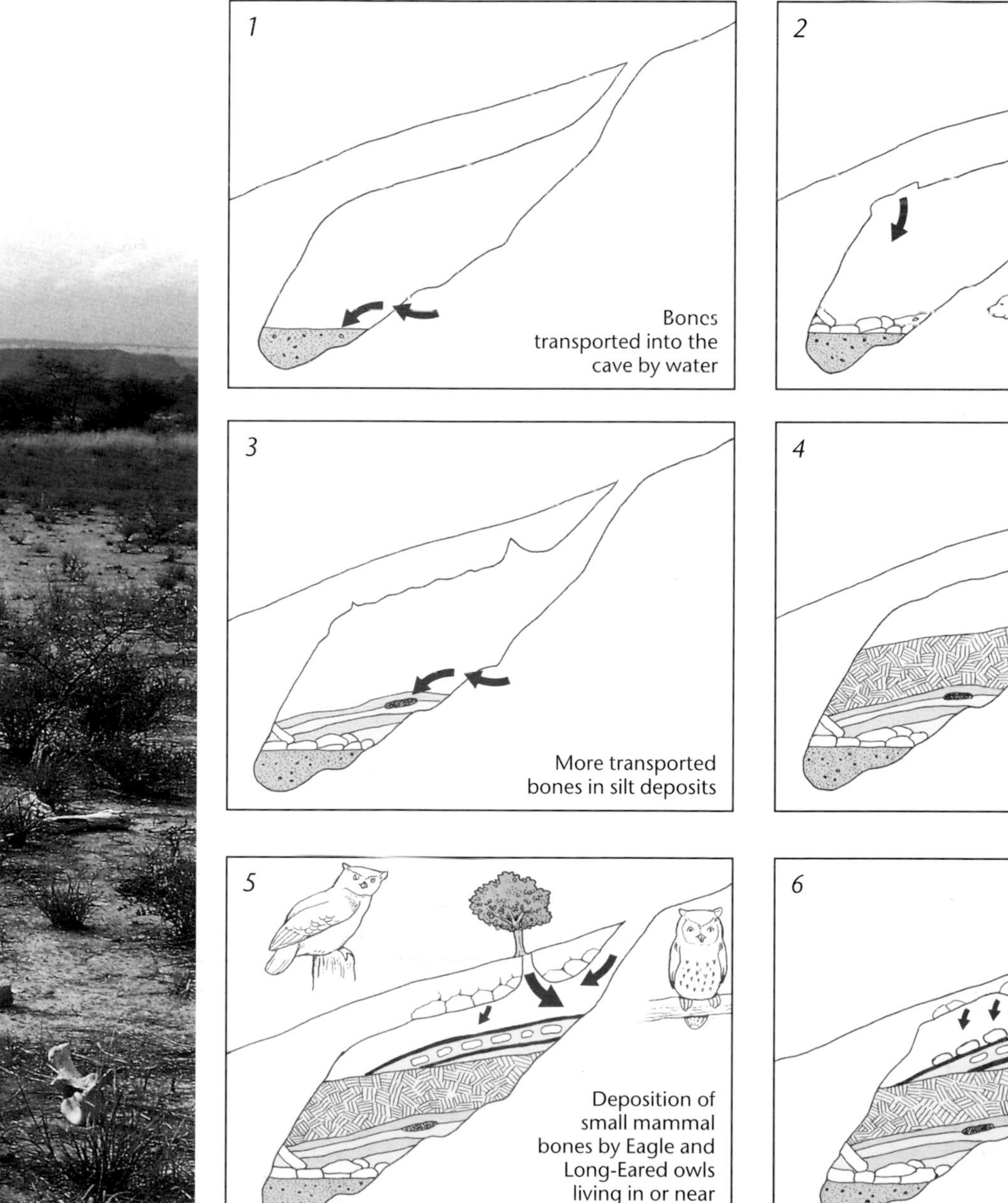

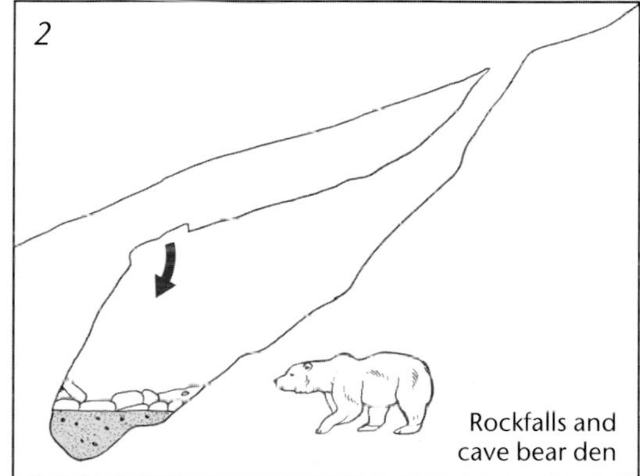

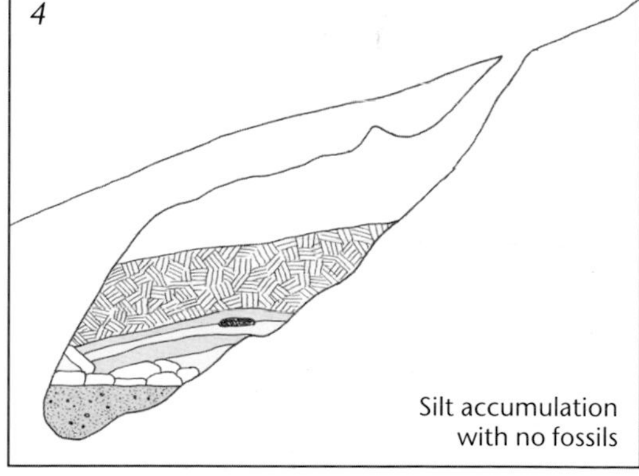

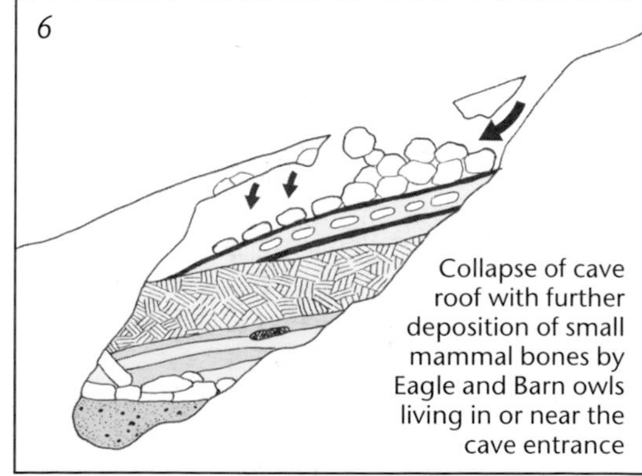

What Fossils Tell us about Ancient Environments

The study of fossil human and ape ancestors can tell us where and when the events making up human evolution took place, but it cannot tell us why things happened. For this we need to know the context in which the evolutionary changes occurred, which means the environmental conditions to which the fossil humans were having to adapt. Only then can we begin to answer questions about the place of humans in the world, both past and present, and their position in world ecology.

Palaeoecology: the study of past interactions with the environment

Ecology is the study of the interactions between animal and plant communities and their environment. By the same definition, palaeoecology is the study of past interactions of communities with their palaeoenvironment. Animal communities interact with the environment, both being modified by it and to a lesser extent modifying it, and the animal and plant communities in any given environment form an ecosystem. Ecosystems are dynamic, with energy flow passing from the sun, fuelling plants through photosynthesis, providing food for animals all the way down to micro-organisms in the soil. The study of these interactions is highly complex. In studying the palaeoecology of fossil humans and fossil apes, the aim is to try and understand something of their position both within the community and in response to their environment. While ecologists can experiment directly with extant ecologies, however, palaeoecologists have to attempt to reconstruct past ecologies on the basis of the fragmentary fossil record.

All methods of palaeoecological reconstruction depend on comparisons of past with present. This is not to say that past ecologies are assumed to be the same as present ones, but the same principles

Frequencies of pollen may provide insights into the vegetation present at a fossil site. These pollen spectra are from the Shungura Formation near the Omo River in Ethiopia compared with modern pollen. The time scale is shown on the left. When drought-tolerant species are abundant (left), moisture-loving trees are rare (right), telling us there must have been a drought in the area around 2 million years ago.

Ancient Pollen – a Record of Past Environments

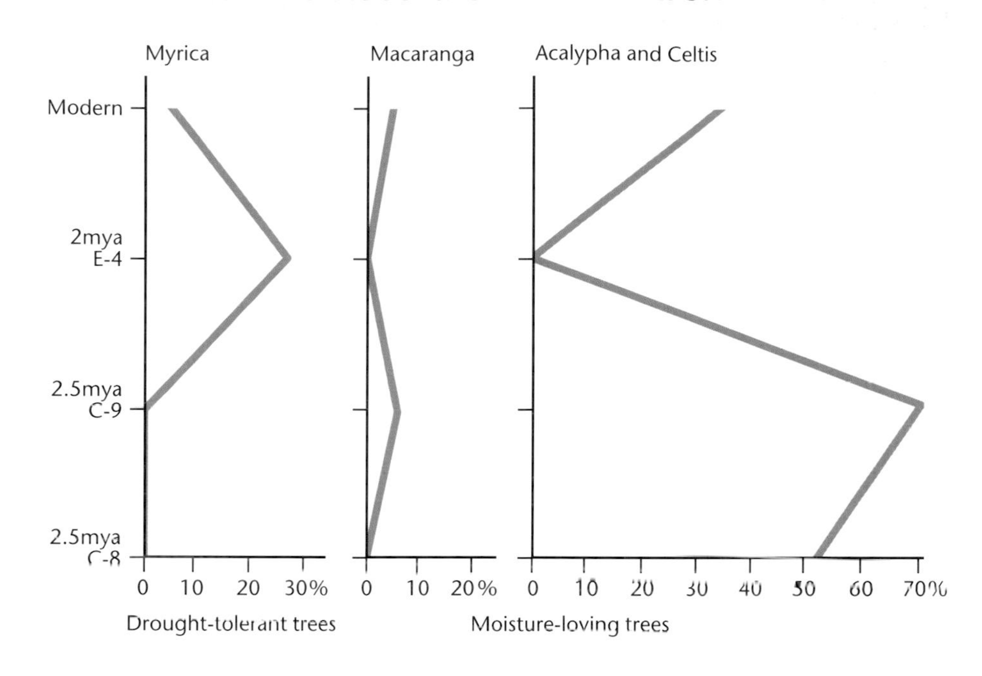

are assumed to apply in the past, so that by understanding these principles and relating them to past evidence it is sometimes possible to build up a picture of past ecology. Fundamental to this is the concept of the niche, which is basically the position of organisms within an ecosystem, what they take from it and what they put back into it; since this cannot be studied directly for fossil animals, it is necessary to find alternative approaches.

Sometimes there is direct evidence for vegetation through the presence of fossil plant remains or fossil pollen. When this is the case, many of the plants can be identified as to species, although in many cases the form of the plant is still uncertain, whether tree, bush or herb, and even more uncertain is the structure of the vegetation association. Evidence may also come from the study of carbon and oxygen isotope differences, since these also differ in different types of vegetation, but by far the most common method of palaeoecological analysis for hominin sites is looking at the fossil mammals.

(Below) Some primates have remained in the forest, and these spider monkeys from South America spend all their lives in the trees.

(Opposite left) Mixed habitat in tropical Africa with grasslands subject to flooding around a small lake with dense woodland in the background. Such mixed habitats are common in Africa and are difficult to interpret palaeoecologically because both woodland and grassland animals occur, as well as aquatic animals.

(Opposite) Primates evolved as arboreal species, but a few species like the baboon shown here have adapted to life on the ground, so that the presence of primates in a fauna does not necessarily indicate the presence of forest or woodland.

Identifying environments from fossil mammals

The oldest and most traditional of these approaches has been to identify fossil animals and equate their past ecology with those of their nearest living relatives. Undoubtedly this sometimes works, for many groups of mammals are ecologically conservative, and to a certain extent this is true of the apes. The problem with this method, however, is that it is not testable because there is no kind of evidence that can be gathered by which to test it. More recently a number of methods have been developed that are based on function, and these enable the behaviour of animals and their adaptations to particular kinds of niche to be reconstructed. These methods become very powerful when the niche adaptations of all the animals in a single fauna are combined in an analysis of the community structure of the fauna.

Jaws and teeth are primarily adapted for processing food, and they come in different sizes and shapes depending on the type of food most commonly eaten. Some of these have been described in the earlier section on function (see pp. 36–37), but the point here is that these different morphologies can be identified in fossil animals, and so widely applicable is the relationship between food and jaw types that it is reasonable to infer diet in the fossils. Vegetarian mammals that eat leaves can be distinguished from species that eat grass on the basis of the height of the teeth, the breadth of

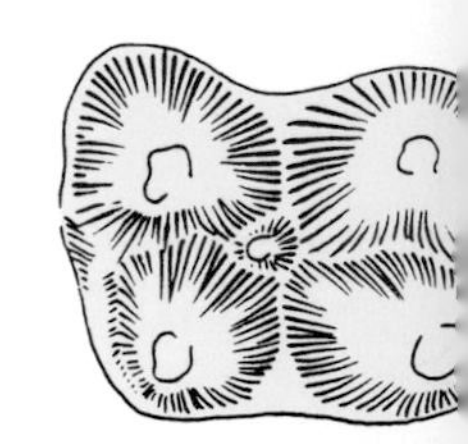

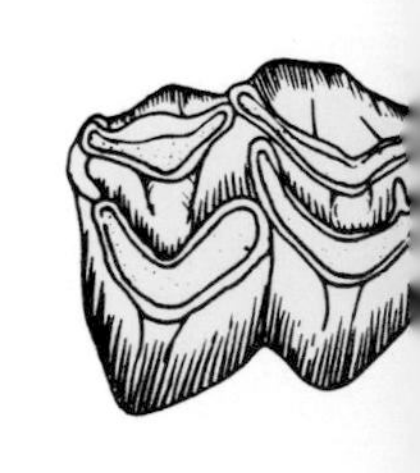

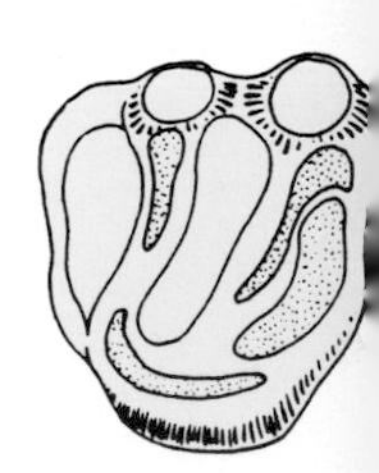

Molar crown types in mammals showing the variation in form. Along the top are low crowned teeth with separated cusps suitable for crushing food objects. The crowns increase in complexity from left to right, and also indicate a change from crushing softer objects on the left to harder objects on the right. The middle row has teeth adapted for slicing through objects such as grass and leaves, with longitudinal and diagonal ridges acting like scissors. The two teeth along the bottom row have strongly ridged crowns with transverse ridges, another way of solving the problem of slicing up tough food items.

Miocene Savannah Habitats and Fauna

Mammal diversity in present-day faunas in East Africa compared with past faunas from the Miocene of North America, showing the variation in form present in both. North America today has lost many of its large mammals, but in the past its fauna had many similarities with those of Africa. The habitat types and associated fauna are: forest environments – small, selective browsers (I); woodland – small to medium-sized browsers/ mixed feeders, territorial (II&III); savannah – medium to large-sized browsers and grazers, and mixed feeders, some herd-forming and territoriality (IV&V); and grassland medium to large-sized grazers, mixed feeders and high-level browsers, herd-forming (VI). Although the fossil species are all different from modern species, there are similarities in their form that show they were adapted for comparable lifestyles and occupied comparable niches to those of living animals.

The species illustrated are: 1 water chevrotain; 2 duiker; 3 gerunuk; 4 bushbuck; 5 greater kudu; 6 impala; 7 black rhino; 8 giraffe; 9 eland; 10 wildebeest; 11 Grant's gazelle; 12 Burchell's zebra; 13 Parablastomeryx; *14* Barbouromeryx; *15* Lambdoceras; *16* Diceratherium; *17* Merycoidodon; *18* Hypohippus; *19* Anchitherium; *20* Moropus; *21* Archeohippus; *22* Merychyus; *23* Parahippus; *24* Oxydactylus; *25* Protolabis; *26* Pseudoceras; *27* Yumaceras; *28* Tapirus; *29* Synthetoceras; *30* Aphelops; *31* Merycodus; *32* Hemiauchenia; *33* Calippus; *34* Aepycamelus; *35* Neohipparion; *36* Pliohippus; *37* Astrohippus.

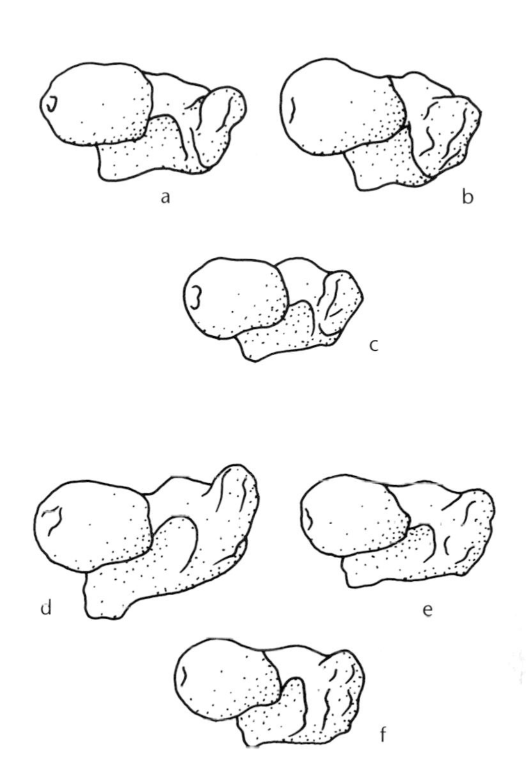

(Above) The head of the femur is a good indicator of type of locomotion of bovids. The cylindrical head of 'a' is from a running antelope, living in a savannah habitat, where movement of the legs is limited to back and forwards; the more rounded head of 'b' is from a leaping species living in closed woodland or forest. The femur head in 'c' is intermediate, indicating intermediate broken cover. The bottom three specimens are fossil femora showing similar adaptations: 'e' and 'f' are similar to 'b', and 'd' is similar to 'c'.

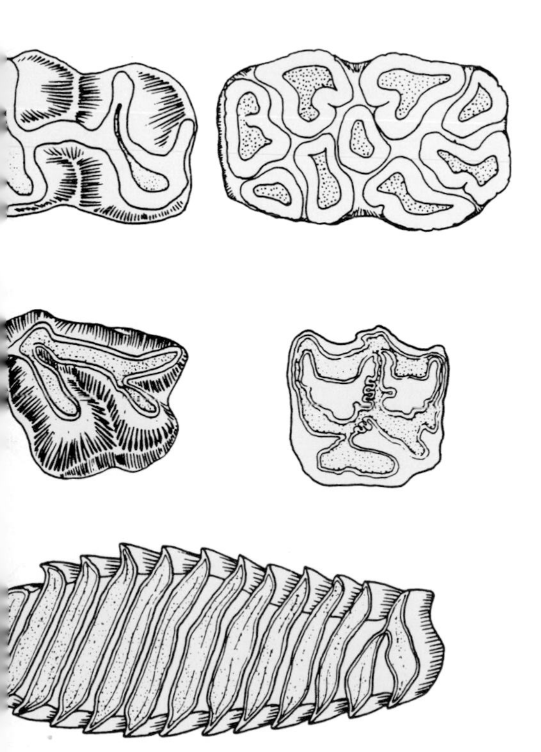

the front of the jaws and from wear on their teeth; fruit-eating species can be distinguished from both by their low crowned teeth with flat surfaces and their large front teeth (incisors); and insect-eating species are distinct because of the high pointed cusps on their teeth and their lightly built jaws. The numbers of different species present in any one place eating these different kinds of food should be the same as the distribution of these food types, so that, for example, if there are many fruit- and leaf-eating animals in a fauna, it may be inferred that the place where the fossil was found was one that had many trees. When this information is combined with other environmental factors, such as latitude and altitude, it may be possible to predict that a tropical forest ecosystem was present.

Adaptations of the limb bones can be analysed in a similar way. Running, leaping and climbing adaptations of the limbs are all different from each other, and even more extreme are adaptations for digging, swimming and flying. Body size has to be taken into account here, for these adaptations do differ between very small and very large species. To elaborate on the example above, we said that many fruit-eating and leaf-eating animals indicated the presence of lots of trees; if there are also many animals in the fauna that had limbs adapted for tree-climbing, together with many small ground-living and flying species, the prediction of a tropical forest ecosystem being present is greatly enhanced.

Changing Climates

Britain is directly influenced by the changing circulation patterns of the adjoining north Atlantic, and this has produced great contrasts in the climate, environment, animals and plants during the Pleistocene. Below is a reconstruction of Three Cliffs Bay in Gower, South Wales, as it may have been about 120,000 years ago. The presence of hippopotamus, hyaena, rhinoceros and elephant make it look like an African scene, but the climate was little warmer than the present day.

The distance of the Earth from the sun is such that our planet has been able to support the evolution and maintenance of life. A major part of this support comes from the stability of climate which allows the presence of liquid water on the surface. However, the Earth's orbit, and its precise orientation in space, are not fixed, and hence the amount of solar radiation it receives is not constant. Also, the circulation of atmosphere or water around the planet may alter because of geological changes such as the position of the continents, or the presence of mountain ranges. At times, our planet becomes colder, and this leads to an accumulation of ice at the expense of liquid water. Thus the Earth experiences a so-called Ice Age.

There have been many Ice Ages in the distant past – for example, the Huronian Ice Age over 2,000 million years ago, and the Ordovician Ice Age over 400 million years ago. The present Ice Age, in which we are still living, began at least 2 million years ago. At the beginning of the last century, based on evidence from the advance and retreat of glaciers in the Alps, it was believed that there had been four major ice advances in recent Earth history, and these became the basis of dating the archaeological and fossil human records from Europe. However, we now know that the sequence of climatic changes and ice growth was far more complex than this.

The Milankovitch Cycles

The main driving force of our present climate was elaborated by a scientist called Milutin Milankovitch, who calculated that there were three main factors behind the ebb and flow of the Earth's ice caps. These concerned fluctuations in the shape of the Earth's orbit (from more circular to more oval), in the tilt of the Earth's axis of rotation, and in the time of the year when the Earth is closest to the Sun. These three factors run through cycles about every 95,000, 42,000 and 21,000 years. When all three signal in the same direction, the Earth's climate swings to an extreme of either glacial (cold) or interglacial (warm) conditions, but most of the time it lies between the extremes. During the last 700,000 years, the longest (95,000-year) cycle has

(Right) The same area in Gower, South Wales is shown as it might have been 100,000 years later, about 20,000 years ago. Now the edge of a massive ice sheet lay only a few miles away, and the sea level had fallen dramatically because of the water locked up in huge ice sheets globally. The cold landscape is now inhabited by lemmings, reindeer and snowy owls.

been dominant, producing a major glaciation roughly every 100,000 years. We are fortunate to be living during a brief interglacial stage, but such stages are the exception rather than the rule in recent Earth history, and about 90 per cent of the

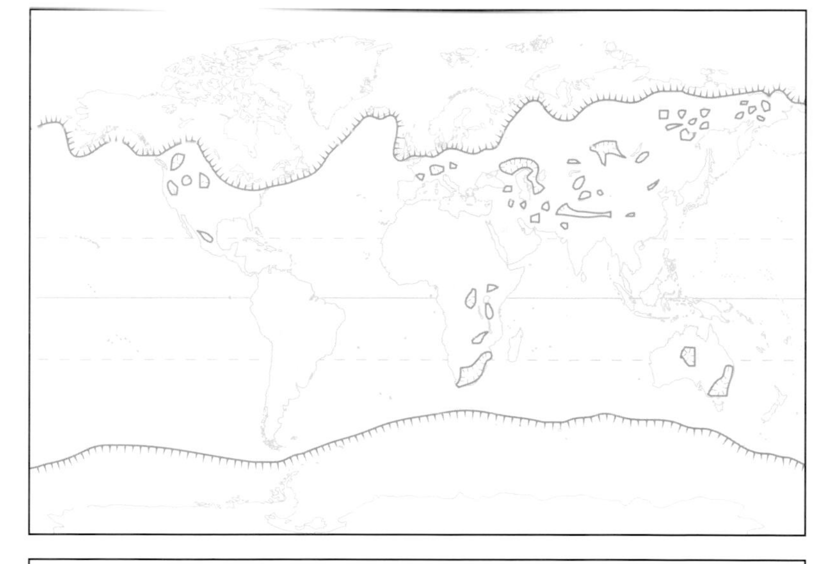

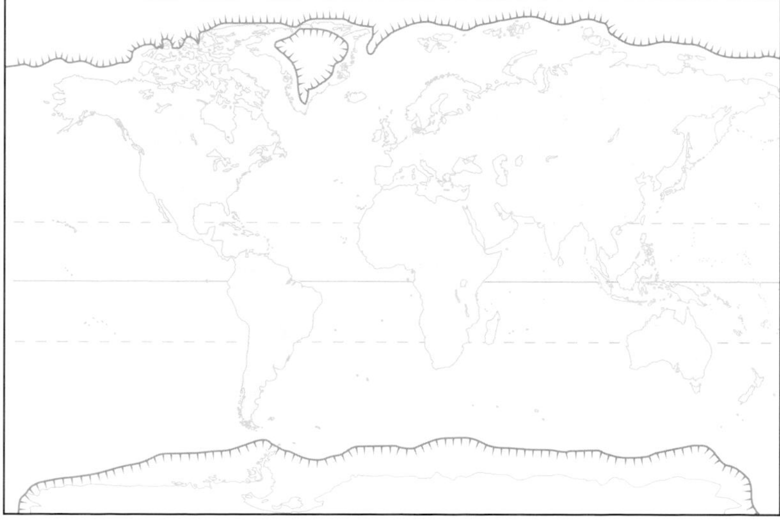

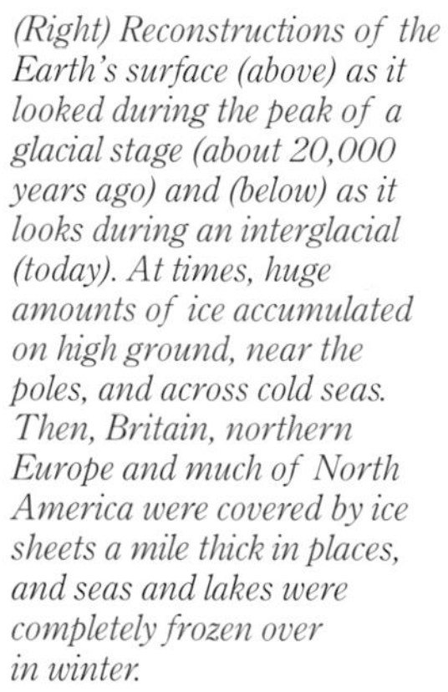

(Right) Reconstructions of the Earth's surface (above) as it looked during the peak of a glacial stage (about 20,000 years ago) and (below) as it looks during an interglacial (today). At times, huge amounts of ice accumulated on high ground, near the poles, and across cold seas. Then, Britain, northern Europe and much of North America were covered by ice sheets a mile thick in places, and seas and lakes were completely frozen over in winter.

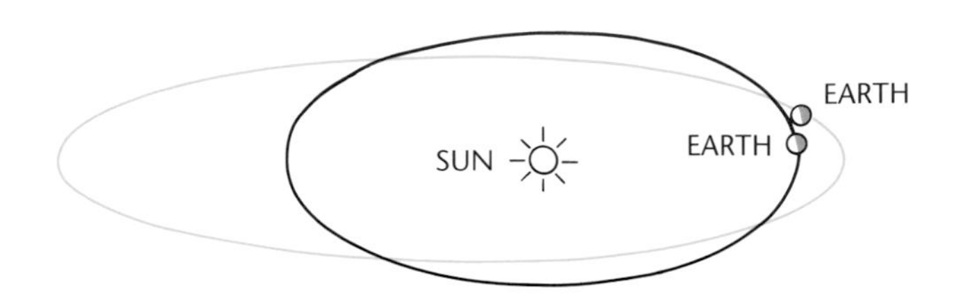

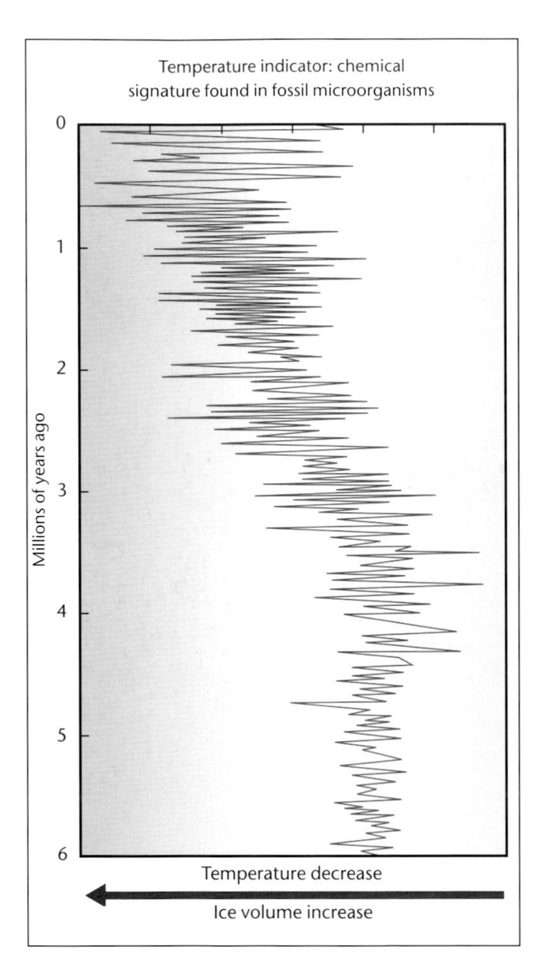

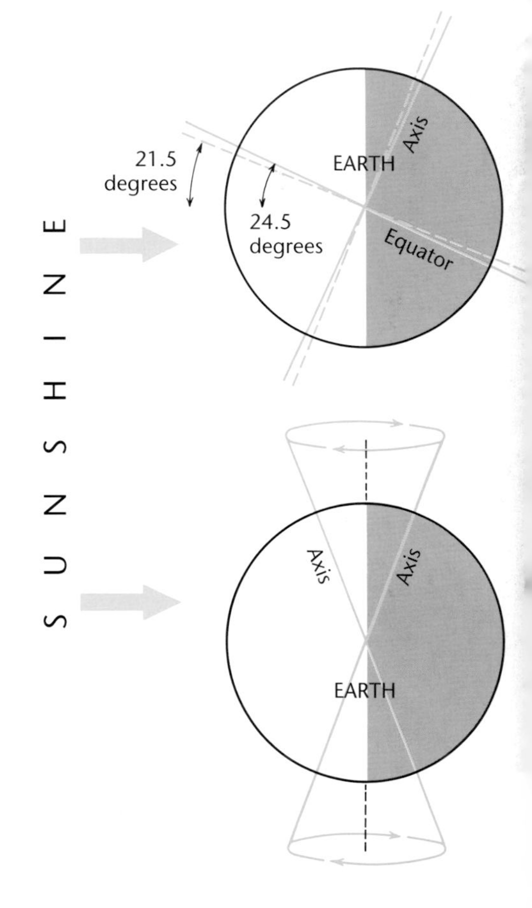

(Top and far right) Some 70 years ago, Milutin Milankovitch described the three main factors that affect how much sunshine reaches different parts of the Earth, and hence also determine our climate. First (top), the Earth's orbit does not describe a perfect circle around the Sun, but oscillates between more and less elliptical states over a cycle of about 95,000 years. Secondly (centre right), the tilt of the Earth's axis fluctuates between 21.5 and 24.5 degrees roughly every 42,000 years. And thirdly (below right), the Earth 'wobbles' as it spins, in the manner of a revolving gyroscope. Each 'wobble' lasts about 21,000 years.

(Near right) Long and nearly continuous sequences of deposits on the floors of the oceans can be used to reconstruct past climatic changes, and in particular the size of the ice caps. This is most commonly done through chemical signatures preserved in the shells of fossilized microorganisms. This deep sea core record shows the major fluctuations of the last 6 million years, and as well as illustrating the reality of Milankovitch Cycles, it illustrates both the overall decline of global temperatures and the growing influence of ice caps through this time.

last 500,000 years has been colder than the present day.

When the ice caps grew, they not only affected the seas or land immediately around them. Sea levels dropped globally by up to 100 m (330 ft) because of the amount of water locked up in the ice caps. Such changes meant that Britain was regularly joined to France, Sicily to Italy, New Guinea to Australia and Asia to Alaska. And the amount of water circulating in the atmosphere also declined at such times, which meant that every glaciation in high latitudes was generally accompanied by the spread of deserts nearer the tropics.

Western Europe, and particularly the British Isles, display some of the most extreme signs of climatic change. This is because the presence or absence of the Gulf Stream, carrying subtropical waters from the western mid-Atlantic towards Europe, is affected by glaciation in the North Atlantic. For the last 10,000 years the Polar Front, which marks the extent of true glacial waters, has lain far to the northwest of Britain. Some 120,000 years ago, the Gulf Stream flowed around Britain much as it does today – in fact it may have been slightly warmer than today. However, 20,000 years ago, at the peak of the last Ice Age, the Polar Front then lay across the Atlantic at the level of the Iberian peninsula. Half of North America was covered by a thick ice sheet and northwest Europe was locked in arctic waters, laden with icebergs. Then, polar bears may have swum in the Thames. The rapidity of climatic change is now known to be even more remarkable than that. Rather than taking 10,000 years to go from glacial to interglacial, it

appears from the records of ice cores, layers preserved on the ocean floor, and cores from lake beds, that some changes from one climatic mode to another could have occurred in only about 10 years. The Milankovitch Cycles have not stopped, and in combination with the effects of global warming our present stable interglacial climate could well disappear in the length of a single human lifetime!

(Above) Glaciers are slow-moving rivers of ice, carving out characteristic valleys as they flow downwards. It was comparisons of the shape of glacier-cut valleys in the Alps with those in regions such as the Lake District of England that showed there must have been major 'Ice Ages' in Earth history.

(Left) Climate history is recorded in rocks, lake sediments and ice cores, as well as on the ocean floors. Here an ice core is being extracted, and such cores may record annual falls of snow over hundreds of thousands of years. Trapped bubbles of gas also record the composition of the atmosphere at the time of deposition.

Site I: Rusinga Island

In order to explain more about how palaeoanthropologists work, we are going to describe a number of fossil sites and how they are excavated. The fossils collected from them, particularly the apes, how they are measured and the way the results are analysed will also be described. First of all we shall consider some Miocene sites spanning the time from about 20 to 10 million years ago, and after that some Pleistocene sites from the last 2 million years.

Rusinga Island lies off the Kenyan shore of Lake Victoria and is made up almost entirely of Miocene fossiliferous sediments. It was one of the sites that made the reputations of Louis and Mary Leakey, who collected there for over 30 years beginning in 1931, but many other people have also worked there. Louis Leakey first mapped and named all the major collecting localities on the island, and it was also the first place that one of us (Peter Andrews) excavated when still a graduate student at Cambridge University, when on an expedition there in 1971 with two other Cambridge students, John and Judy Van Couvering.

The sediments on Rusinga are mostly early Miocene in age, with the main fossil beds dated radiometrically by potassium-argon dating to 17.8 million years ago. There are also some earlier deposits nearly 20 million years old, and some lake

(Above) The alignment of a fossil being measured by Glenn Conroy at an excavation at Kaswanga Point, Rusinga Island.

(Right) Geologists at work on Rusinga Island. John Van Couvering was the first to accurately map and date the Rusinga deposits, and he is shown here with Cary Madden.

(Right) Map of Miocene sites in western Kenya.

(Below right) Lake Victoria from space. Rusinga Island is just visible in the northeast corner of the lake. The position of the lake is marked by the lack of cloud cover, but in fact the strong evaporation of lake waters during the day, frequently gives rise to a build up of clouds during the day leading to afternoon rain on the land surrounding the lake.

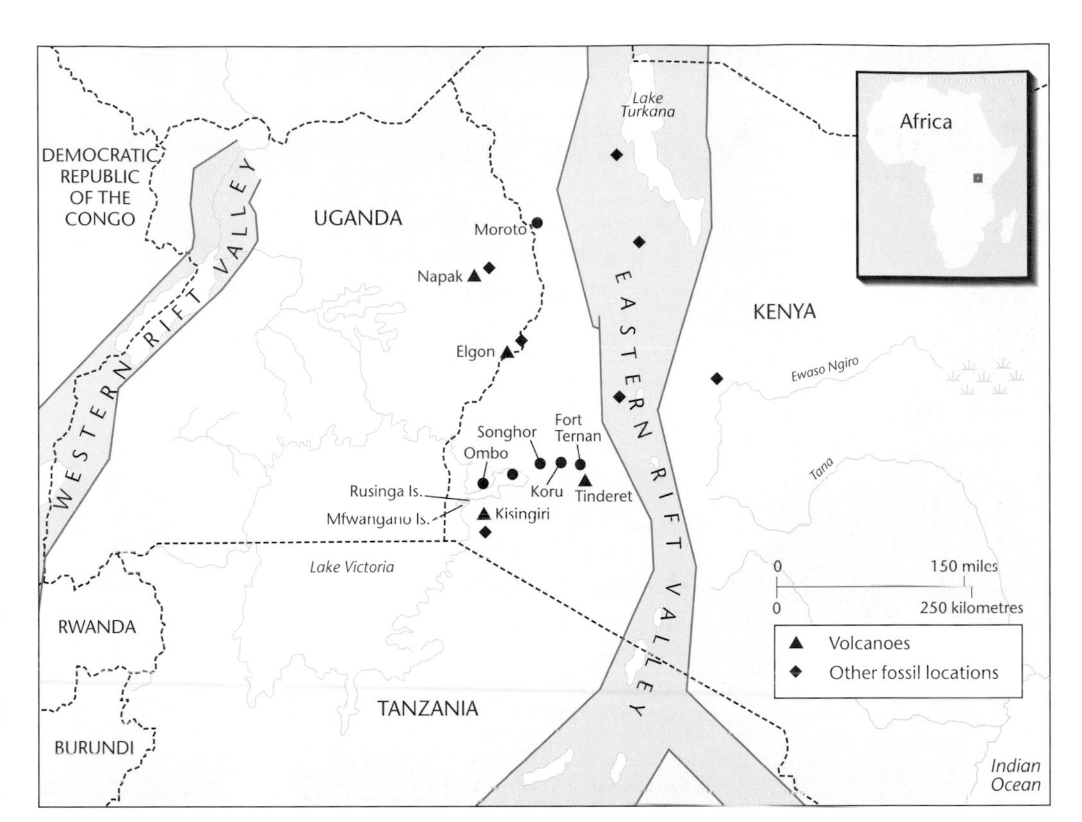

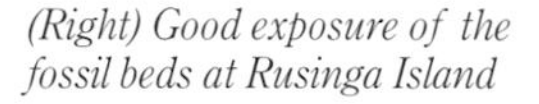

(Right) Good exposure of the fossil beds at Rusinga Island

(Below) The fish beds of the Kulu Formation in the upper levels of Rusinga Island has abundant remains of fish skeletons, being indicated here by Judith Harris.

beds at Kulu with abundant bird and fish fossils that are about 17 million years old. All these sediments come from an ancient volcano called Kisingiri at a time when there was no lake there but a large river which deposited some of the Rusinga sediments. This was also a time before the development of the East African rift system, so that the present highlands that break up the relief of much of East Africa were not yet present and the area would have been flatter and probably continuous geographically with the central African lowlands.

Well-preserved fossils

One of the peculiarities of the fossils that have been found on Rusinga Island is their excellent state of preservation. This has arisen because the sediments in which they were buried came from a volcano with strongly alkaline emissions. Kisingiri was what is known as a carbonatite volcano, producing an abundance of ash with a high component of carbonates that quickly replace the original structure of bones so that they fossilize extremely quickly. Similar processes can be seen today when bones are fossilized within a few years in highly calcareous environments, for example in limestone areas. When this happens the bones do not have time to decay, for their organic parts are replaced so rapidly by the carbonates, and even soft parts may be preserved if the replacement is rapid enough. This has happened in some cases on Rusinga Island, with the preservation of the skin of some small reptiles, and the presence in other places of intact woodland floors, with all the variety of leaves, flowers, seeds, twigs and fruits.

Animal fossils are also common in the Rusinga deposits, and in many cases they are present in the fossilized remains of the soils in which they were first buried. Many of the animals have complete or nearly complete skeletons, with some remarkable finds of early bovids, rodents and apes. Recently a group of partial skeletons of the early fossil ape *Proconsul* was found, and the only known skull of *Proconsul* came from here. Another remarkable find of *Proconsul* together with many other animals

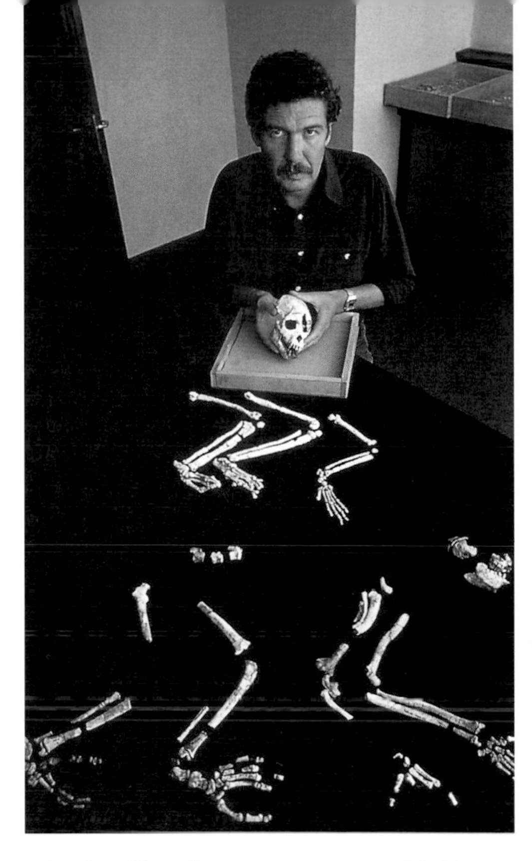

came from an area only a couple of metres across, known as Whitworth's pothole, and this was formed by the decay of an ancient tree trunk. The tree was evidently large and hollow, and many of the animals probably lived inside it, particularly the large pythons whose bones are mixed with those of other animals, and they were buried when the tree was covered over by sediment.

So far, preservation of fossils in soils and in the hollow tree have been mentioned, both particular kinds of taphonomic preservation. Most commonly on Rusinga Island, however, the fossils were preserved in river sediments. A large river flowed through the area in the Miocene, and most of the sediments were laid down in water, including the volcanic deposits, which were washed off the slopes of the volcano and redeposited where they are now. In most cases, the sediments were exposed on the surface long enough for soils to form on them, with trees growing on the soils, and the fossils preserved mostly came from animals living in these habitats, either in the trees or on the surface of the soil.

The two species of fossil ape, known as *Proconsul heseloni* and *P. nyanzae*, are found at many different sites on Rusinga Island, together with several species of smaller ape, particularly species of *Dendropithecus* and *Limnopithecus*. These will all be described in later sections of this book. They all lived in trees, and they were found with many other tree-living species, such as lorises and flying squirrels, and animals adapted to life on forest floors, such as elephant shrews and water chevrotains. When we excavated at Rusinga in 1971, we found several layers with these kinds of animal associations strongly suggesting tropical forest environments during the times these deposits were laid down. We also found other layers where these forest species were absent and instead there were larger animals like those living today in more open woodland. Excavations of the woodland floor also demonstrated the presence of non-forest environments, and it became apparent in the course of our work that conditions changed rapidly as the Rusinga deposits accumulated, with forest faunas in one time and place being succeeded by non-forest faunas within short periods and replaced by other faunas only a few tens of metres to either side. Probably there was a very mixed type of habitat, perhaps with small patches of forest big enough to support forest faunas in an overall wooded area.

(Above) Fossil fruits of forest trees found on nearby Mfwangano Island: (left) two fruits of a species of Entandrophragma, *a large forest 'mahogany' species, and (right)* Sterculiaceae, *a large emergent forest tree. These taxa are diagnostic of forest conditions during the early Miocene of Mfwangano Island.*

(Above) The partial skeletons of Proconsul *(reconstructed below in its woodland to forest environment) found by Alan Walker on Rusinga Island.*

Site II: Paşalar

In 1976 one of us was working with a team in southwestern Turkey on deposits of middle Miocene age, about 13 million years old. Our first choice of sites was not very satisfactory, for we did not find any fossil apes, and so at the end of the season, we decided to try and find a somewhat older site further to the north where some fossil apes had been found in deposits exposed in the side of a road cut. All we knew was that this site was near a small village named Paşalar in a remote country district. It consisted of a small pocket of sediment no more than 40 m (130 ft) in extent, but it was supposed to be rich with fossils.

Survey and excavation

Having found a site, the next stage was to obtain permission to work there; this took several years to accomplish. When work finally began, we established a fixed point set in concrete from which to take measurements, and then we surveyed the whole area, using standard surveying equipment. Maps were drawn of the area and of the excavation site in particular. Finally, we laid out a metre grid to control the excavation and started working progressively through the outcrop. At the same time, we cut trenches through the deposits to determine their depth and extent. Both the excavation material and the sediment dug out of the trenches were screened with water to separate the fossils missed during the digging from the sediment. The information from all this work has told us that the fossils at Pasalar were transported to the site with the sediments in which they are now found. The sediments themselves can be easily identified as coming from a source very close to the site, from the adjacent hillside, and the weathering and abrasion of the fossils indicates that they were exposed for a short time on this hillside before transport.

Over the years we have accumulated hundreds of thousands of specimens, including about 1,700 specimens of fossil ape. As work progressed, we were joined by many specialists who each contributed to a particular aspect of the study. Sedimentologists, geomorphologists, geologists and geochemists have analysed the sediments and stratigraphic setting to show where the fossil deposits came from. Geophysicists have analysed the carbon and oxygen isotopes to determine the soil chemistry and the type of food eaten by the fossil animals. Palaeontologists have identified all the animals and this has enabled us to date the site by comparisons with other sites with similar faunas, giving a date of about 14–15 million years, early in the middle Miocene. We have also investigated the functional morphology of most species

(Below) Location and simplified geological map of Paşalar.

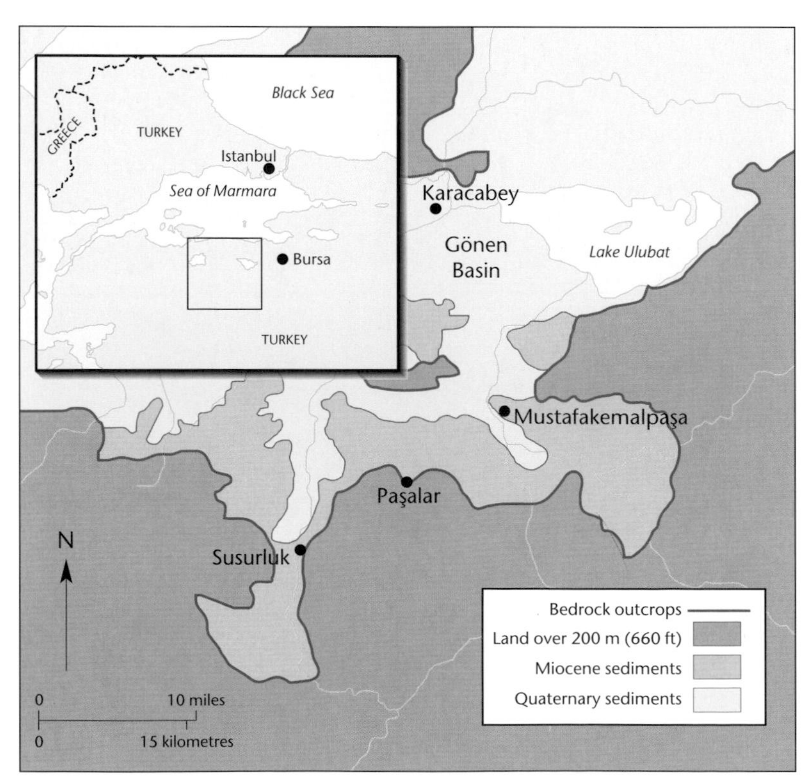

present at the site, showing that most animals were ground-living and ate vegetation, but there were also many tree-living animals that ate fruit, and this suggests seasonal forest being present at that time. The analysis of the soils also indicates a highly seasonal climate in Turkey because there was evidence of both water-logging and drying of the soils. As a result of this work we have been able to reconstruct the climate and vegetation in some detail for this part of Turkey in the middle Miocene. There was a seasonal monsoon climate, with a long dry season during the winter but heavy rains in the summer. The vegetation was probably deciduous forest with many openings in the canopy and rich ground vegetation.

Analysing the fossils

Taphonomists have analysed the fossils found at the site. Some bones showed signs of digestion, and many of the small rodents and insectivores had been eaten by owls. Regarding the large mammals like rhinos and giraffes, there were large numbers of young individuals compared with numbers of adults, and this indicates that carnivores had selected the young of large animals and probably there was a carnivore den close by where these

(Above) The fossil site at Paşalar. The sediments are exposed on a small hill, originally extending to a line running from the small tree in the left foreground to the trees in the right background. The excavation (left) has cut into the hill by between 9 and 12 m (30–40 ft) and has produced many thousands of fossil specimens.

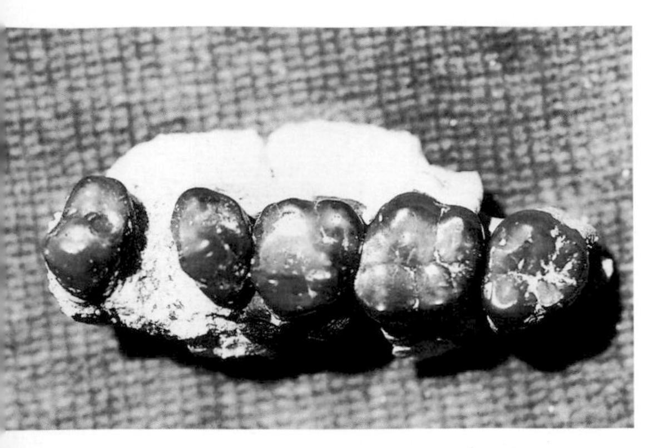

(Top) Specimen of Griphopithecus alpani, *the common fossil hominid from Paşalar.*

(Above) Mandible of the giraffid Giraffokeryx, *one of the more common species at Paşalar.*

bones accumulated. Most bones showed signs of weathering, and the weathered bones were abraded later on by transport, while others were buried in soil, in most cases before transport and in some cases afterwards. All the bones were transported to the site over a distance of 2–3 km (1.25–1.9 miles) together with the sediment, which was then reworked by spring action on their new resting place.

Much of the analytical work has concentrated on the large sample of fossil hominoids. Variation in the fossil apes has been investigated, as has their tooth structure. By sectioning the teeth and examining them under a scanning electron microscope (SEM), it was found that they all had very thick enamel on the teeth. Other students have worked on the wear patterns of the teeth, and on the micromorphology of the teeth using sophisticated image analysis techniques to compare the morphology with other apes. One scientist used the SEM to describe the microwear on the teeth and found that the fossil ape was a hard fruit feeder as suggested by the enamel thickness. Another worked on the small number of limb bones found (all finger and toe bones) and found that the Paşalar ape was just like the earlier and more primitive apes from Africa. All these techniques provide a consistent picture of the Paşalar fossil apes as arboreal climbers eating fruit in a seasonal forest environment very much like the monkeys living in similar habitats in India today.

Many Turkish students (opposite) were trained in excavation techniques and palaeoanthropology at Paşalar. Insaf Gençtürk (right) was one of these students, and she has now progressed to complete her doctorate in the subject.

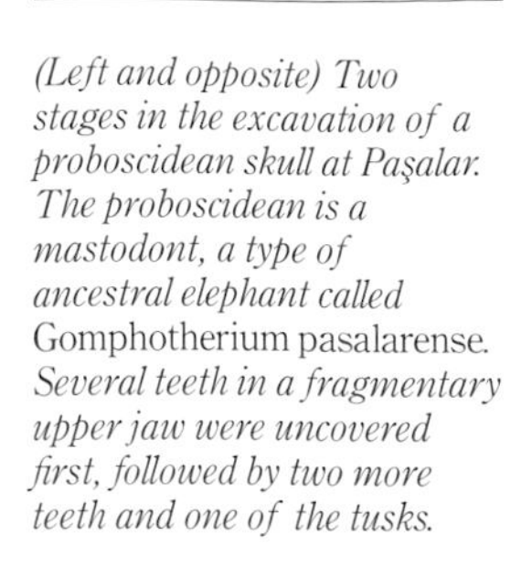

(Left and opposite) Two stages in the excavation of a proboscidean skull at Paşalar. The proboscidean is a mastodont, a type of ancestral elephant called Gomphotherium pasalarense. *Several teeth in a fragmentary upper jaw were uncovered first, followed by two more teeth and one of the tusks.*

Site III: Rudabánya

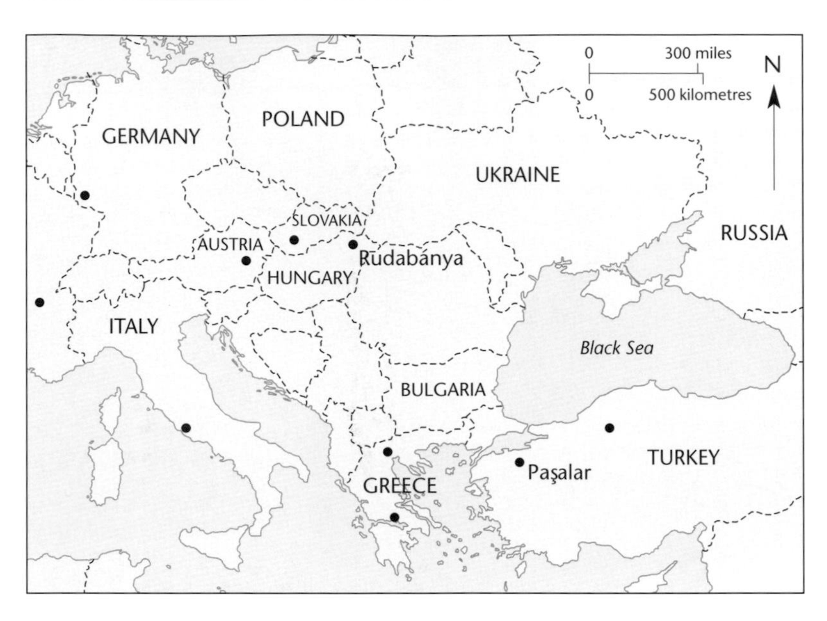

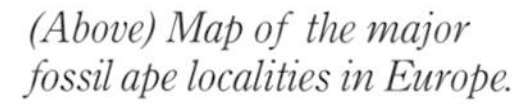
(Above) Map of the major fossil ape localities in Europe.

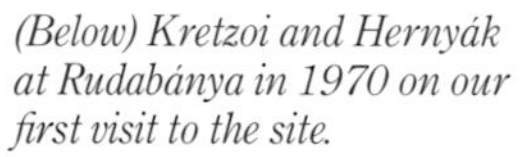
(Below) Kretzoi and Hernyák at Rudabánya in 1970 on our first visit to the site.

Rudabánya is a small town in Hungary in an area of considerable industrial and mining development. A few miles outside the town is a massive quarry, the digging of which has exposed extensive deposits of Miocene sediments which were laid down in and along the shores of a large Miocene lake. This lake was so large it was more of an inland freshwater sea, occupying what is called the Pannonian basin which covered large parts of present-day Hungary. The sediments cannot be dated directly, but from the age of the lake and from the animals found in the sediments they have been dated to the late Miocene between 9 and 10 million years ago.

The significance of the site

The fossil site at Rudabánya was discovered in 1967 by G. Hernyák, a Hungarian geologist working with the mining company, and the early fossils from there were described by an Hungarian palaeontologist, Miklós Kretzoi, who quickly realized the significance of the site. What is so important is that some of the best specimens of European fossil ape from this period have come from Rudabánya together with abundant plants and animals that provide a detailed knowledge of the area. During much of the middle and late Miocene, from about 1 million years ago until slightly later than the time of Rudabánya, apes were common in Europe, extending as far north as Poland, and although the climate was different then, with much of southern Europe being subtropical, it is still very different from present-day ape habitats, and the question remains how did apes survive.

Earlier collections at Rudabánya had paid scant attention to the environment of early apes, and nothing was known of the taphonomy of the site. The taphonomic excavation took some time to set up, but eventually we were able to start work there in 1992. We brought together a group of people based around specialists in excavation so that we could combine descriptions of the geology of the sediments and the surrounding rocks with the taphonomy of the fossils and their palaeontological identification.

Revealing the deposits

Excavation procedure was much the same as described earlier for Paşalar (p. 62). We excavated with hand tools in metre squares, recording the positions of all fossils in three dimensions so that the distribution of all finds could be plotted. Fossil wood and other plant remains are extremely abundant, because the deposits consisted of two superimposed swampy soils, and fossil bones were locally abundant as well. The deposits were formed in a shallow valley sloping to the north and leading into the Pannonian Lake. The lake level was evidently variable, and at the beginning of the depositional sequence the lake level was low but gradually rising up the valley. As a result of this the sediments got wetter and wetter, culminating in a

(Left) Excavation at Rudabánya, Hungary. Excavation was by metre grid, but for greater precision the position of each specimen was measured in three dimensions from a fixed point.

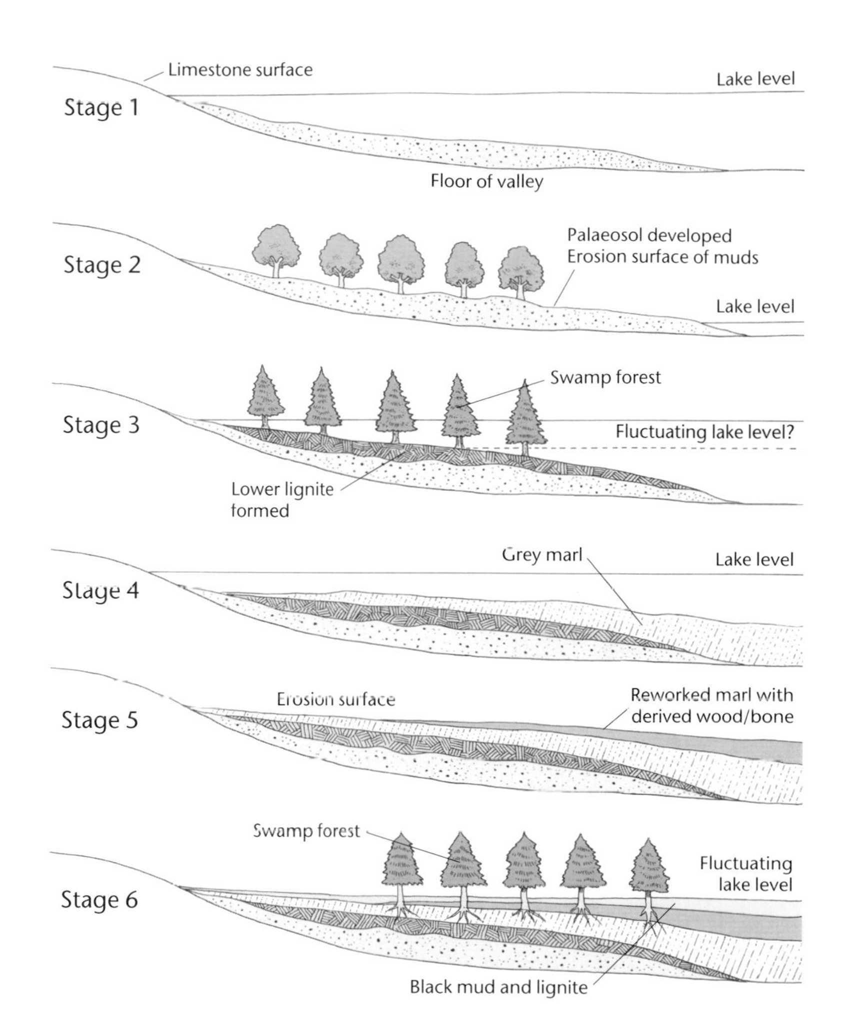

(Above) Reconstruction of events during the accumulation of the Rudabánya deposits. Rise and fall of the Pannonian Lake level led to deposition of lake marls when the lake was high and organic rich swamp deposits when the lake was low. There are some fossils in the marls, washed in from the surrounding land, but most are found with the swamp vegetation, which consisted of Taxodium *forest, and in small pools left behind by the receding water.*

lignite, which is formed in swamp conditions where there is so much water that dead vegetation does not rot down but accumulates into thick deposits over time, like peat does today. With time and pressure, lignite eventually turns into coal, but this had not happened in the Rudabánya deposits. What did happen was a further major increase in lake level, so that on top of the lignite a thick deposit of lake sediments was laid down. At times the lake dried up, leaving the surface of the sediment exposed as lake flats with local ponding for just a few years and producing localized patches of black clay in the ponds. Finally, gradual rise in the lake produced thick black muds succeeded by another lignite at the top of the sequence.

The fossils

Fossils occur everywhere except in the lignites, which would have been very acid because of the anaerobic (lacking oxygen) conditions (plants preserve well in acid conditions, but bones are dissolved and destroyed by the high acidity). Fossils in the lake deposits were relatively complete, with several partial skeletons probably from animals that were washed into the lake from the surrounding land. In the black clay, formed by local ponding on barren lake flats, there was an abundance of small mammals including the small ape-like primate *Anapithecus hernyaki*, extremely broken up and probably the leftovers from predators' meals.

Finally, the upper black mudstone was rich in fossils again, very broken up but including all the specimens of fossil ape that we found. This is named *Dryopithecus hungaricus*, and this fossil ape is of special interest because of its advanced morphology (see pp. 110–113), and it is interesting that it was found in conditions that showed the presence of subtropical forest, for the abundant tree remains found at this level indicate that subtropical *Taxodium* forest was present there at the time. *Taxodium* is the swamp cypress, common today in the Florida Everglades, and the exclusive link between it and *Dryopithecus* suggest strongly that this fossil ape inhabited this type of environment.

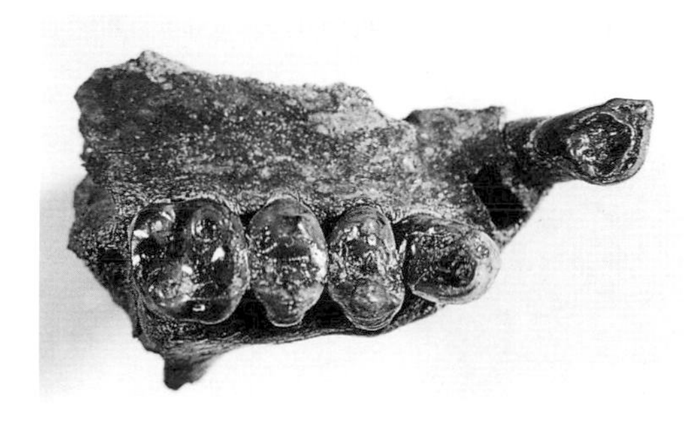

(Left) Maxilla of Dryopithecus hungaricus. *This specimen provides details of the teeth, palate and the incisive fossa, which can be seen as a notch cut into the top edge of the bone.*

Site IV: Olduvai Gorge

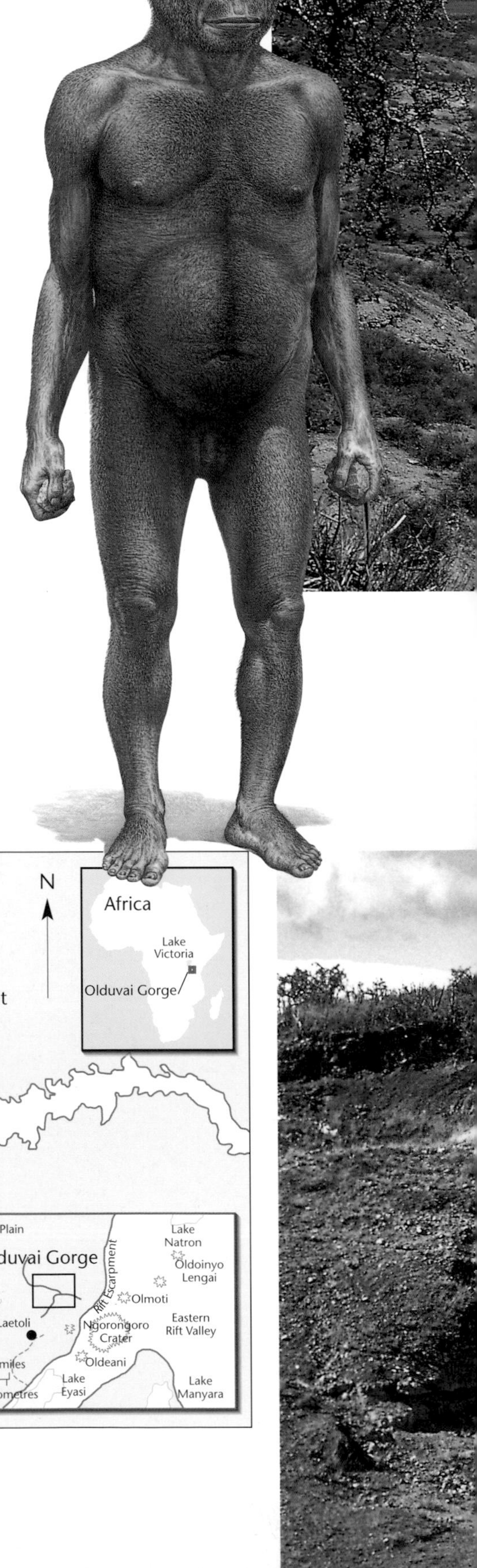

(Right) The first finds of Homo habilis *included only a few from the body skeleton. However, in 1986 a partial skeleton was discovered, on which this reconstruction is based. It was a small biped, with rather ape-like proportions.*

Olduvai Gorge is cut into the eastern Serengeti Plain of northern Tanzania. It is about 50 km (30 miles) long and up to 100 m (330 ft) deep, and its exposed sequence of lake, river and volcanic sediments records portions of the last 2 million years of East African prehistory. It was first identified in 1911 by a German scientist, who made a small collection of fossil bones from its slopes, and took them back to Berlin (Olduvai was then in German East Africa). Important species such as *Hipparion*, the three-toed horse, were recognized amongst the material, so an expedition led by the geologist Hans Reck returned to the Gorge in 1913. Reck's team collected more than 1,700 fossils, confirming the great antiquity of the site, and these included a human skeleton which he claimed was an ancient fossil (it is now known to be a relatively recent burial dug into much more ancient deposits).

The Leakeys at Olduvai

Serious work only resumed at Olduvai when the Kenyan-born anthropologist Louis Leakey invited Reck to join his 1931 expedition to the region. This expedition discovered the first ancient stone tools at Olduvai, and Leakey developed the idea that the site recorded a gradual sequence leading from crude pebble tools (of the so-called Oldowan Culture) in

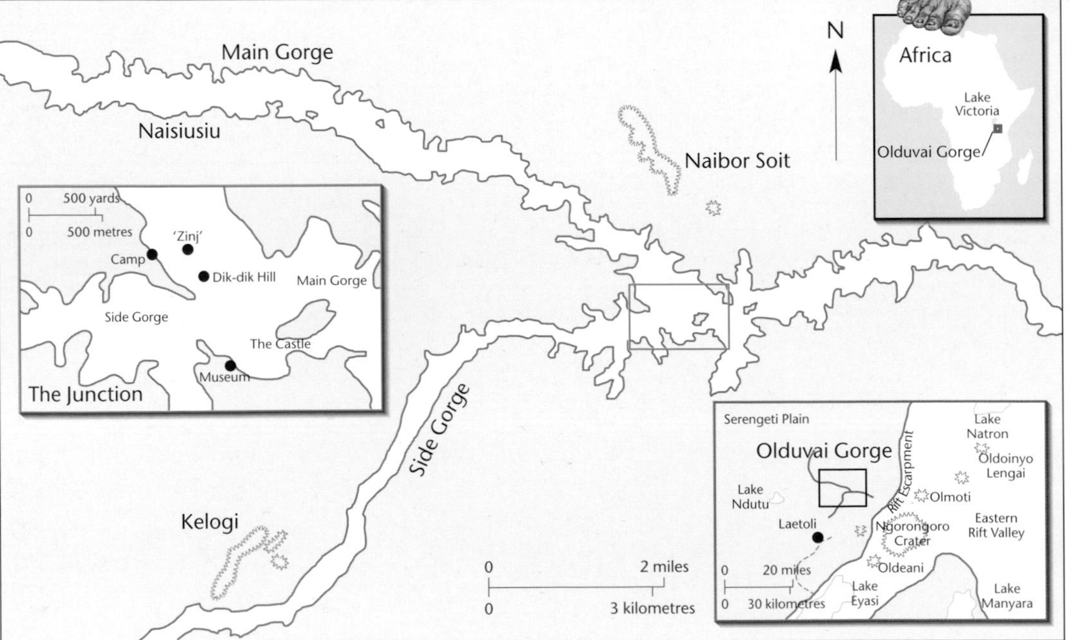

(Below) Olduvai Gorge in Tanzania is one of the classic localities of African prehistory. This plan view shows its basic Y-shape, and the position of many of the most important sites where fossils have been discovered, close to the junction of the main and side gorge.

the earliest Bed I to the much more sophisticated handaxe tools of the later Bed IV. Leakey was joined by the archaeologist Mary Nicol on a further expedition to Olduvai in 1935, and she married him the following year.

The record of fossil animals and stone tools continued to increase as they expanded their work over the next 20 years, and the changing patterns of animal fossils and artifacts indicated a great time depth for the Olduvai sequence. At the base of the sequence, deposits were laid down by a large salty lake, the size and depth of which fluctuated considerably. The lake eventually disappeared, and the region became a flat plain crossed by seasonal streams. Later still, as the present gorge began to open up because of geological changes, wind-blown

(Above) Olduvai Gorge.

(Below) The 'Zinj' site at Olduvai Gorge in 1972. The plinth in the foreground shows the position where the skull of Zinjanthropus *(now* Paranthropus*)* boisei *was found by the Leakeys in 1959.*

(Right) Excavations at Olduvai over a period of nearly 100 years have produced a wealth of finds. Early work often consisted only of surface collections of material washed out of the sediments, but later, more careful excavations have produced many thousands of finds from small areas.

sand and dust were added to the top of the lake and stream deposits. Periodically, nearby volcanoes erupted their lava or ash into the layer-cake of accumulating sediments. Animals represented as fossils include large forms of hartebeest, water-buck, pig and springbok. Early on there were unusual elephants with downward-pointing tusks in their lower jaws, called deinotheres, with modern forms appearing later in the sequence.

(Below) This skull-cap was found by Louis Leakey at Olduvai in 1960, and is believed to date from over 1.2 million years ago. It was originally called 'Chellean Man' because of its supposed association with 'Chellean' hand-axes found at Olduvai. However, in its massive brow ridge and long low shape, it bears close resemblance to skulls of Homo erectus, *with which it is often grouped.*

(Above) The skull of Olduvai Hominid 5 was nicknamed 'Nutcracker Man' from its massive back teeth. First assigned to 'Zinjanthropus' boisei, *and considered a possible human ancestor by Louis Leakey, it is now commonly classified as a robust australopithecine* Paranthropus boisei.

Discovering the hominin fossils

The Leakeys were not rewarded with a spectacular fossil of a possible primitive human until 1959, when Mary Leakey discovered the skull of a robust australopithecine (see pp. 126–129) associated with Oldowan tools (see p. 208) in Bed I. At first the Leakeys named it *'Zinjanthropus boisei'* ('Boise's East African man', after Charles Boise, one of their benefactors), but the fossil is now assigned to *Paranthropus boisei*. Louis Leakey at first assumed he had found the early human who was manufacturing the Oldowan tools, but over the next two years remains of a more human-like form began to be found at the same levels in Bed I, and higher up, in Bed II. These remains, including parts of a skull, jawbones, hand, leg and foot bones, were assigned to a new human species, *Homo habilis* ('Handy man'), in 1964. This species, with its smaller back teeth and larger brain size, was now assumed to be the tool-maker of Olduvai Beds I and II, rather than its contemporaneous australopithecine cousin.

The potassium-argon dating technique has shown that the oldest human remains and stone tools at Olduvai date from about 1.8 million years ago, near the end of the Pliocene period. Remains of the more advanced species *Homo erectus* have also been excavated from the top of Bed II Bed IV at Olduvai, sometimes associated with handaxe tools. Louis Leakey died in 1972, but Mary continued her meticulous excavations at Olduvai until 1984. These have yielded much important evidence about early human behaviour, including stone tool production, possible stone structures, and evidence of scavenging, hunting and butchery. Two years after Mary finally left the site, a partial skeleton attributed to *Homo habilis* was discovered in Bed I deposits by an American team, and in 1996, another American-led team found a further upper jaw attributed to this species. So Olduvai Gorge continues to figure prominently in the story of human evolution.

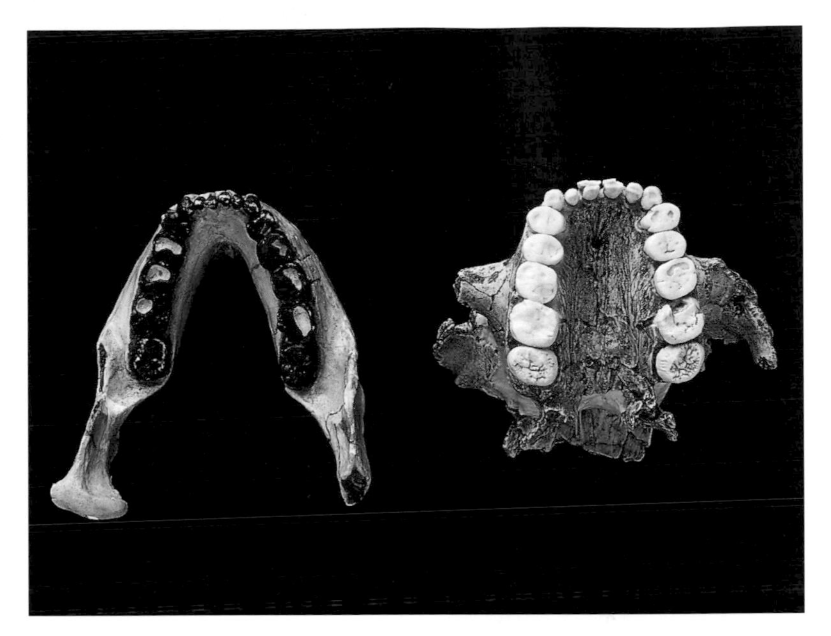

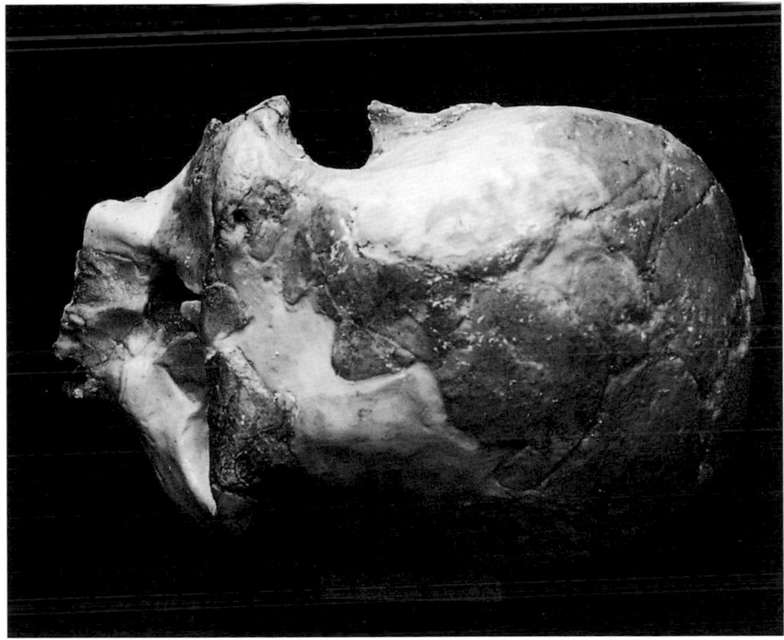

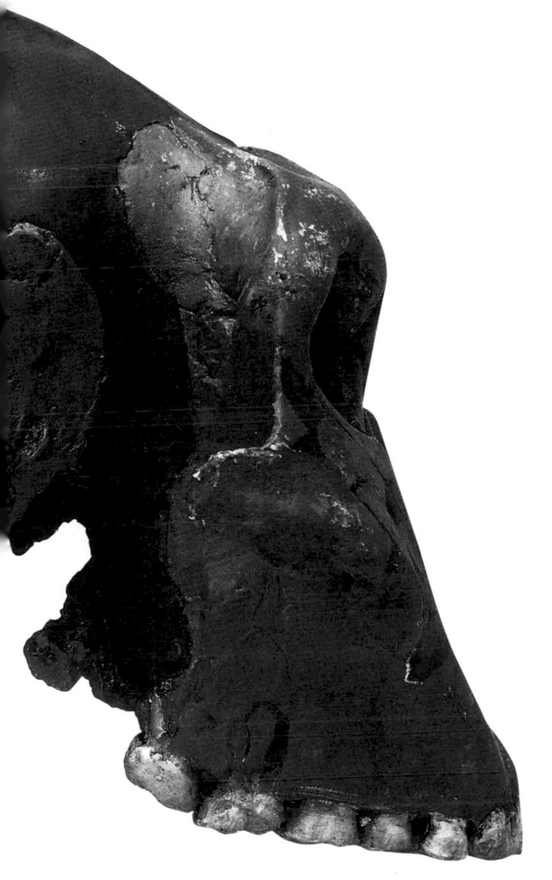

(Top) The unusual dental form of Paranthropus boisei, *as shown in (left) a lower jaw found at Peninj in Tanzania, and (right) the upper jaw of Olduvai Hominid 5. The front teeth are relatively small while the premolars and molars are huge, and often heavily worn.*

(Above) A view from above of the Olduvai Hominid 24 skull, often assigned to Homo habilis. *It is one of the most ancient fossils from Olduvai, about 1.85 million years old.*

Site V: Boxgrove

(Right) This reconstruction of a male individual from Boxgrove is based on the large and robust tibia found there, as well as material of Homo heidelbergensis *from other sites. In keeping with evidence of the impact of a projectile point on a horse shoulder blade, as well as preserved spears from younger deposits in Britain and Germany, he has been equipped with a wooden javelin.*

The site of Boxgrove is located in a quarry 10 km (6 miles) north of the current shoreline of the English Channel near Chichester in southern England. Detailed excavations at Boxgrove began in 1985 and gradually developed into a large project, involving over 40 specialists and dozens of excavators. But it was not until the end of 1993, with the discovery of a human shin bone (tibia) about 500,000 years old, that the site became world famous. That discovery was followed in 1995 by the excavation of two human teeth. These finds represent the oldest humans known from the British Isles, and the site has yielded a wealth of data on the behaviour of the Boxgrove people.

Over half a million years ago, the coast was 10 km (6 miles) north of its present location, and the sea cut huge cliffs into the chalk hills of what is now Sussex. Then the sea level fell slightly, and salt marshes and coastal grasslands developed over the sandy beaches which it left in its wake. Herds of game grazed on this new coastal plain – animals such as red deer, bison, horse and even elephant and rhinoceros, but there were also the animals which preyed on them, such as lion, hyaena and wolf. And people roamed here too, living off the land and the game.

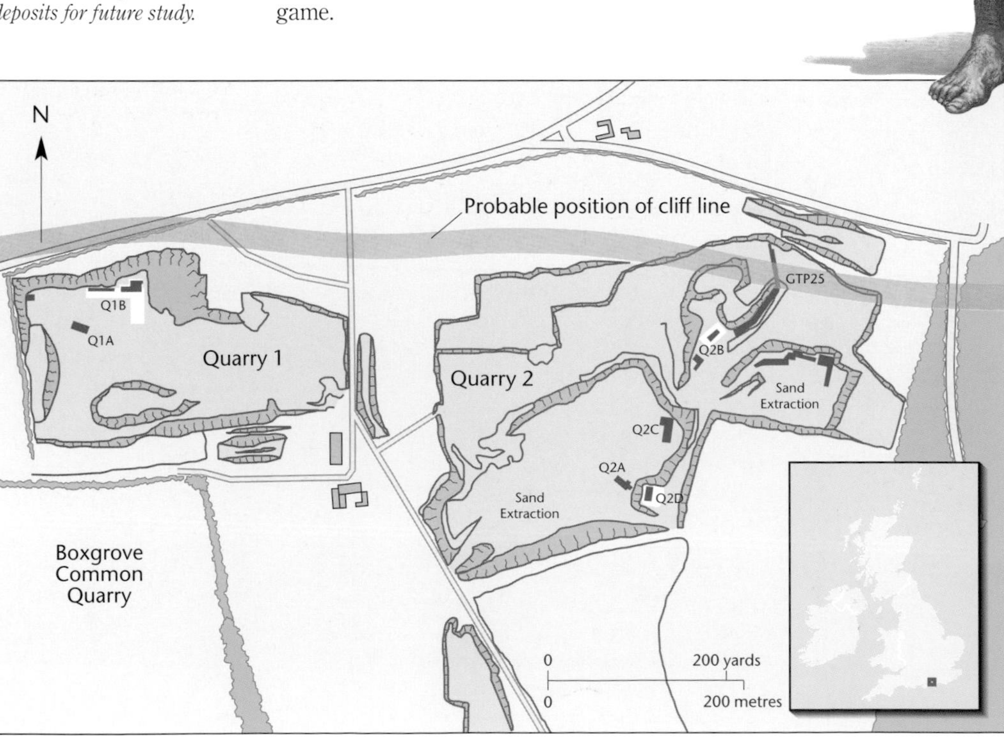

(Below) The Boxgrove archaeological site is located near Chichester in England, and was discovered in a working quarry. Quarry 2 contained the butchered remains of a horse, while Quarry 1 was the locality that contained butchered rhinoceros remains and the human tibia and teeth. The site is now protected and covered over to preserve its deposits for future study.

(Opposite above) One of the main sequences of deposits at Boxgrove. Analysis of these deposits has allowed a detailed history of the site's past environment to be built up, from chalk sea cliffs to ecologically rich coastal plain.

(Opposite) An excavation in Quarry 1 in 1995. This locality produced the two human lower incisors found that year, as well as dozens of handaxe tools. Some of the deposits at Boxgrove had been barely disturbed in 500,000 years.

Tools and bones at Boxgrove

People were also drawn by the presence of flint in the chalk cliffs – an excellent source of raw material from which they could produce the most characteristic stone tool found at Boxgrove – the handaxe, of which over 300 examples have been excavated. Because the land surfaces at Boxgrove were repeatedly covered over by gently flowing water, covering them with a fine silt, those ancient surfaces have been preserved with only minimal disturbance. The preservation is so good that the exact places where people crouched down to make their stone tools have been preserved, so every flake of flint they struck off is still lying where it fell some half a million years ago. Not only that, but the bones of the animals they ate are also there, surrounded by tools, and often covered in butchery marks.

The handaxes, which are predominantly oval or almond-shaped, were used to butcher large carcasses of giant deer, red deer, bison, horse and rhinoceros. There are very few traces of cut marks on any of the bones from smaller animals such as roe deer, suggesting that smaller carcasses were either ignored, or carried elsewhere for

(Below) One of the beautifully preserved handaxe tools from Boxgrove. The handaxes at Boxgrove are generally well made from flint blanks that must have been collected locally from the chalk cliffs. Most have this characteristic almond or teardrop shape.

butchery. It is clear that the humans at Boxgrove had access to complete carcasses, since most parts of the animals are represented at the butchery sites.

The site is dated from its mammal remains to a warm stage, or interglacial, in the Middle Pleistocene, at the end of what is known as the Cromerian Complex, about 500,000 years ago. The molar teeth of water voles changed by evolution during this time, and the type of water vole found at Boxgrove, called *Arvicola terrestris cantiana*, as well other associated species, match those present at the Mauer sand pit near Heidelberg in Germany, the site which produced the lower jawbone of *Homo heidelbergensis* (see pp. 148–151) in 1907. The Boxgrove tibia has also been placed in that species, and it is one of the most massive early human leg bones ever found. The individual concerned must have been about 1.8 m (5ft 11in) tall, and the great thickness of bone in the walls of the tibia suggest that this person (probably a man, because of the size of the bone) was heavy and muscular – probably over 90 kg (200 lb) in weight. The strength of the bone must also reflect the physically demanding life-style that these people had to endure.

The two teeth found at Boxgrove are from the front of a lower jaw, and are not so exceptional in size, although heavily worn. Under a microscope, they reveal a mass of scratches and pits on their surfaces (see p. 45). Many of these must have been made as the individual concerned sliced with stone

(Right) Another view of Quarry 1 in 1995, showing the large number of excavators involved in work at the site. The sediment that has been carefully scraped away is kept for further screening in sieves, revealing even tiny pieces of flint and bone.

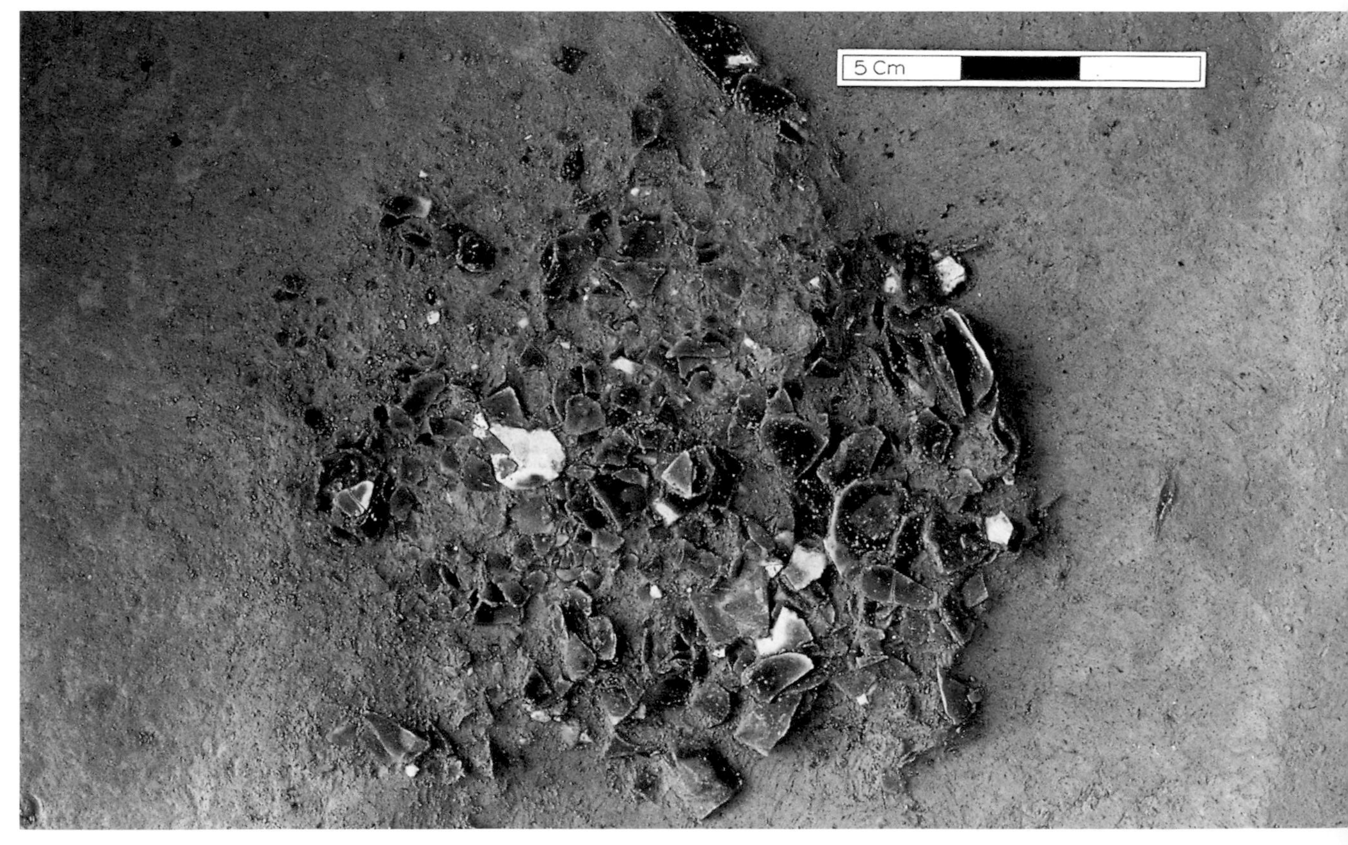

(Below) In some cases at Boxgrove, the remains of stone toolmaking lie just as they were left by a flint knapper. Here a profile of flint debris shows where someone squatted to make a handaxe some 500,000 years ago. The pieces can be put together to show the whole process of manufacture.

(*Below*) *The human fossils from Boxgrove have been assigned to the species* Homo heidelbergensis, *based on their estimated age and resemblances to other material assigned to that species. Here, one of the Boxgrove incisors is compared with a replica of the jawbone from Mauer near Heidelberg in Germany, found in 1907, from which the species is named.*

(*Right*) *The Boxgrove left tibia, found in 1993. It is from a tall and heavily built individual, and microscopic studies of the bone structure suggest its owner (probably a male) was relatively old when he died.*

tools through meat or vegetable materials clenched in their jaws. The direction of the slices can even be determined, and indicate that the tools concerned were apparently being held in the right hand. The bases of the teeth are covered in tartar deposits, and these even extend down the roots at the front. This means that the roots must have been partly exposed during life, and that the teeth were probably being regularly forced back and forth while chewing or clenching the jaws.

Site VI: Gibraltar

The Rock of Gibraltar has been a landmark for the peoples of the Mediterranean for countless millennia, and must have provided a marvellous vantage point for some of its early inhabitants 50,000 years ago, the Neanderthals. Gibraltar was one of the first places to yield up evidence of these ancient people. It may also have been one of their last refuges before extinction.

The accidental discovery of a fossilized human skull, blasted out during work at Forbes' Quarry, nearly placed Gibraltar in the forefront of prehistoric studies over 150 years ago. But the find was neglected for 50 years, and the 1856 Neander Valley skeleton from Germany got most of the scientific attention. A second significant Neanderthal find was made in Gibraltar in 1926, at the Devil's Tower site, surrounding a cleft in the North Face limestone, not far east of Forbes' Quarry. This find was

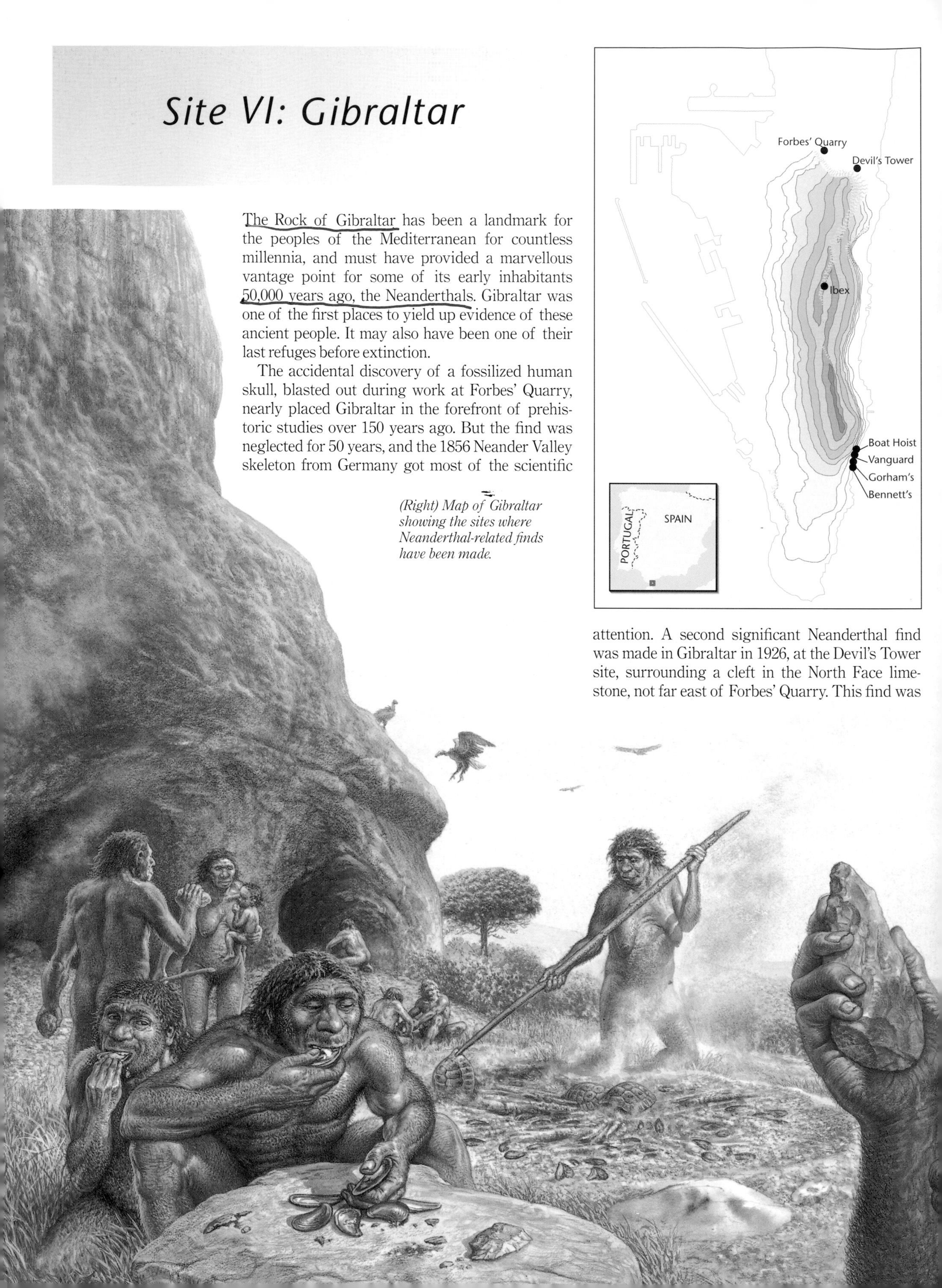

(Right) Map of Gibraltar showing the sites where Neanderthal-related finds have been made.

(Above) Gorham's Cave, with Bennett's Cave visible to the left.

(Above left) This fossil skull, probably of a woman, was found in 1848, eight years before the skeleton from the Neander Valley (Neander Thal) that gave its name to the species. Unfortunately its scientific importance was only recognized in 1863, and it was not properly described for another 40 years after that.

excavated systematically, and had associated animal bones, stone tools, and charcoal from ancient fires. The fossil remains consist of parts of the upper and lower jaws and braincase of a Neanderthal child (see also pp. 42–44).

Gorham's and Vanguard Caves

Although the fossil finds from Forbes' Quarry and Devil's Tower continue to attract scientific interest, and they remain two of the best preserved of Neanderthal skulls, the sites from which they came are now virtually empty of deposits. However, there are also a series of caves near the sea on 'Governor's Beach', on the southeast side of the Rock, and two of these caves – Gorham's and Vanguard – still contain rich evidence of Neanderthal occupation. Gorham's Cave, in particular, contains an immense record of human occupation, spanning much of the last 100,000 years. An ancient raised beach, which dates to about 120,000 years ago, lies at the base of the site. There is then at least 10 m (33 ft) of evidence of Neanderthal occupation, marked by Middle Palaeolithic tools, and on top, about 3 m (10 ft) of deposits containing Upper Palaeolithic tools, which elsewhere in Europe are characteristic of occupation by modern humans, after about 35,000 years ago. Capping the whole sequence are historic levels containing pottery and

This reconstruction shows a Neanderthal group outside one of the Gibraltar caves about 50,000 years ago. At this time the Earth's ice caps were large and a low sea level globally allowed the formation of a fertile coastal plain below the caves. Evidence from the caves show that the Neanderthals' diet included tortoises and shellfish.

(Above) This view of Gorham's Cave shows the depth of sediments preserved there, spanning over 100,000 years.

(Left) Vanguard Cave, on Governor's Beach, has recently revealed clear evidence that the Neanderthals exploited marine resources there over 100,000 years ago. This included baking mussels in a fire and consuming marine mammals such as seals and dolphin. The mussels must have been systematically collected, but the marine mammals might have been scavenged from carcasses washed up on the shore.

metal artifacts of the last few thousand years. Ancient hearths are preserved at various levels in the cave, especially in the Upper Palaeolithic, and animal bones are common throughout the sequence.

In 1995, a new series of excavations were begun in Gibraltar. The first season at Gorham's produced rich collections of stone tools, bones, burnt nuts, seeds and charcoal, mapping areas where the Neanderthals had built fires and processed meat and vegetable foods. A radiocarbon date for some of the charcoal came out at close to the limits of the method – about 45,000 years, and indicated that there would be even younger Neanderthal levels above. By 1997, a series of radiocarbon dates from charcoal and fossil bones for occupation levels at both caves had been obtained. They showed that Vanguard, the smaller of the two sites, must have been virtually full of sand and sediment by about 40,000 years ago, and thus was only likely to contain significant evidence of Neanderthal occupation.

At Gorham's, however, there were several Upper Palaeolithic levels which were dated to between 26,000 and 30,000 years ago, and these showed that the arrival of modern people in Gibraltar could have occurred while Neanderthals were probably still living in the more mountainous interior of Iberia. A particularly remarkable find from the Vanguard cave site is the first well-dated and unequivocal evidence that Neanderthals utilized marine food resources – a topic which has been a source of debate for many years. There, a discrete Middle Palaeolithic layer consisting of mussel shells of a large and consistent size, mixed in with ash and stone tools, some of which showed edge damage that might have come from opening or scraping out the shellfish, was excavated. This level has been dated to about 115,000 years ago by the thermoluminescence method, and this is some of the earliest evidence of the use of marine foods by humans anywhere in the world.

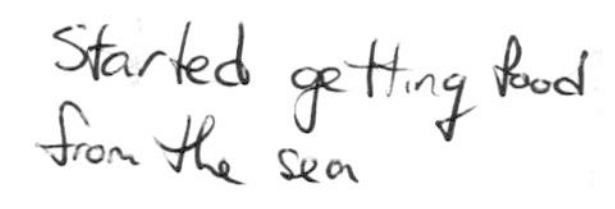

How often the last Neanderthals in the Gibraltar region encountered the first modern people there, and the manner of their interactions, is something we can only guess at. There may have been violent encounters, they may have avoided each other, or they may have made relatively peaceful contact. From the Gibraltar and southern Iberian evidence, it seems that the Neanderthals did not change their technology when the new people arrived – unlike their counterparts further north, they maintained their ancient stone toolmaking traditions essentially unchanged until they disappeared.

(Below) A possible Neanderthal spearpoint from Gorham's Cave. It is not known for certain how these points were mounted on wooden shafts.

(Left) A selection of Neanderthal stone tools found in Gorham's and Vanguard Caves. In most cases they could have been made from material found locally on beaches or in river beds, but some rocks appear to have been transported from sources in what is now Spain.

The fossil evidence for human origins can be traced back to the origin of the primates, the taxonomic order to which apes and humans belong. We can trace the acquisition of more and more human characters during the evolution first of the anthropoids, the group that comprises the monkeys and apes, and then the hominoids, the apes and humans. Anthropoids probably diverged from other primates about 30–40 million years ago, and the hominoids are known to have diverged from the monkeys slightly over 20 million years ago. The early fossil hominoids are the first group that we discuss in any detail, and they include a range of small to large species from East Africa 17–22 million years ago. However, there is some doubt that these species were apes at all, and in fact they may have been ancestral to both monkeys and apes. Later fossil apes about 15 million years ago were definitely on the hominoid line, and soon after that fossil apes are known for the first time outside Africa. At around 12 million years ago there is evidence for the divergence of the pongids and hominids, the former being the orang utan lineage and the latter the line leading to the African apes and humans.

At this stage there is something of a gap in the fossil record. Fossil apes continue in Eurasia until 7–8 million years ago, but in Africa in the period 12–8 million years ago the fossil ape record dries up. Some have taken this to indicate that the hominids evolved outside Africa and re-entered Africa 7–8 million years ago, but this is to misunderstand the nature of the fossil record: absence of evidence does not necessarily mean evidence of absence. The African record resumes after 7–8 million years ago with hominid fossils from Chad and Kenya that are controversially claimed to be hominin, that is, exclusively on the line leading to modern humans. After 4 million years ago the hominin record starts to fill out with fossils that provide good evidence of bipedal (upright two-legged) locomotion. Following on from the first recognized stone artifacts, the earliest species of our genus, *Homo,* appear around 2 million years ago, distinguished by a larger brain size and evidence of meat eating. Shortly after this, one or more of the early *Homo* species leaves Africa for the first time. To a certain extent thereafter, evolution began to follow separate paths in Africa, East Asia and Europe, until the last major event in human evolution, when our species *Homo sapiens* evolved in Africa and began to disperse from there to populate the world.

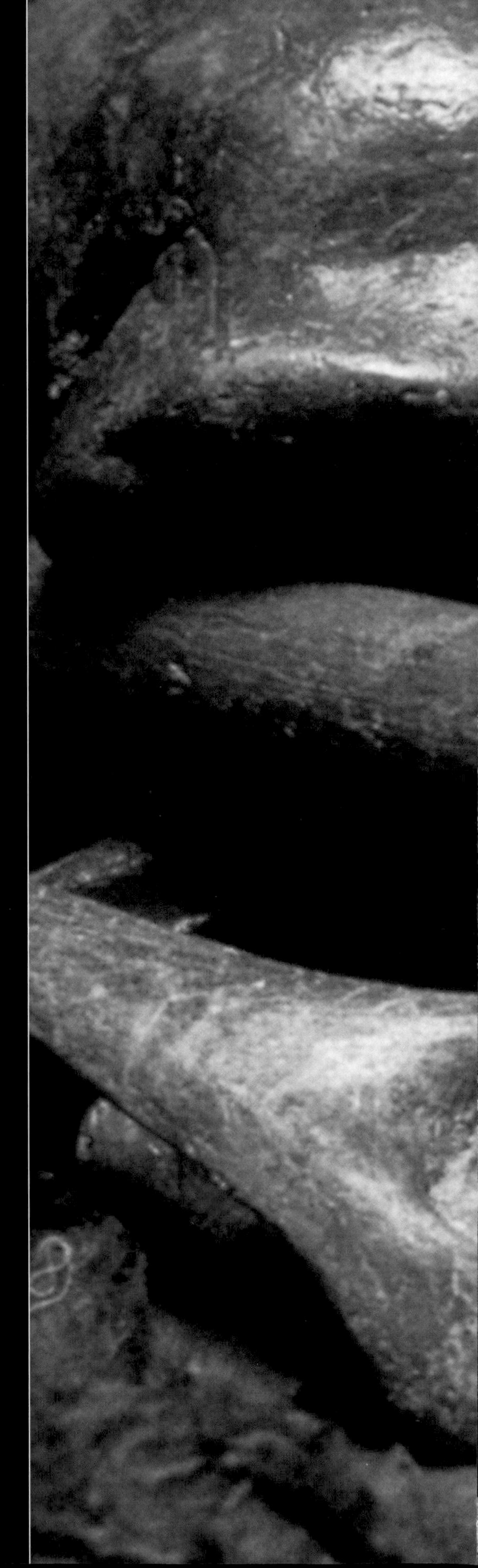

The discovery of a skeleton at the Feldhofer Grotto in the Neander Valley, Germany, in 1856 was a watershed in human evolutionary studies. It led to the recognition of a distinct kind of ancient human, and the fossils were named as the first extinct human species, Homo neanderthalensis, *in 1864.*

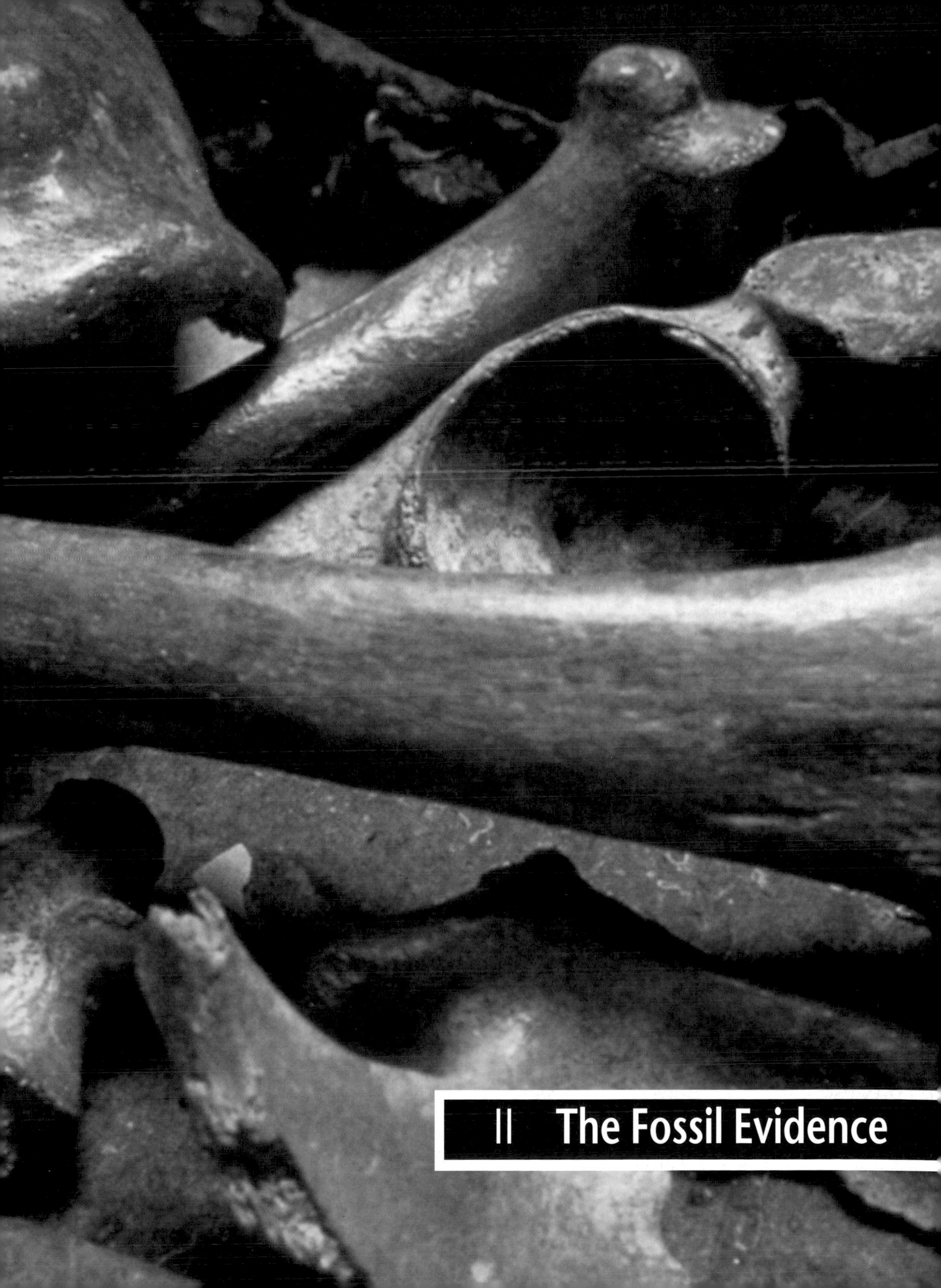

II The Fossil Evidence

Origin of the Primates

(Above) Tiny Moholy bushbabies. Notice the forward directed eyes, the grasping hands, and nails at the finger tips rather than claws. The large size of the eyes and ears are characteristic of nocturnal species.

The primates are mammals comprising mainly tree-living species. They are the group to which apes and humans belong, but when they first appeared they were very different – generally small and often nocturnal animals. Some primates survive to the present time that are equally small and nocturnal, for example the bushbabies of Africa and the lorises of both Africa and Asia, although many aspects of their anatomy are different from ancestral primates and they cannot be considered representative of them except in the most general terms.

The primates have often been considered to lack any clear defining characters, and even today there is some uncertainty about which group of animals primates are most closely related to. There is also disagreement as to which groups, both living and fossil, should be included. For years the tree shrews of Southeast Asia were included, but it is now recognized that they share only primitive characters with primates. These conflicts arise out of the nature of the characters used to define primates, and hence we will list these characters (see box) before discussing primate origins. This is an abbreviated list of characters present in all or most living primates, and it can form the basis for interpreting the fossil evidence for primate origins.

Primate Characters

1 Grasping foot with divergent big toe (hallux)
2 Presence of nail (rather than claw) on the big toe (hallux)
3 Elongation of the heel (calcaneus)
4 Dominance of hindlimb during locomotion
5 Frequent occurrence of grasping hands with opposable thumbs
6 Nails commonly present on most digits; may be secondarily lost
7 Some degree of forward rotation of the eyes
8 Eyes set close together, directed forwards for stereoscopic vision
9 Increased brain size, especially brain size before birth
10 Gestation length is long relative to body weight
11 Foetal growth is slow relative to mother's body weight
12 Life history prolonged
13 Loss of one incisor and one premolar from the tooth row

The early fossil record

The fossil record is patchy in the extreme, with certain times and places well represented and others not at all. Today there are at least 194 species of living primate, and the number is going up as

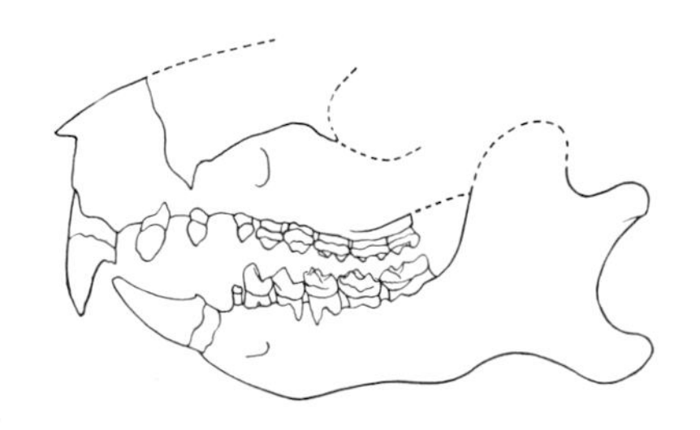

(Centre right) The rodent-like dentition of Plesiadapis, *showing the teeth in position of occlusion.*

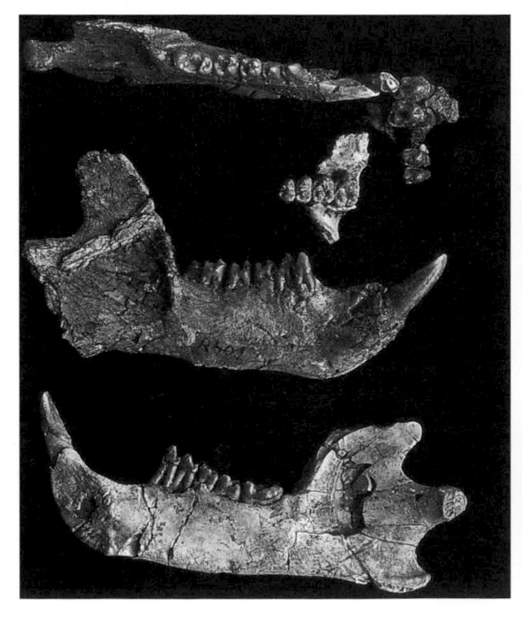

(Right) Mandibles of Plesiadapis, *long thought to be an early primate, but now generally excluded from the order. It has forward pointing rodent-like incisors but primate-like molars.*

subspecies of groups like bushbabies are made into separate species. Even at its most complete during the Eocene, the fossil record has less than half of the expected numbers of primates, assuming comparable diversity to today. At other periods and in large parts of world, no fossil primates are known at all, and even where they are known the fossils are usually extremely fragmentary.

The plesiadapiformes were a Palaeocene group of at least 14 species. They are often considered to belong to the primates, but the fossils share almost none of the characters listed in the box. The same is true of other Palaeocene and earlier species (before 55 million years ago), and the reason they have even been considered likely candidates for early primates is that they may have slightly enlarged brains (character 9 in the list) and they have primate-like molar teeth. They share these characters, however, with many other early mammals, and their front teeth were highly specialized for cutting up vegetation. Many of these Palaeocene animals are now recognized as related to dermopterans, or flying foxes.

The first true primates

It is in the Eocene epoch, spanning 55 to 35 million years ago, that primates appear in something like their present form. The beginning of this time was a period of climatic warming, with tropical conditions extending into northern Europe and the south of England. The warm climate may have been connected in some way to primate origins, although exactly how is not known, and several major groups of primates appear spread over the whole world (excluding Australia and Antarctica on present evidence).

The most diverse group of Eocene primates are classified in the family Adapidae, with two rather distinct groups in North America and Europe. None of these still exists today, nor do any of their relatives, but all of the basal primate characters listed in the box are present in the adapids. They are present also in the second major group, the Tarsiiformes, which may be related to living tarsiers of Southeast Asia. Both groups diversified at the same time at the beginning of the Eocene, and some of the earliest fossils are known from China, along with Eocene representatives of tarsiers. There is also controversial evidence for the presence of higher primates in China from this time. It is possible that the tarsiiforms were more closely related to the monkeys and apes than were the adapids. This is indicated by characters of the eye socket, the shorter face in tarsiiforms, and the proportions of the brain that are linked with change in sensory behaviour from smell to sight. This implies that not only the primates as a group, but the higher primates (Haplorhini encompassing tarsiers, New World monkeys, Old World monkeys and apes) also emerged at the same time, but their place of origin cannot be firmly identified at present.

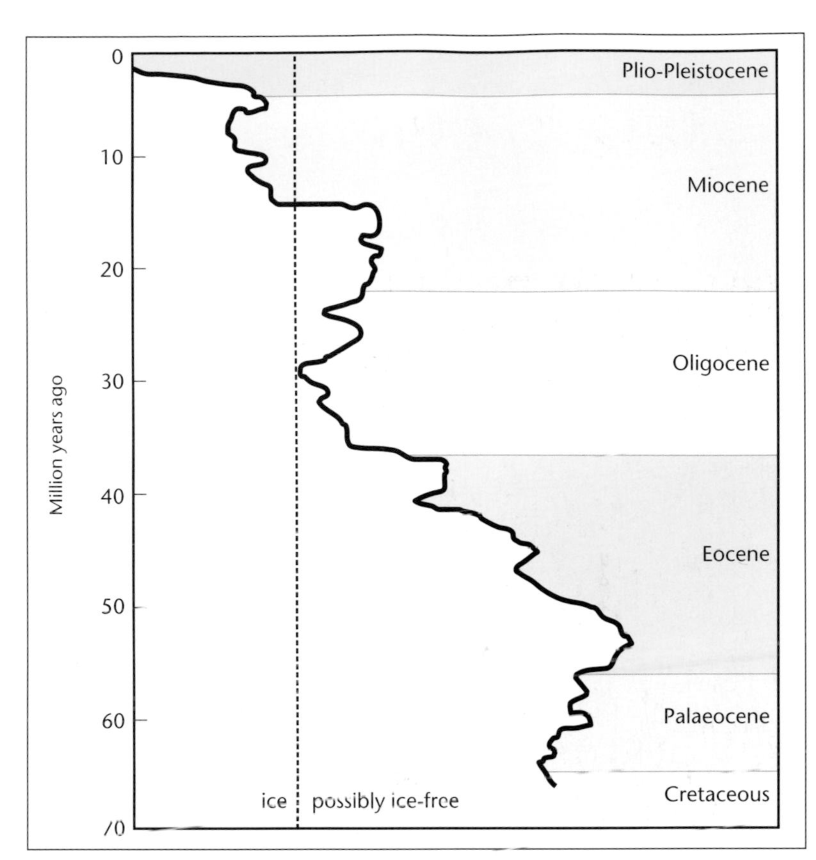

(Above) Oxygen isotope curve based on evidence from ocean-floor foraminifera. These provide a good proxy for determining global annual mean temperature, which can be seen to have fallen more or less continuously since the beginning of the Eocene. This coincides with the time when the first recognizable primates such as Adapis *are known from the fossil record.*

Both adapids and tarsiiformes were fast-moving and agile jumpers and branch runners. They were primarily insectivorous in their diet and almost totally arboreal in habit, many of them also being nocturnal as well. They are found exclusively in association with forest habitats, of which there was great abundance in the warm conditions of the Eocene, and these forests were very like the tropical forests of today, especially in Africa. Towards the end of the Eocene temperatures got cooler and the forests contracted, resulting in major changes in the primate faunas at the end of the Eocene and into the Oligocene.

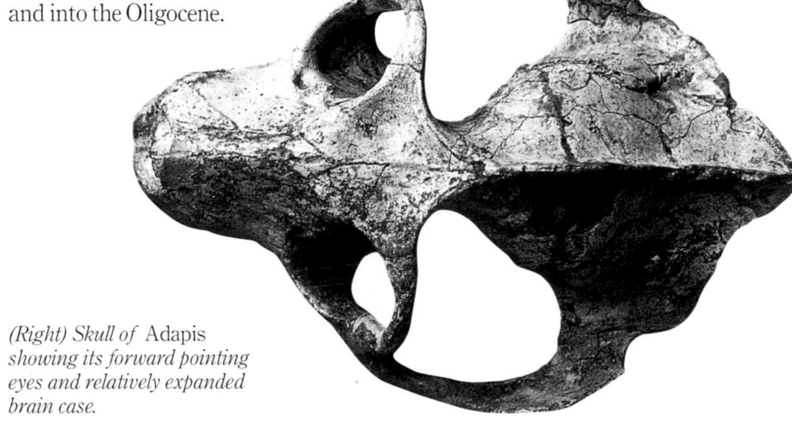

(Right) Skull of Adapis *showing its forward pointing eyes and relatively expanded brain case.*

Early Anthropoids

(Right) A team of palaeontologists search the Fayum Depression in Egypt for fossils.

We have seen how the early primates diversified during the warm conditions in the Eocene epoch. They lived in the tropical forests that covered much of the Americas, Eurasia and Africa. Climatic cooling at the end of the Eocene around 40 million years ago brought about major changes in the mammalian faunas, particularly the primates, which become much less widespread during this time and the adapids and tarsiiforms were nearly exterminated, with the latter hanging on in tropical forests of eastern Asia. In the later part of the Eocene, however, there was a proliferation of anthropoid primates in North Africa, and there are controversial anthropoids known from Burma and Thailand.

The Fayum fossils

The primate fauna from the Fayum in Egypt was very rich and provides most of our knowledge of late Eocene and Oligocene primates. At least five species of primate were present in the earliest of the Fayum deposits, basal anthropoids living about 36 million years ago. These were still very primitive, but slightly later in time there were eight species of propliopithecids with ape-like teeth, parapithecids with monkey-like teeth, such as *Parapithecus* and *Qatrania* and several species of uncertain affinities such as *Afrotarsius* ('African tarsier'), *Arsinoea*, and from Algeria, *Algeripithecus*. These groups expanded in the Oligocene forests of the Fayum depression forming a rich primate fauna lasting until about 31 million years ago.

The fossil primates in the Fayum were first discovered early this century by a professional collector, Richard Markgraf, but it is the collections of Elwyn Simons during the past 30 years that have yielded most of the specimens known today. There are several quarries producing fossils from the Fayum, and the longer the field work continues, the more fossils are found, partly with increasing knowledge of the site but partly also because as areas are cleared in the course of searching for the fossils, the rate of wind erosion increases on the exposed surfaces and so more fossils are uncovered by the next season's collections. This is an unusual but very effective way of collecting, and it does not often happen that a fossil site gets richer the longer it is collected.

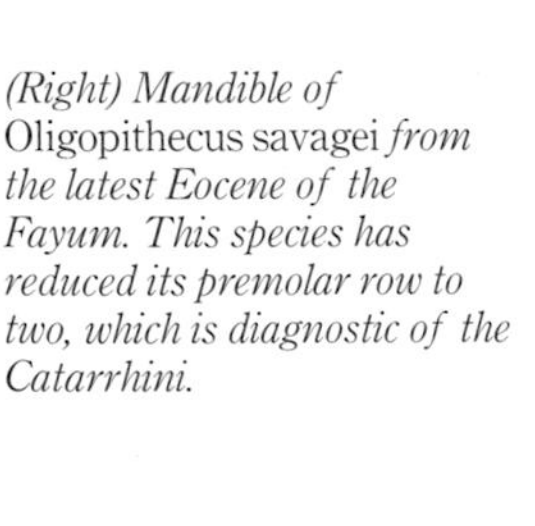

(Right) Mandible of Oligopithecus savagei *from the latest Eocene of the Fayum. This species has reduced its premolar row to two, which is diagnostic of the Catarrhini.*

The Propliopithecids

The best known fossils from the Fayum are the two ape-like species of *Propliopithecus*, *P. haeckeli* and *P. zeuxis* (formerly *Aegyptopithecus*). They were small,

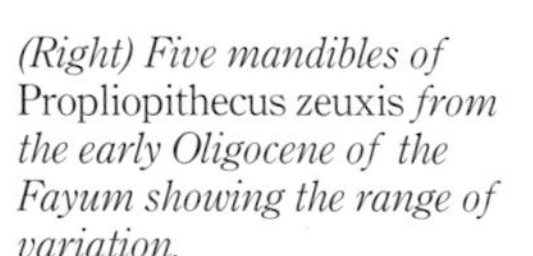

(Right) Five mandibles of Propliopithecus zeuxis *from the early Oligocene of the Fayum showing the range of variation.*

with a size range in the order of 2–6 kg (4.4–13.2 lb), like small- to medium-sized dogs, and like dogs they had rather projecting faces with long slashing canines. Their teeth were ape-like, which led to their being classified as apes when first discovered, but it is now known that their teeth are primitive. As will be seen later, this makes it difficult to identify apes on the basis of their teeth alone. Their skeletons are quite distinctive, with rather robust bones, particularly their forearms. The upper arm was robust with an elbow joint adapted for stability and incapable of being fully extended (see pp. 88–89 for discussion on the critical nature of the elbow joint).

(Above and right) Two skulls of Propliopithecus zeuxis *from the early Oligocene of the Fayum Depression, probably a male on the left with a strong sagittal crest, and a female on the right.*

(Left) Reconstruction of Propliopithecus zeuxis. *This early anthropoid was a rather heavily built and probably slow- moving climber in trees, living high up in the forest canopy of the tropical forest indicated as the probable environment in Egypt at this time.*

The fingers and toes were also stout and adapted for powerful grasping, and overall the anatomy suggests slow quadrupedal (four-footed) climbing in trees as the main way of life of these primates, probably very like howler monkeys in the tropical forests of South America today.

The adaptations of the skull and teeth of the propliopithecids indicate that they lived on a diet of fruit, probably relatively soft fruits. This is consistent with the type of environment indicated by the fossil evidence. Many trunks of large trees are found with the fossils, indicating that this area was covered in high forest, probably with a tropical to subtropical climate. This is also indicated by the remarkable fauna: in addition to the two propliopithecids, there were several other primate species, notably the monkey-like parapithecids mentioned above, with at least four genera and five species all

(Right) Based on the postcranial bones of Propliopithecus zeuxis, *this basal catarrhine primate had a morphology very like that of the howler monkey shown here (except for the prehensile tail) and probably had a similar form of locomotion.*

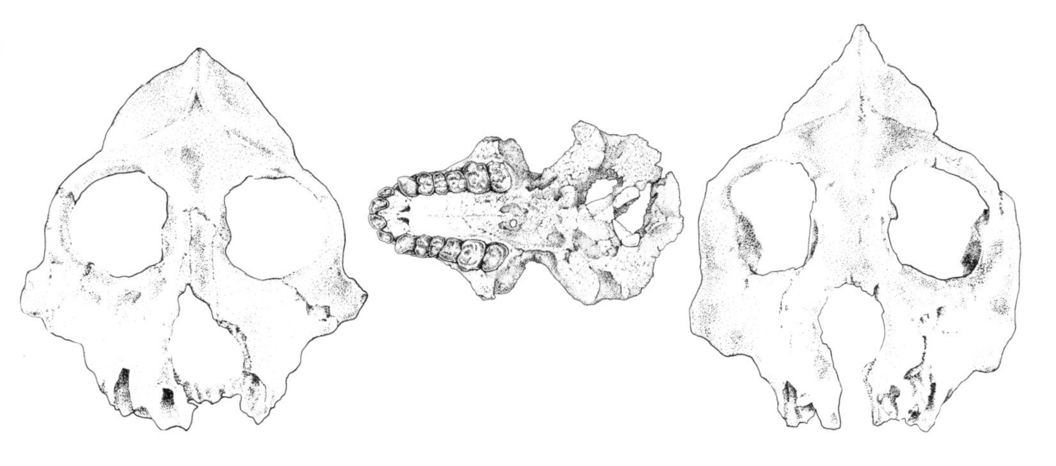

(Left) Three drawings of Propliopithecus zeuxis *showing the morphology of the skull.*

much like the living squirrel monkeys of South and Central American tropical forests. The largest species approached 2 kg (4.4 lb), the size of capuchin monkeys today. Other mammals include a remarkable diversity of hyraxes and related proboscideans (elephant family), the large rhino-like *Arsinotherium* and small elephant shrews and rodents as well as some marsupials. Sirenians were also present, living in the lagoons, so that the picture is built up of low-lying lagoons next to the sea, with coastal and riverine forests interspersed with areas of water, the whole in a hot and steamy tropical climate.

The propliopithecids and probably also the parapithecids are related to the monkeys and apes, and they provide the strongest evidence for the origin of this group. This indicates an African origin for the group, since this evidence comes only from northern Africa, Egypt and Oman (in the Oligocene, Arabia was part of the African continent). In contrast the few fossils from Burma have ambiguous relationships. The African connection is continued when the first apes are considered, for they also are known only from Africa, although it is hard to distinguish early monkeys and apes. Fossils such as *Dendropithecus* ('Wood ape'), *Limnopithecus* ('Lake ape') and *Micropithecus* ('Small ape') belong in this category, and most anthropologists accept that although they are ape-like in their skull and teeth, as indeed is *Propliopithecus*, they actually belong to a considerable radiation of small primates from the Oligocene through to the middle of the Miocene, lasting some 24 million years (36–12 million years ago) but leaving no living descendants.

(Below) The hot and wet tropical habitat of the Fayum primates is based on evidence from both the types of animals living in the area during the Oligocene and the large forest trees found in the deposits.

What Makes an Ape?

One of the difficulties in identifying the earliest apes, let alone describing them, is that they were little different from a whole range of ape-like primates living at the same time. At its simplest level, monkeys have tails and apes do not, but using the lack of something as a distinguishing feature is not straightforward for fossils. The section on taphonomy (pp. 46–49) showed that the preservation and fossilization of animal remains is often a destructive process, so that we rarely find complete skeletons. When a partial skeleton is found with no tail bones, can we be sure they were never there, or did some hungry animal eat or carry off the tail? For example, the claim has been made by one scientist that a tail could be shown to be absent from one of the species of *Proconsul* ('Before Consul' – Consul was a well-known chimpanzee in London Zoo) from Rusinga Island (see pp. 58–61): this evidence is based on the shape of the sacral vertebrae, which form the bottom end of the backbone and which are very small in *Proconsul*. On the other hand it has recently been reported that there are several tail vertebrae associated with one of the *Proconsul* skeletons found recently on Rusinga Island which would suggest it did have a tail. If this proves to be true, it would cast doubt on the ape affinities of this whole group. This matter has still to be resolved, but it does demonstrate the difficulty in answering an apparently simple evolutionary question.

Female gorilla with young. Apes are distinguished by their lack of a tail.

Despite this difficulty, the principle on which the question was based is sound. This is that we look first to the living representatives of a group to see what characters they share in common to the exclusion of other groups: apes lack tails, other primates have tails, therefore lack of tail identifies apes. What we are doing in addition is imparting direction to the evolutionary process, because by looking at other higher primates and indeed to mammals in general, we see that most of them also have tails, so that the character 'presence of tail' is a general character of mammals. For the monkeys and apes, therefore, the presence of a tail is an ancestral character, and its retention in monkeys is the retention of an ancestral character of no evolutionary significance; in apes, however, absence of tail is an evolutionary novelty and is highly diagnostic of the group. Since all living apes (and humans) lack a tail, it is assumed that this condition was present in the last common ancestor of the apes (and humans) and that this condition also defines the group, known as superfamily Hominoidea. To identify a fossil as a hominoid, therefore, it should have this condition, for if it does not, it must precede the last common ancestor. This is why the issue of the presence or absence of a tail in *Proconsul* is so important and why the uncertainty about it raises doubts as to its hominoid relationships.

The issue of the presence or absence of a tail has been used here to establish a procedural point, but the situation is not quite as black-and-white as this implies. Living apes share a variety of characters in common, ranging from behavioural similarities to points of detail about their anatomy, but clearly these did not all arise at once and so there is a sort of halfway house where fossil primates that were indeed ancestral to the apes had some – but not all – of these characters. The fossil species in this halfway house are given a special name, stem apes, which implies that they are in fact related to the apes, and grouped in the same superfamily Hominoidea, but they only have a few of the ape characters.

The importance of the elbow region

It was mentioned in the section on living apes (pp. 16–19) that the structure of the elbow region is a significant character in hominoid evolution. Apes (and humans) have highly mobile arms: if you extend your arm out straight, palm uppermost, and twist your forearm, you will find you can rotate it by nearly 360 degrees, and we use this ability in all sorts of ways during our daily lives. On the other hand we can support our weight on our hands or even hang from an overhead support by our arms, so that as well as having great mobility at the elbow joint, we also have great strength and stability. What we can do, sometimes with considerable effort, apes can do even more readily. This, therefore, is a general character of apes, and it is based on the structure of the elbow joint whereby the articulation between the humerus (upper arm) and the radius and ulna (lower arm) has dual function; the surface of the humerus where it articulates with the radius is round so that the radius can rotate around it; the part of the humerus that articulates with the ulna, however, is deeply keeled and ridged, so that the ulna is held in place as it were by tram lines and can only move in one direction. The humerus thus has a highly diagnostic shape in apes, as do also the heads of the radius (round) and ulna (ridged), and finding these features in fossils would indicate relationship with living apes, for the other primates do not have this combination of characters. It is the presence of this functional complex in the elbow region that has led to *Proconsul* being recognized as a hominoid while earlier ape-like primates such as *Propliopithecus*, or similar aged primates such as *Dendropithecus*, both of which lack it, are excluded.

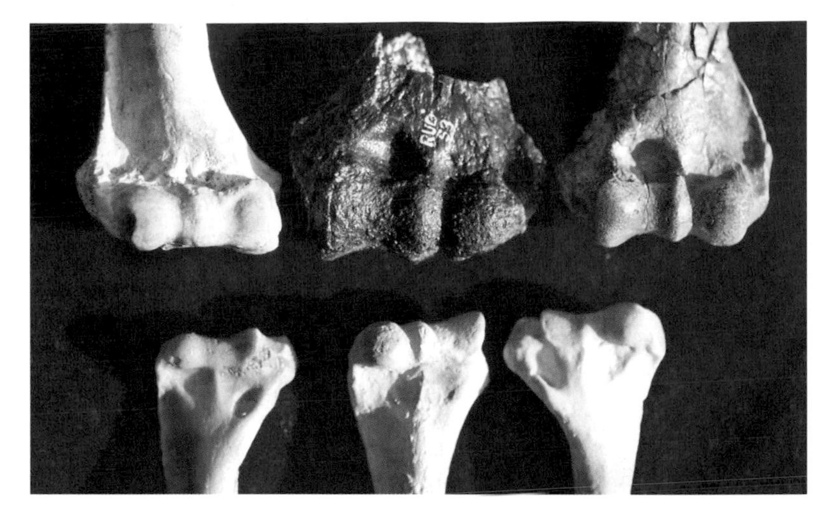

(Above) The elbow joint of Propliopithecus *compared with other fossil anthropoids, showing the high degree of similarity between them. Clockwise from top left:* Equatorius, Dryopithecus, Sivapithecus, Pliopithecus, Proconsul, *and* Propliopithecus.

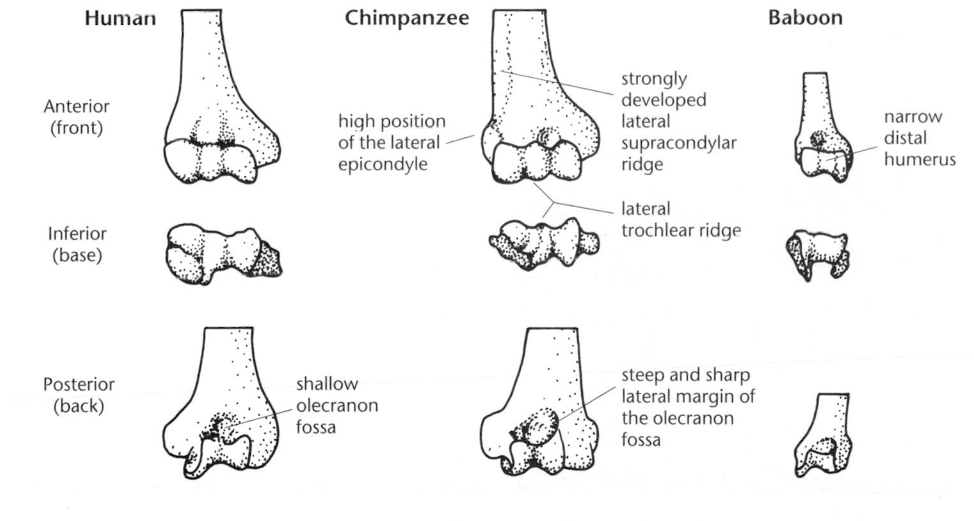

(Left) Comparison of the humerus of humans (left) chimpanzee (middle) and baboon (right), showing their differing morphologies.

(Below left) Arm bones of Dendropithecus macinnesi. *These bones are long and gracile, indicating a suspensory form of locomotion which was once thought to be gibbon-like.*

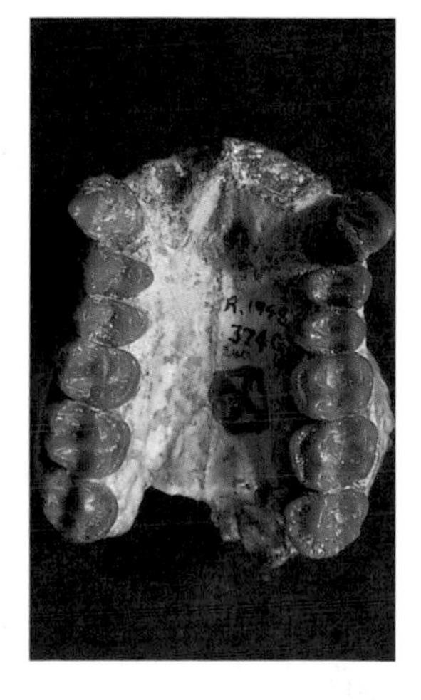

(Below) The maxilla of Dendropithecus macinnesi, *which has ape-like teeth.*

Ancestral Apes

A baboon's upper canines have blade-like crowns that are long and laterally compressed, providing an example of the honing mechanism seen in Dendropithecus macinnesi. *This condition is rare in apes, where the effect tends to grind the tooth down, not to sharpen it.*

The earliest direct evidence for the existence of apes is seen in the structure of the elbow joint on a fossil from Rusinga Island, already mentioned several times (pp. 58–61). Several partial skeletons of *Proconsul* species include the elbow, and they all have an ape-like condition, but they also have a great many other characters that are not ape-like at all, and combined with the equivocal interpretation of the presence or absence of a tail, there is still doubt as to the status of these early Miocene primates.

In addition to *Proconsul*, and coming from the same place and time, there was a remarkable radiation of ape-like primates in the early part of the Miocene about 20 million years ago. There was a species similar in size to *Proconsul* but with rather different adaptations. This was *Rangwapithecus gordoni*, about the same size as *Proconsul africanus* and known from the same sites in western Kenya, particularly Songhor and Koru. Its jaws and teeth indicate a different feeding adaptation from *Proconsul*, for it had more pointed cusps on its teeth and greater ridge development for piercing and cutting tough food objects. This is an adaptation commonly seen in living mammals that eat tough vegetation, where they need to cut up tough leaves to release their nutrients much as gorillas do today. It seems likely that *Rangwapithecus* ('Rangwa ape') was a leaf-eater in contrast to *Proconsul*, all the species of which were adapted for a soft fruit diet similar to that of living chimpanzees. What little is known of the rest of its skeleton suggests that *Rangwapithecus*, like *Proconsul*, was tree-living, but the evidence for it being an ape is even less than that for *Proconsul*.

A point of interest about the diets of these fossil apes is that comparisons with living species show a different spectrum of adaptations. When all the Miocene fossils are considered together and

(Left) The upper canine of Dendropithecus macinnesi *is elongated and blade-like, finely grooved and laterally compressed. During life, this tooth was honed by the lower third premolar, maintaining a sharp edge on the distal surface. Most fossil and living apes lack this honing mechanism.*

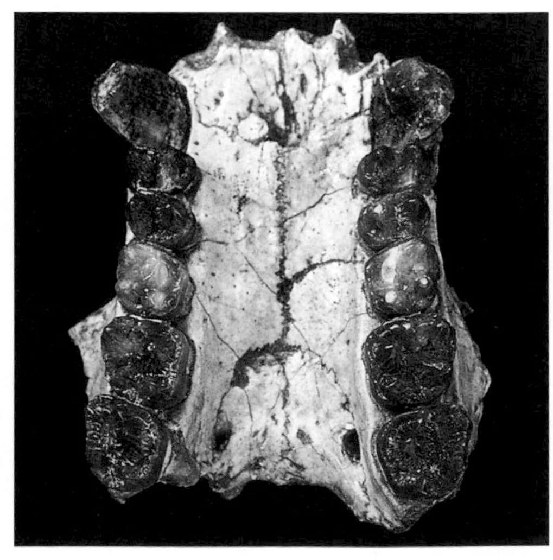

compared with the observed pattern in living monkeys and apes, the Miocene apes are shifted downwards towards the fruit-eating end of the spectrum. There are no species in the Miocene that have as strong leaf-eating adaptations as any of the living apes and much less than the leaf-eating monkeys. It would appear, therefore, that there has been an adaptive shift in the course of primate evolution, with some species adapting more strongly to eating tough objects like leaves and all species adapting to increasing toughness in fruit.

Small-bodied apes

Other primates present at the same time as *Proconsul* were what are commonly called the small-bodied apes. Some of these are similar to *Proconsul* except for their much smaller size, for example *Limnopithecus legetet* and *Limnopithecus evansi*. These were small primates the size of *Propliopithecus*, and with very similar jaw and tooth morphology but lacking its slashing canines and high pointed premolars. They also had broader incisors, the front teeth used for eating and processing fruits, as, for example, when we eat an apple whole. There were two other genera present in the early Miocene that did resemble *Propliopithecus* even in this respect, namely *Micropithecus* and *Dendropithecus*. Their back teeth were similar to those of *Limnopithecus*, but in addition they had high pointed canines, extremely blade-like in the case of *Dendropithecus* and very similar to baboons. Male baboons use their long pointed canines for display, to scare off other males, and for fighting, and they make fearsome weapons when they actually do use them. Baboons are a common prey of leopards throughout Africa, but if a group of males combines an attack on a leopard they can turn the tables and tear it to pieces. On this basis, it is probable that *Dendropithecus* had similar behaviour to baboons, with highly structured social systems in multi-male groups and very strong dominance hierarchy.

Reconstruction of Dendropithecus macinnesi *depicting it as gibbon-like. There is no indication that it brachiated like gibbons, nor that it even looked like gibbons as depicted here, but it is evident from its morphology that it had long slender limbs suitable for suspension in trees.*

Most of these stem apes are known only from jaws and teeth, but *Dendropithecus* also has some partial skeletons from Rusinga Island again. Remains of four individuals were found by Louis Leakey at a site where little else has been found. Both front and back legs are known, and they are very long and slender. Animals that live on the ground tend to have strongly built and robust bones to support their weight, and this may be true also of tree-living animals if they move around in the trees on the tops of branches rather than hanging beneath them. Bones as slender as those of *Dendropithecus* would be poorly adapted for weight support, and it is likely that it suspended itself beneath branches as do living apes like gibbons, which have similarly long and slender leg and arm bones. This similarity led early workers to suggest that *Dendropithecus* was ancestral to gibbons, but since it lacks any of the distinctive characters of the apes, particularly of the elbow joint, it seems more likely that the slender bones indicate only a similar life-style, and not any evolutionary relationship.

The species discussed so far are all known from early in the Miocene, about 18–20 million years ago. The 'small-bodied apes' are also known later in time, and of particular importance is *Simiolus* ('Small ape') from a site in northern Kenya called Kalodirr, because it is adaptively similar to *Dendropithecus*, while being about 1–2 million years later in time. Several of these later small species changed but little from *Dendropithecus*, the best known of them being *Pliopithecus* ('Primitive ape') from western Europe, *Anapithecus* from eastern Europe and *Laccopithecus* from China. Some of these will be discussed further, but for now we shall just note that they all appear to have slender limbs and probably a similar way of arboreal life.

(Opposite) Maxilla of Rangwapithecus gordoni, *a species related to* Proconsul *but distinguished from it by having sharper cusps on its molars, indicating that it ate leaves, unlike the fruit-eating* Proconsul.

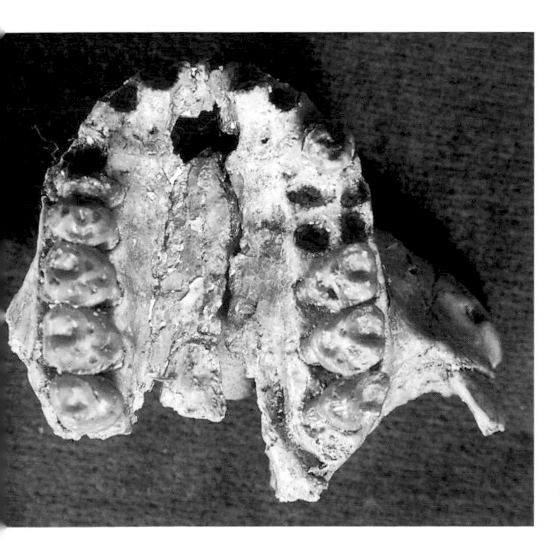

(Left) Maxilla of Micropithecus bishopi, *a small species probably related to* Dendropithecus macinnesi.

Proconsul and its Contemporaries

(Left) Partial skull of one of the earliest hominoid species known, as yet unnamed. It comes from a site in the Koru area in Kenya called Meswa Bridge.

(Right) Palaeontologists search for fossils on Rusinga Island.

The earliest fossil attributed to the Hominoidea is *Kamoyapithecus hamiltoni* ('Kamoya ape' – named in honour of the fossil hunter Kamoya Kimeu) from northern Kenya. It has not been well dated, but on the basis of radiometric dates and the animals associated with it, it appears to be Oligocene in age, about 26–24 million years old. There are only a few fragments of jaw known so far, and it has been identified as hominoid solely on the basis of similarities in its teeth, but since these are similar to most other higher primates living at that time, including *Proconsul*, they are not very diagnostic. The same is true of rather more complete material from a site called Meswa Bridge in western Kenya, 22 million years old; these have been attributed to *Proconsul*, but in both cases the identifications were made on the basis of primitive characters shared with that genus and hence are not reliable.

The earliest good evidence for *Proconsul* is from a group of western Kenyan sites all dated to between 18 and 20 million years ago. One of these is Rusinga Island, from where *P. heseloni* and *P. nyanzae* are known (body weights 11 kg/24 lb and 36 kg/79 lb respectively, based on estimates from 18 individuals), but notable additions are Songhor and Koru, where two different species are known, *P. major* and *P. africanus* (the latter about 11 kg/24 lb again, but

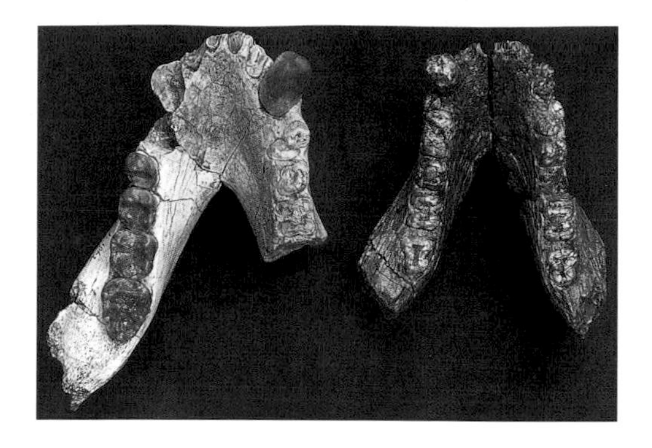

(Right) Two mandibles of Proconsul major *from Songhor. The one on the right has lost many of its teeth, but judging from the root sizes and the dimensions of the jaw bone, it came from a female individual, while the bigger mandible on the left was a male.*

the former up to 76 kg/168 lb based on three individuals). *P. africanus* was the first species found and named by A.T. Hopwood from the Natural History Museum in London in 1933, but the major collections came from the work of Louis Leakey in the two decades after the Second World War. As a result of more than four decades of collecting and research, *Proconsul* is now one of the best-known fossil apes from anywhere in the world.

Monkey-like features of *Proconsul*

Unfortunately, there is only one skull of *Proconsul* known, despite all the effort put into finding more, and that is the fossil found in 1948 by Mary Leakey on Rusinga Island. This skull is very monkey-like, with a narrow nose, lightly built face and brow ridges and short muzzle. This skull is probably from a female individual, and males were somewhat larger. This evidence is based largely on the size of the teeth, and in the smaller species of *Proconsul* males were 1.3 times the size of females, which is similar to the size difference between male and female chimpanzees. For *P. major* this ratio is larger and is more like the difference between male and female gorillas.

(Below) The type specimen of Proconsul heseloni, *the 1948 skull found by the Leakeys on Rusinga Island.*

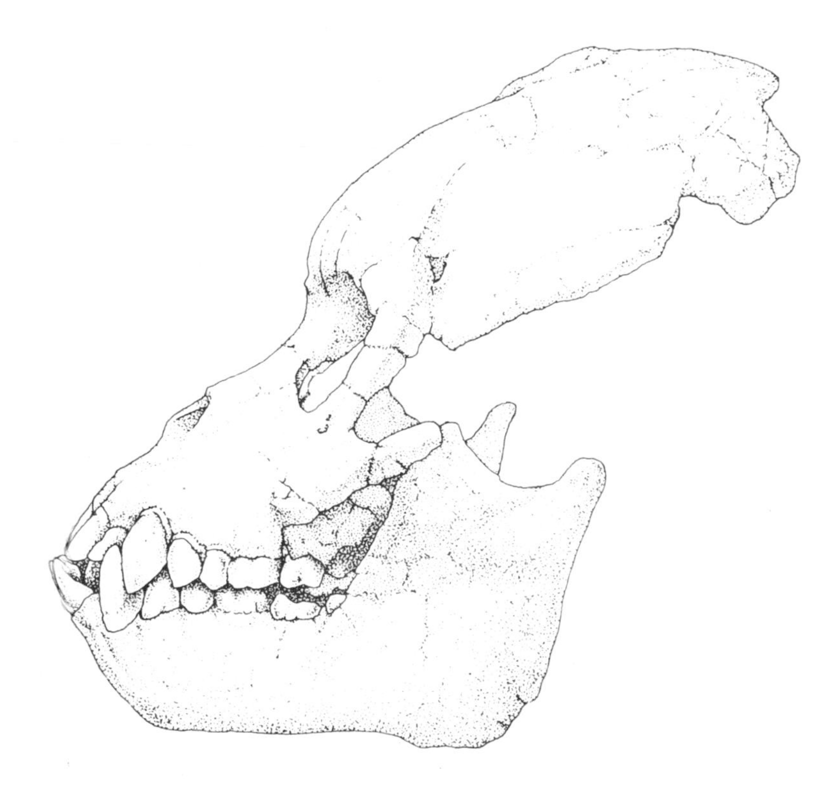

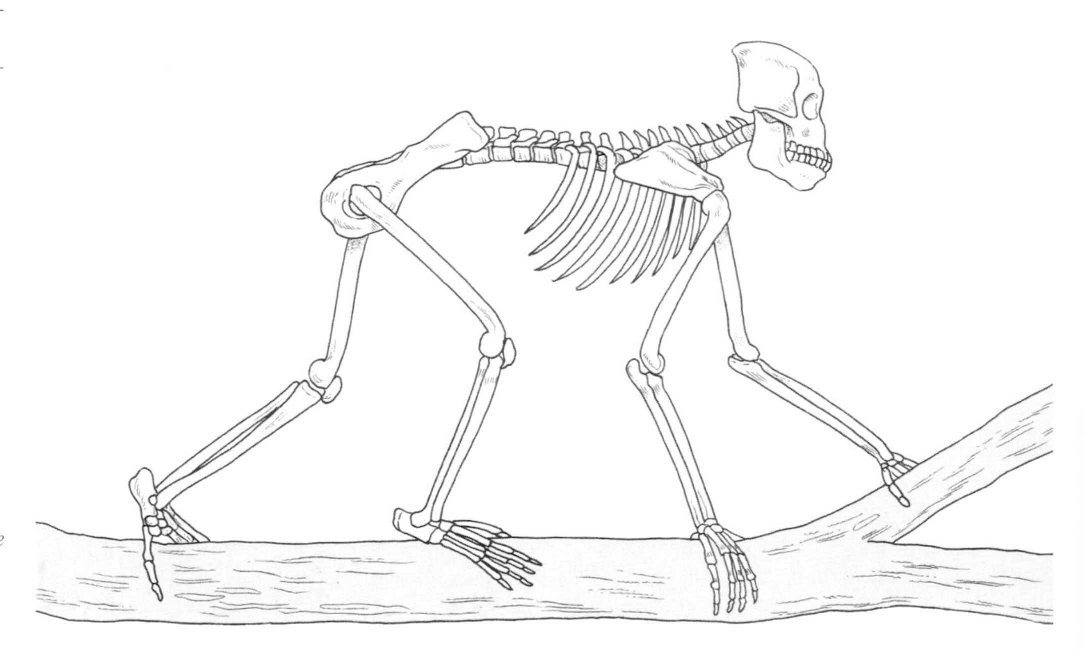

Reconstruction of the Proconsul heseloni *skeleton. Most bones of the skeleton are known for this species, and this is a composite reconstruction.*

Proconsul is now known to have had a monkey-like torso which was deep and narrow like that of cats and dogs and other four-footed animals. This contrasts with the torso of living apes where the chest is broad from side to side and shallow from front to back, as in humans. The back was long in *Proconsul*, particularly the lower back (lumbar region with six lumbar vertebrae), the shoulder joint was directed backwards (rather than laterally as in living apes), and *Proconsul*'s scapulae (shoulder bones) were on the side of the body rather than on the back as in apes. All these characters show that *Proconsul* moved around on all four legs in prone position just like monkeys and other mammals, such as cats and dogs.

Hindlegs, feet and hands

The hindlegs of *Proconsul* confirm that it was adapted for a limited range of movement, notably weight bearing during four-footed walking and climbing. In particular the head of the femur, where the upper leg articulates with the hip bone, was not rounded as in apes but was extended laterally, which limits the movement at the hip to more of a forwards-and-backwards motion rather than laterally as in living apes. On the other hand, it was not as limited in this respect as are the monkeys, so that in this character, *Proconsul* was intermediate between monkeys and apes.

The foot of *Proconsul* was powerful and grasping, with a widely divergent big toe and powerful grasping muscles, as in the living apes, and so it would have been able to grasp branches with its feet as well as its hands. All this adds up to *Proconsul* being an active tree-climber, although it was more of a slow climber than an agile leaper. It may be that some of these characters may have been precursors to the primary adaptation seen in the living apes, that of suspensory locomotion, whereby the body is supported in trees suspended below branches rather than being supported quadrupedally on the tops of the branches. This will be discussed further in relation to *Dryopithecus* and *Oreopithecus* (pp. 110–113).

One interesting aspect of the *Proconsul* hand is that it is proportioned very like a human hand. The hands of the living apes are extremely elongated

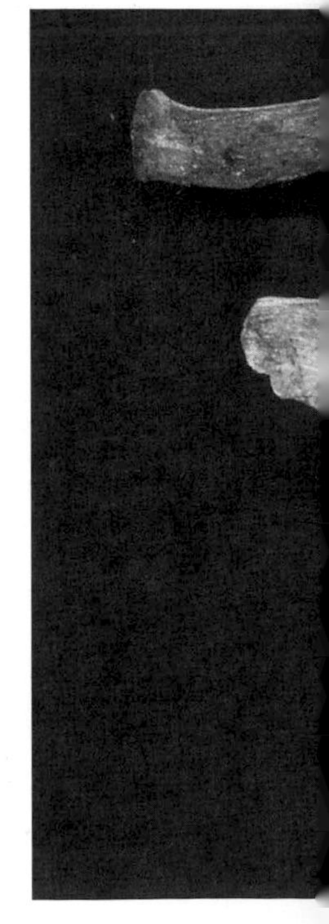

Four of the limb bones from the original skeleton of Proconsul heseloni *found on Rusinga Island in 1951. They are two humerus fragments, a distal ulna (bottom left) and a nearly complete radius at the top.*

Flesh reconstruction of Proconsul heseloni. *This shows it as an active above-branch climber in trees, moving on all four legs, which are equal in length, but probably it was not able to leap from tree to tree as some monkeys do today. Since there is some evidence that it lived in more seasonal environments, it may have come down to the ground when moving from tree to tree, or one patch of woodland to another.*

and the thumb is short, so that when a chimpanzee tries to pick up a small object, it has to hold the object between the thumb and the side of the hand rather than between finger tips as we do. *Proconsul* had hand proportions similar to the human hand, and the thumb joint was also similar to that present in the human hand, so it might well have been able to grasp small objects very precisely.

At this stage it cannot be said with any confidence that *Proconsul* either was – or was not – an ape. What can be said is that if it is identified as an ape, it is certainly the most primitive member of the superfamily, and in overall appearance and behaviour it had little in common with any of the living apes. It lived in a mixture of environments, in tropical forest in some sites, particularly *P. major* and *P. africanus* at Songhor, while some of the Rusinga sites indicate drier, more seasonal environments, although still closed woodland for such species as *P. heseloni* and *P. nyanzae*.

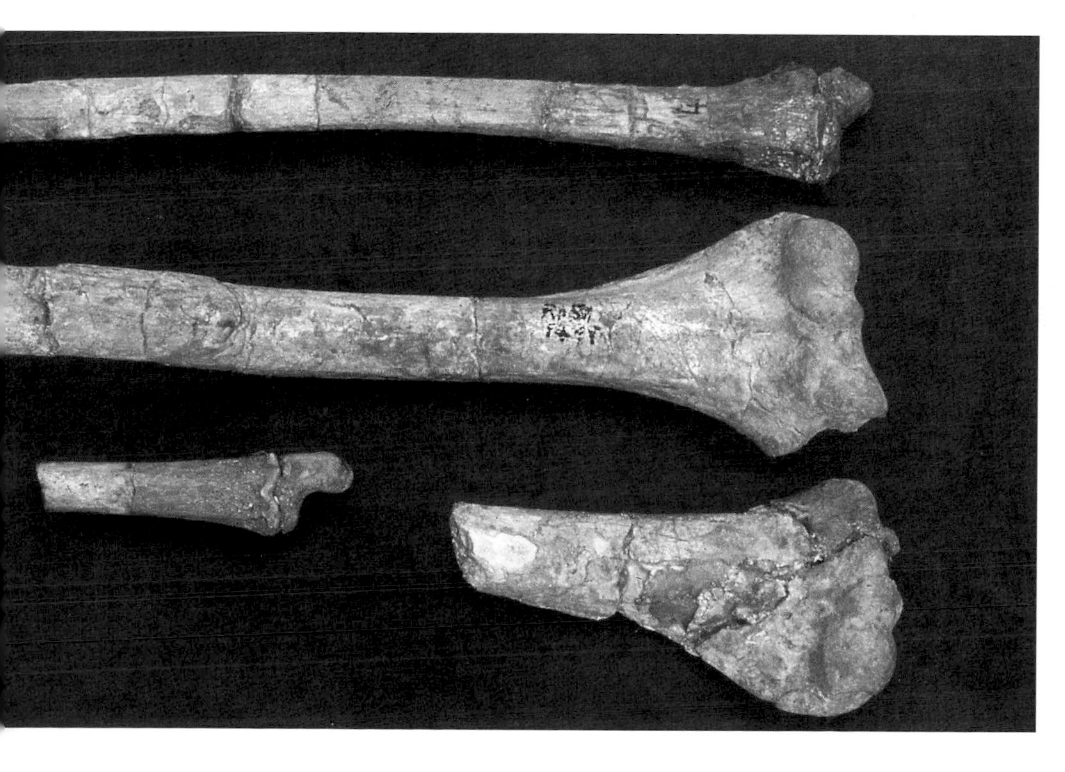

Middle Miocene African Apes

(Right) This Morotopithecus bishopi *lumbar vertebra is in some respects similar to those of living hominids. It is more advanced than other early Miocene apes such as* Proconsul.

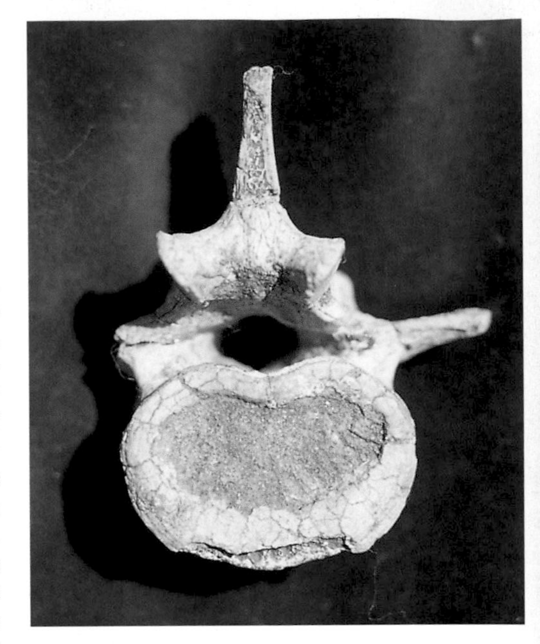

Two groups of fossil ape are known from around 17–15 million years ago in Africa. One of these is still very similar to *Proconsul*, a recently described partial skeleton from northern Kenya, *Nacholapithecus kerioi*. It is similar also to another recently described partial skeleton from the same part of Kenya, *Equatorius africanus*, and together they may belong in the same group as, although some 2 million years later than, *Afropithecus* ('African ape'), known from Kalodirr in northern Kenya. This site has already been mentioned as the place of discovery of *Simiolus*, and a third species is also known from the same site, *Turkanapithecus* ('Turkana ape'), all dated to about 17 million years ago.

(Below) The heavily buttressed skull of Afropithecus. *The ridge along the top of the skull, called the sagittal crest, was for the insertion of the large jaw muscles that were associated with the enlarged teeth, especially the incisors and premolars.*

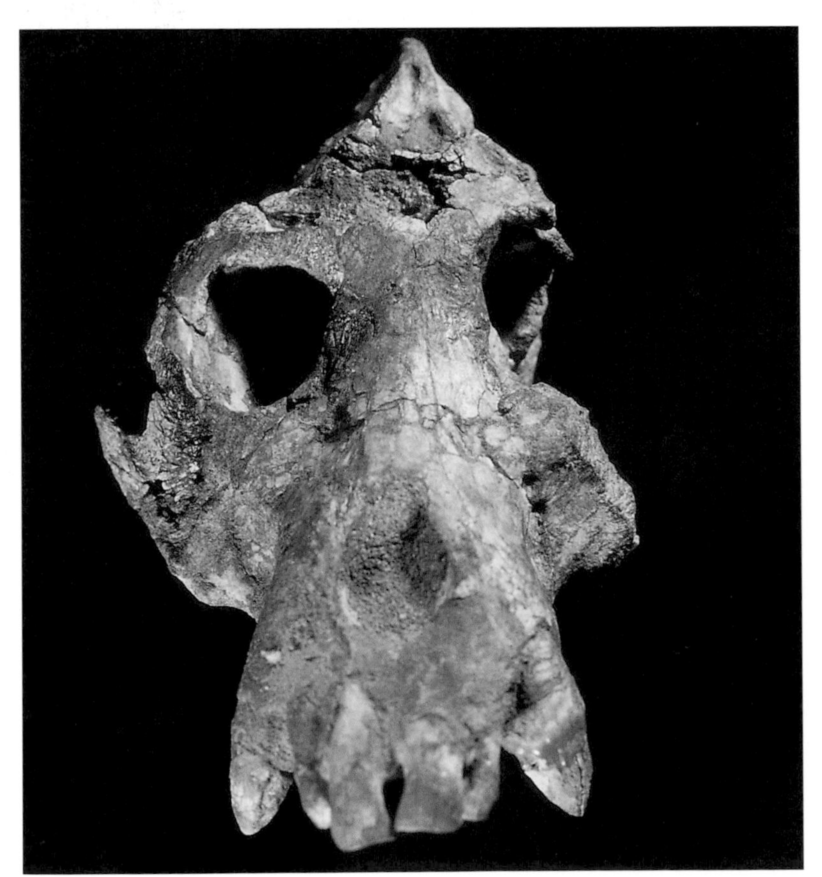

Afropithecus and *Morotopithecus*

The *Afropithecus* skull is one of the most complete specimens of fossil ape known, with complete face and upper jaw and the front part of the skull looking rather gorilla-like and about the size of a small gorilla. The skull had a very long and projecting face as in modern apes, but the shapes of its nose and eyes were quite different, being narrower than in living apes. It had large front teeth, and the bone in which the incisors are implanted, the premaxilla, was also very long and projecting. All this, together with the thick enamel on the teeth suggest that *Afropithecus* was adapted to an extremely hard diet that required large and robust teeth protected by thick enamel.

Other species are known that are similar to *Afropithecus*, both in Kenya and in Uganda. One of these was first described in 1996 as *Morotopithecus bishopi*, based on a specimen originally attributed to *Proconsul major*. This is from slightly older deposits in Uganda but is considered here with *Afropithecus* because it has similar adaptations of the skull and teeth. Like *Afropithecus*, *Morotopithecus* ('Moroto ape') has massive front teeth and premolars and a long face, and it also appears to be adapted for a diet of extremely hard food except again for its relatively lightly built jaws. It is also remarkable for having part of the backbone (vertebra) that is similar to that of living apes and different from monkeys.

Kenyapithecus wickeri

Equatorius and *Nacholapithecus* mentioned above are 2 million years later than the afropithecines, and they are considered by some to be derivatives of this group. They retain many primitive characters that they share with *Proconsul*, and the evidence is still insufficient to be sure of their relationships. There is also controversy over their relationship with a later fossil ape, *Kenyapithecus wickeri* (Kenya ape found by (Fred) Wicker), but since they

(Left) The fragmentary remains of the femur of Morotopithecus bishopi *compared with a chimpanzee femur.*

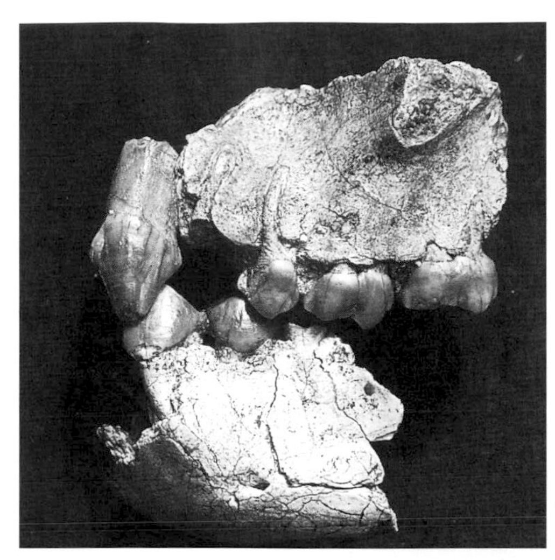

(Right) Maxilla and mandible of Kenyapithecus wickeri. *The association of these specimens cast new light on the interpretation of the phylogenetic significance of this species, formerly thought to be ancestral to humans.*

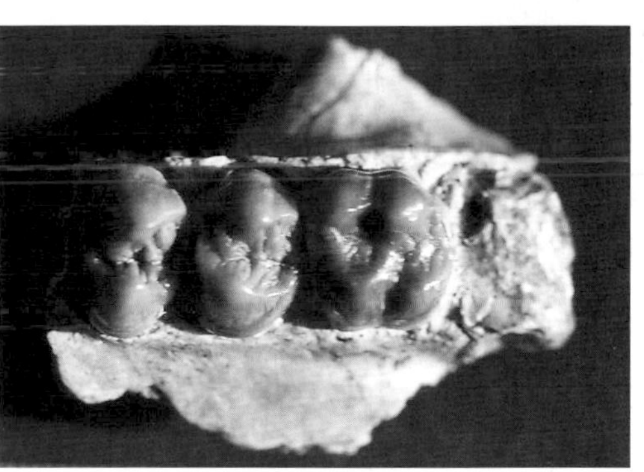

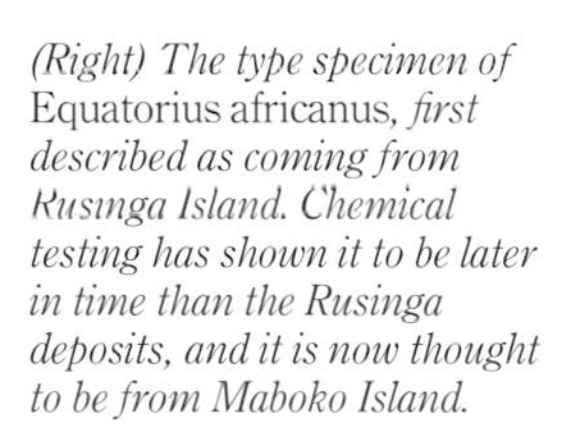

(Right) The type specimen of Equatorius africanus, *first described as coming from Rusinga Island. Chemical testing has shown it to be later in time than the Rusinga deposits, and it is now thought to be from Maboko Island.*

lacked key specializations (synapomorphies) present in *Kenyapithecus* there is no case for this. Louis Leakey named this species in 1962 at a time when Elwyn Simons was active in promoting the affinities of an Asian ape called *Ramapithecus* ('Rama's ape', after the Indian god) with human ancestry. A sort of competition developed between Leakey and Simons, each hoping to find an earlier human ancestor than the other. Since both were articulate and active researchers, their rivalry did much to generate interest in ape evolution in aspiring students. Each had their adherents, and at times the rivalry became very intense.

Simons' most eminent student was David Pilbeam. Together they revolutionized the study of fossil apes with probably the most influential paper ever written on the subject in 1965. Simons and Pilbeam referred *Kenyapithecus wickeri* to their Asian species, *Ramapithecus punjabicus*. *Ramapithecus* in those days was thought to have been an early hominin ancestor (a view now discounted – see p. 106), and *Kenyapithecus wickeri* was also very human-like with large teeth that were covered with a thick layer of enamel so that the crown appeared rounded and smooth, more like human teeth than apes'. In addition the jaws were extremely robust, the front teeth appeared small, and the face appeared to be short, again all similarities with humans. On the other hand, the lower jaw was more ape-like, particularly the presence of a massive and shelf-like buttress across the front where the two halves of the jaw meet. Using this evidence together with the already known upper jaws, the face could then be reconstructed, which showed that it was in fact ape-like in all characters.

(Below) Reconstruction of the lower face of Kenyapithecus wickeri *based on the associated mandible and maxilla shown above. It is very different from the longer faces of the living great apes.*

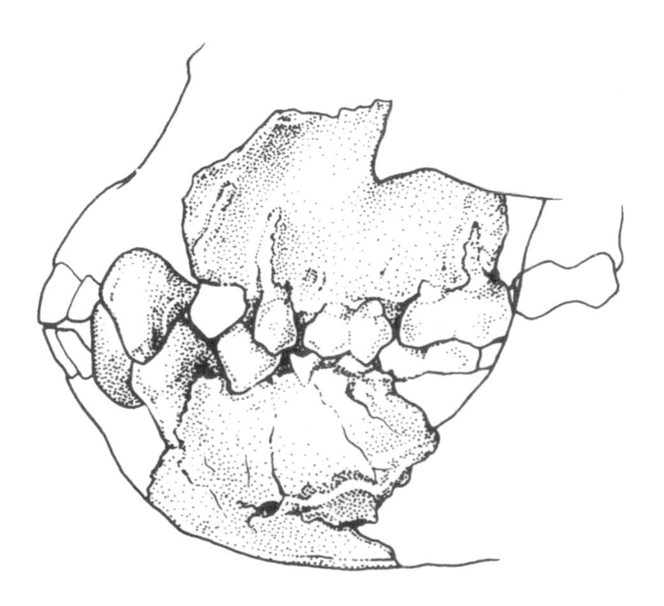

Fort Ternan, Kenya, showing (above) the 1974 excavation at the back of the massive trench cut out by previous excavations by Louis Leakey. A platform (right) was cut out behind the trench to sample the fossil-bearing channel-like feature that runs across the site.

There is also a single limb bone from Fort Ternan in Kenya, the upper arm bone or humerus, and this belongs to *Kenyapithecus wickeri*. It is very like the fossils from Maboko, indicating that the Fort Ternan ape was also partly ground-living, moving about on all-fours. None of these apes would have been entirely ground-living any more than any of the living apes are (except for the mountain gorillas), but it is likely that they moved around their territory on the ground since they were not living in tropical rain forest, with its dense tree canopy, but in more open woodland conditions where it would have been difficult for a large-bodied species to move from one tree to another. This evidence for some degree of ground-living, however, occurring as it does so early in ape evolution, will be seen to be of great importance in understanding about human evolution.

In addition to Fort Ternan, there are several other sites in Kenya that may provide additional evidence on *Kenyapithecus*, but unfortunately these have not been adequately published yet. There is also an interesting but tantalizing fossil from southern Africa, the only Miocene African ape known beyond the limits of East Africa. This is *Otavipithecus namibiensis* (Ape from Otavi in Namibia), which seems to have affinities with *Afropithecus*. The critical question with regard to these new fossils is whether they will prove to belong with *Kenyapithecus* or with *Equatorius*, or in other words do they have advanced characters like the former, or primitive like the latter. This becomes critical when the evidence from outside Africa is considered, and this will be discussed next.

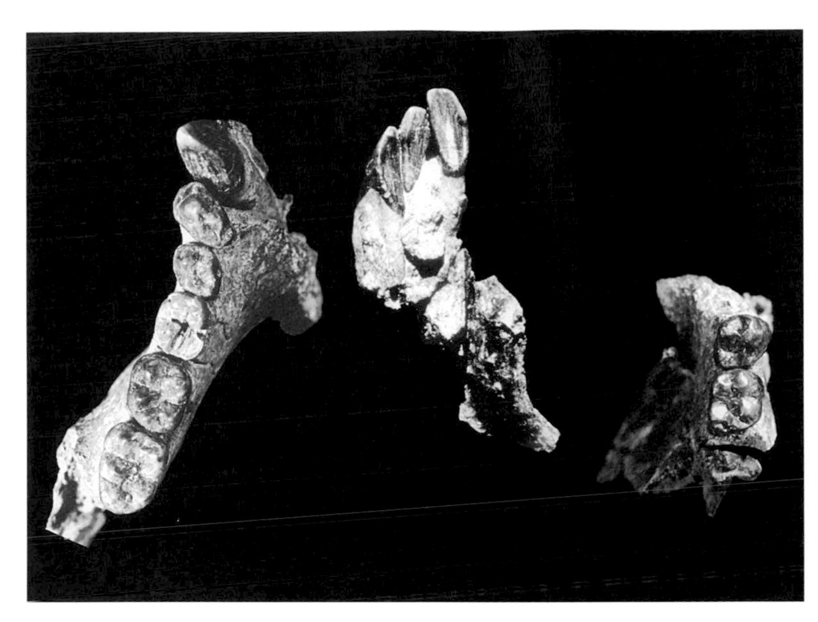

(Above) The recently described specimens of Equatorius africanus *from Kipsaramon in Kenya.*

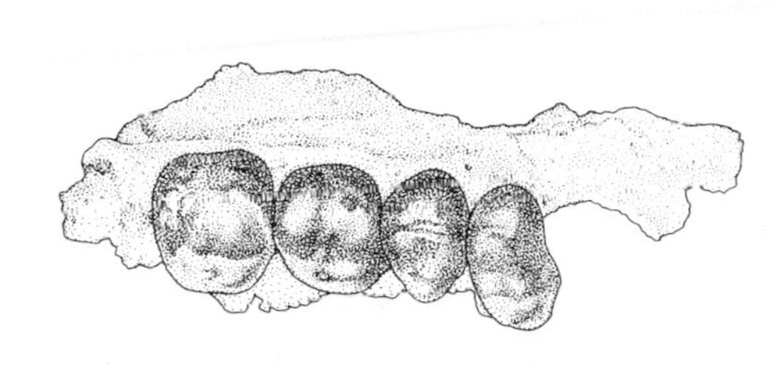

(Left) Drawing of Heliopithecus leakeyi, *a fossil ape from Ad Dabtiyah in Saudi Arabia. It is similar to* Afropithecus *and is the northernmost fossil ape from the African continent at a time when fossil hominids were still restricted to Africa.*

(Right) Reconstruction of Heliopithecus leakeyi. *Little is known of the Saudi Arabian environment during the early Miocene, but indications are that it was forested.*

The Exit from Africa

Up to this point, all the fossil apes described have come exclusively from Africa. From the end of the Eocene, temperatures across the world fell (see p. 83) and the tropical forests of North America and Eurasia receded. As a result, primates had almost completely disappeared from these northern continents by the end of the Oligocene and into the early Miocene. During most of this time, Africa was completely cut off from Europe and Asia, but with movements of the continents Africa was drifting north and made first contact with Europe in the late Oligocene to early Miocene. About 20 million years ago there is evidence of a land connection separating the Arabian Gulf and the eastern Mediterranean: formerly these two seas were connected as part of a larger sea called the Tethys Sea which ran between Africa and Europe. This

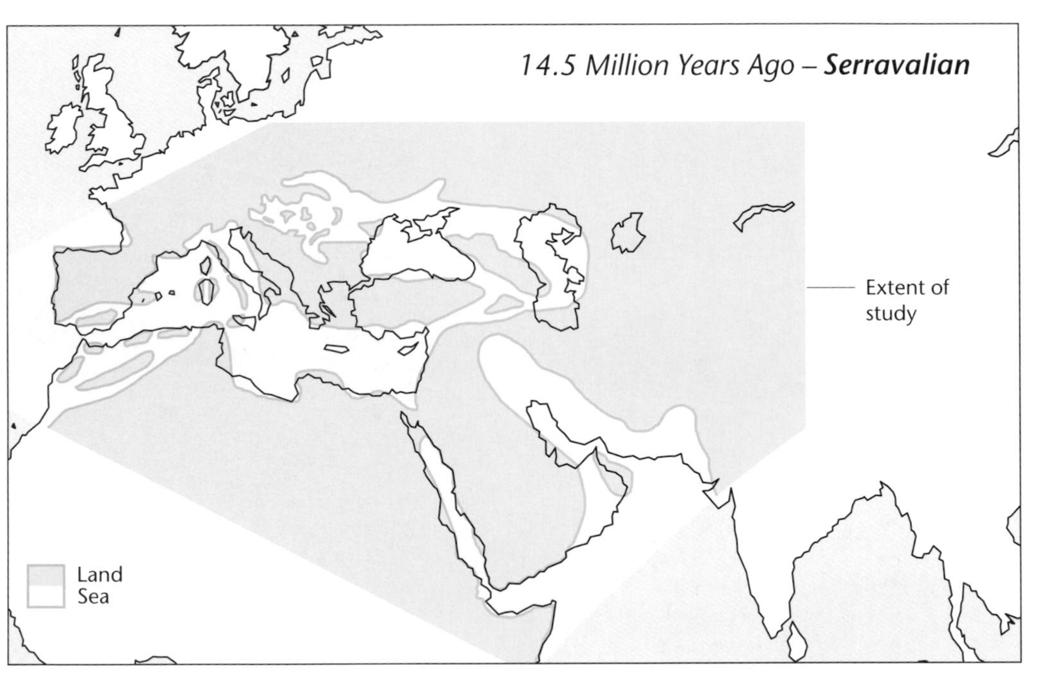

Three stages in the joining of the African plate to Eurasia. No contact was present in the Aquitanian, 21–23 million years ago; there was a narrow contact in the late Burdigalian, 17–18 million years ago, across which some groups of mammals migrated, including Dionysopithecus, *but no hominoids at this stage; and broader contact had been made by about 14–15 million years ago during the Serravalian, which is the time of first appearance of hominoids in Europe.*

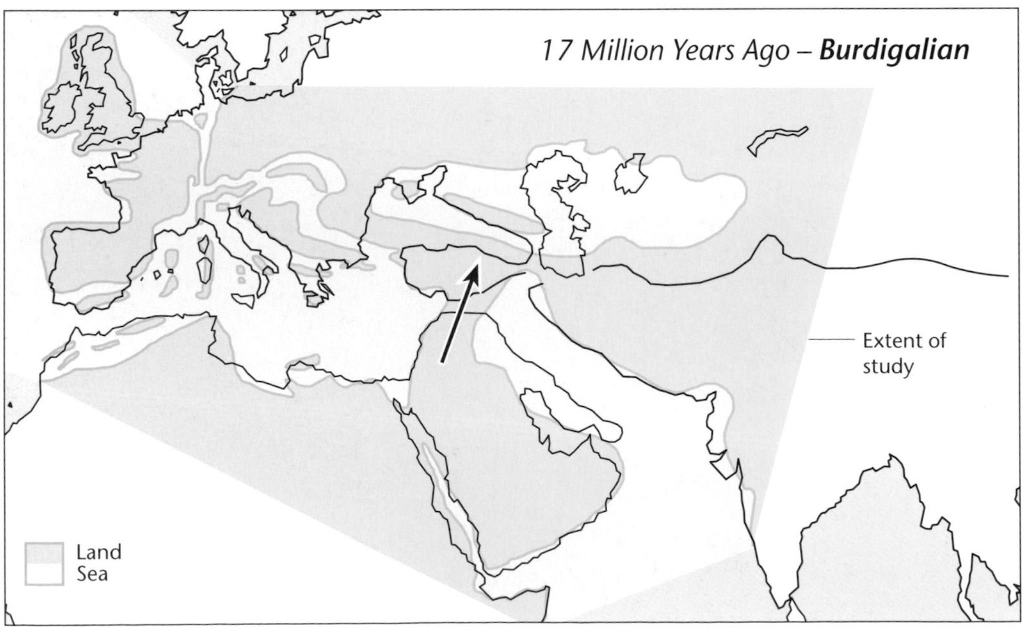

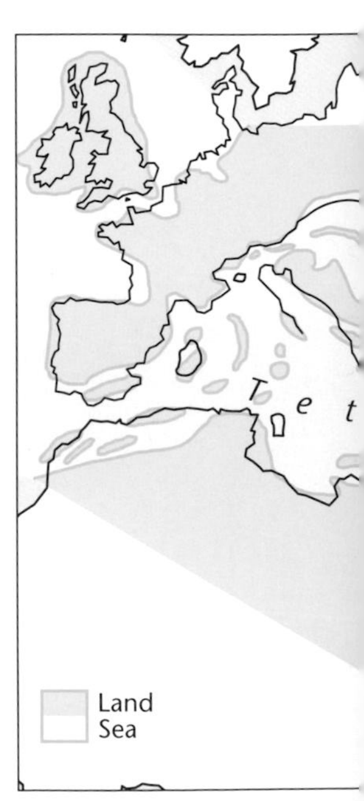

period coincides with the emigration of proboscideans (elephants) from Africa into Europe and Asia, and it was probably also the time when *Dionysopithecus* ('Dionysus ape'), a small-bodied ape very similar to the African genus *Micropithecus*, moved from Africa to Asia. It is known from several places in China and Vietnam and from deposits aged about 17 million years ago.

Griphopithecus

At around 14.5 million years ago, the geographic isolation of Africa ended for the final time, and fossil apes appear for the first time in Europe. At this time the land connection was further east, connecting Africa with the Middle East, and there were major migrations of rodents (mice and rats), bovids (antelopes) and primates. The earliest fossil apes known from Eurasia come from central and eastern Europe. A single tooth is attributed to the genus *Griphopithecus* from southern Germany, and slightly later is the material from the Czech Republic and Turkey. At the same time another group of ape-like primates appears in central Europe, the pliopithecines, a very primitive group of unknown origin. Here we are concerned with *Griphopithecus*, for which a large sample now exists in Turkey from the sites of Paşalar and Candir, the former being one of the three Miocene sites described earlier. This site is at least 14–15 million years old, similar in age to Fort Ternan.

Griphopithecus from Paşalar was originally known from a small collection of teeth made by Heinz Tobien in 1970. Since that time nearly 2,000 specimens have been found, including 15 upper and lower jaws and 17 hand and foot bones, but by far the most common element continues to be single teeth. As described earlier, this is due to the highly destructive taphonomic factors active at this site. It is not likely that a more complete skull or skeleton will ever be found at Paşalar unless by a stroke of complete luck, but there is enough now to give a good idea of what *Griphopithecus* was like. The lower jaw of *Griphopithecus* is very robust (it is not known for the second species described below), like that of the Kenyan fossils, and the front of the lower jaw is similar in having a huge buttress, but the upper jaws are less robust.

The molar teeth of *Griphopithecus* are almost identical to those of *Kenyapithecus wickeri*, so much so that when Tobien's collection was first described it was actually assigned to the African species. It is now recognised that they are also similar to *Equatorius africanus*, mentioned above (p. 97). The anterior teeth, incisors, canines and premolars, are also similar to those of *Equatorius*, but there is a second species at Paşalar that is different. This has a suite of characters in the anterior teeth that are recognized as being derived characters shared with *Kenyapithecus* to the exclusion of the shared primitive characters in *Equatorius* and *Griphopithecus alpani*. There is thus good evidence of relationship between the new species at Paşalar

(Left) Two incisors from Paşalar, Turkey, showing examples of two morphologies of fossil hominid present at the site. On the right is Griphopithecus alpani, *a primitive but thick-enamelled species, and on the left is a new species of* Kenyapithecus, *indicating the earliest link between African apes and a new radiation of fossil apes in Eurasia.*

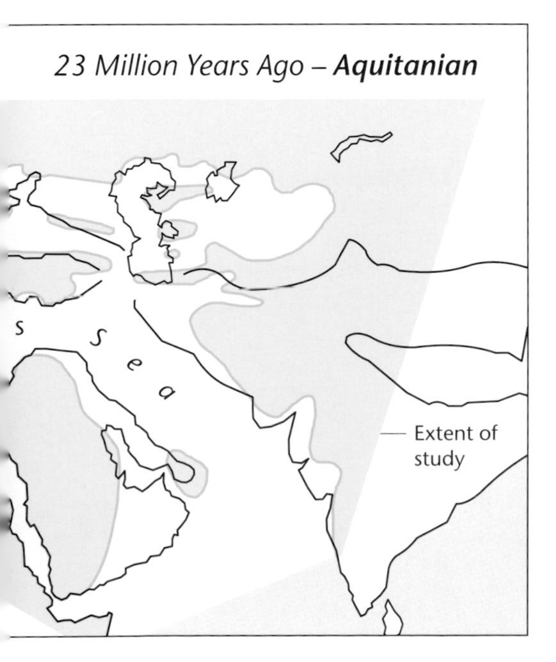

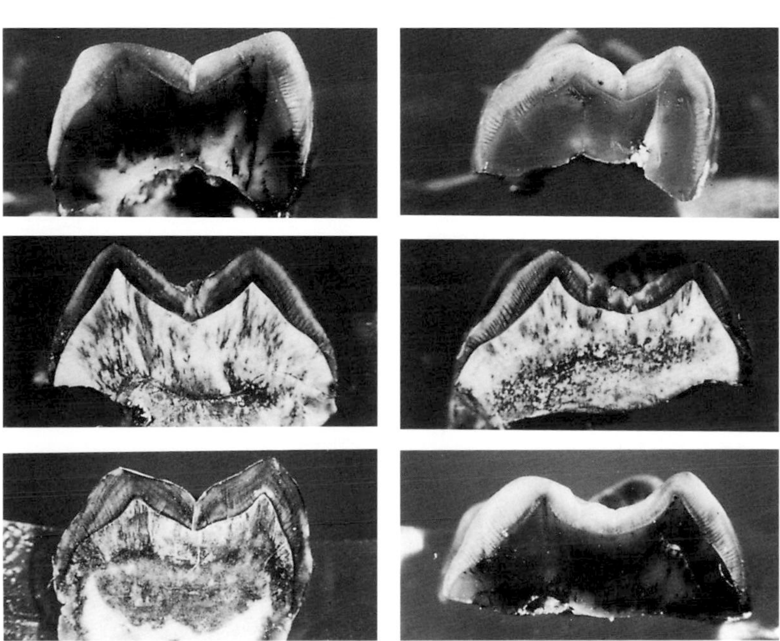

(Below) Many of the early Miocene fossil apes and ancestral catarrhine primates had thin enamel on their teeth, as do chimpanzees and gorillas today, but many of the middle Miocene species, including the ones shown here from Paşalar, had thickened enamel, which is considered an adaptation for strengthening the teeth to withstand a harder, coarser, diet.

and *Kenyapithecus*, so much so that they are being put into the same genus. Conversely, the great similarity between *Griphopithecus* and *Equatorius* is not evidence of relationship other than the fact that they both conform to the generalized and probably widespread ape morphology of the middle Miocene.

One of the great advantages in having access to a collection like that of *Griphopithecus* from Paşalar is that it can produce so much information on that particular window of time. Work on the microwear of *Griphopithecus* has shown that it was adapted to a diet of hard fruits, different from the African apes and most similar to the diet of the orang utan. The postcranial anatomy shows that *Griphopithecus* was partly tree-living and partly ground-living, but with adaptations still little different in general from the pattern already described for the earlier African fossil apes.

Both of these kinds of behaviour fit well with the reconstruction of the environment at Paşalar in the middle Miocene. The sediments indicate that the climate was seasonal, with alternate wet and dry periods. The mammal fauna was similar to present-day faunas from subtropical to tropical woodlands. These earliest fossil ape species living in Europe during the middle Miocene thus lived in subtropical environments similar to the monsoon woodlands that exist today in central and northeast India. These areas today consist of a mosaic of wooded patches interspersed with open tall grass meadows, and *Griphopithecus* occupied a similar ecological niche in these subtropical forests as the earlier Miocene apes in tropical East Africa. The climate during the middle Miocene was warmer than today, so that much of the land around the Mediterranean was subtropical and offered habitat for apes where none is possible today.

Reconstruction of the environment of Griphopithecus alpani *based on the abundant faunal evidence from Paşalar supplemented by carbon isotope and stratigraphic evidence. This fossil ape was almost certainly partly terrestrial, for although it lived in subtropical forest environments, the forests were not dense enough to allow movement through the trees for an animal of this size.*

Pierolapithecus catalaunicus

Another fossil ape has recently been described from the middle Miocene of Spain. This is *Pierolapithecus catalaunicus* ('Ape from Pierola in Catalunia') dated to 12.5–13 million years ago. It is slightly later in time than *Griphopithecus alpani*, and it is very different in morphology. The molar teeth are elongated, an uncommon condition in hominoid primates, and while the skull shows similarities with *Afropithecus* with its long sloping face, it also has some advanced great ape characters such as the deep palate and high zygomatic (cheek) region. More importantly, however, the skull was found associated with parts of the postcranial skeleton (although no limb bones), and these also show a mixture of primitive and derived characters. The hand has primitively short fingers, similar to those of other fossil apes and different from living apes (and *Dryopithecus*, see p. 110), and this indicates a lack of suspensory abilities. On the other hand, characteristics of the clavicle, ribs and vertebrae indicate a broad shallow thorax like those of living great apes, indicating upright posture very different from that of monkeys and most other Miocene apes. *Pierolapithecus catalaunicus* appears therefore to have been an upright climber, much as chimpanzees are today, but lacking the suspensory ability seen in all living apes. This calls into question many of the currently accepted ideas of ape evolution, which assume that the development of upright posture in apes was associated with the ability to suspend the body beneath branches of trees. However, the presence of so many great ape characters places this species in an ancestral position to the living apes and humans (Hominidae), and in so doing, it excludes earlier fossil apes such as *Proconsul*, *Afropithecus* and *Griphopithecus* from that position.

Ankarapithecus – a Fossil Enigma

(Below) View of the excavation in the Sinap deposits in Turkey. This is the site where the earlier partial skull of Ankarapithecus meteai *came from, and a large excavation was started in 1994. The skull shown on the right was found the following year.*

In 1996 a new fossil ape skull was described in Turkey, and it changed many things in palaeoanthropology. The skull was found in 1995 during excavations in 9–10-million-year-old deposits together with many fossil mammals. The skull has been assigned to the species *Ankarapithecus meteai* ('Ape from Ankara') on the basis of its similarity with a larger but less complete partial face described in 1980. A lot is known about its way of life (which will be described later in the book, pp. 204–205), but here we are going to show how the discovery of this single skull has changed our outlook on ape and human evolution.

Looking both east and west

The skull preserves much of the face and the front of the skull. The body size of the individual to which it was once a part has been estimated to be 29 kg (64 lb), approximately the size of a female bonobo (pygmy chimpanzee). The lower jaw is very robust and the enamel on the teeth appears to be thick, and the shape of the lower jaw is the same as in living female apes. The skull is from a female individual whereas the 1980 face is from a male, and it is interesting that the size of the canine tooth is relatively small in both male and female compared with the living apes. This is a feature that used to be thought of as human-like compared with the apes, the males of which have large projecting canines.

Differences in size between the front teeth (incisors) are extreme, which is like the condition in the living orang utan but different from the living African apes. On the other hand, the eyes are about as broad as they are tall which is more similar to the African apes than the generally narrow and tall orbits seen in the orang utan and *Sivapithecus* ('Siva's ape', after the Indian god, see next section). In side view the profile of the face is straight but has well-developed ridges above the eyes. Overall this profile is similar to *Graecopithecus* ('Greek ape', see pp. 111–112) and *Dryopithecus* and differs from the concave face of the orang utan and *Sivapithecus*.

Ankarapithecus has further similarities with the orang utan and *Sivapithecus* in the connection between the floor of the nose and the mouth through the hard palate. It is smooth in *Ankarapithecus* as in the orang utan and *Sivapithecus*, rather than stepped as in *Dryopithecus* and the African apes. The flat, forward-pointing cheek region is shared with *Dryopithecus* in addition to

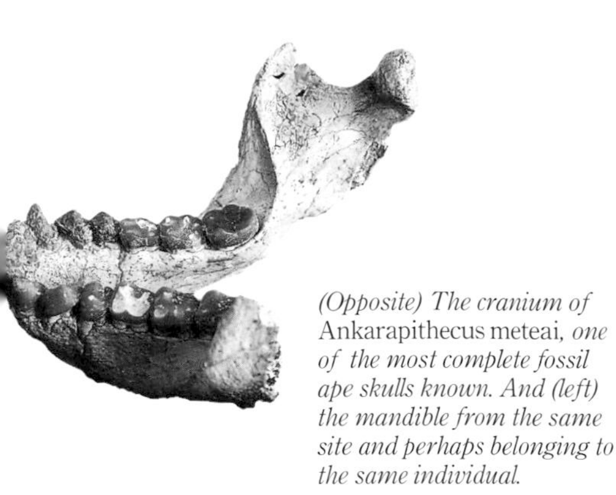

(Opposite) The cranium of Ankarapithecus meteai, *one of the most complete fossil ape skulls known. And (left) the mandible from the same site and perhaps belonging to the same individual.*

Sivapithecus and the orang utan. These characteristics, plus a narrow distance between the eyes, have all been linked with the orang utan-*Sivapithecus* evolutionary group. The presence of some but not all characteristics in *Ankarapithecus*, and its similarities in other respects with *Dryopithecus* and the African apes, suggests that these characters are not as strongly linked together in hominoid evolution as was formerly thought and make it difficult to place *Ankarapithecus* in its evolutionary context.

There are also characteristics shared between *Ankarapithecus* and *Graecopithecus*, such as the broad, low front teeth, eye shape, the small canines mentioned above, and the presence of brow ridges. These characters differ, however, from those present in the African apes and humans, and even the brow ridges are different from that seen in the African apes. Visually striking differences between *Ankarapithecus* and *Graecopithecus*, such as a narrow versus broad distance between the eyes, are no more marked than between the closely related bonobo *Pan paniscus* and *Gorilla gorilla*.

The complex of skull characters that individually are found in various configurations in other fossil apes are found together in *Ankarapithecus*. One set of characters links it with forms found to the east, *Sivapithecus* and the orang utan. A second set of characters links it with forms found to the west, *Graecopithecus* and *Dryopithecus*. A full consideration of the exact relationships among what is now a diverse and better represented sample of late Miocene hominoids will require a clear and detailed definition of characters, but what it shows for now is that there is no such thing as an eastern morphology linked with the orang utan and a western one perhaps linked with the African apes. Parts of both sets of characters are present in *Ankarapithecus*, so that if it were to be linked with one or other descendant group, the non-fitting characters for that group would have to be interpreted as convergent, that is not evolutionarily significant, and if that is the case these same characters cannot be used for the other fossil apes.

For example, the close-set eyes have been used to propose an evolutionary link between *Sivapithecus* and the orang utan, and as this is also present in *Ankarapithecus*, it also has been linked with the orang utan. If, however, *Ankarapithecus* is not related to the orang utan, the narrow face would be better interpreted as an ancestral ape character. Similarly, the presence of brow ridges on *Dryopithecus* and *Graecopithecus* has been used to suggest relationship between these fossil apes and the living African apes; the presence of the same character on *Ankarapithecus* could suggest the same relationship for it, but if it is in fact related to the orang utan then this character also would have to be convergent. In fact many of the characters of *Ankarapithecus* can be looked at more in terms of their function than their phylogenetic or evolutionary significance, and it appears that this fossil ape is related to neither group of living ape, in which case the argument for convergence applies even more strongly. We will use the case of *Ankarapithecus* later on in this book as an illustration of the interplay between function and evolutionary history (phylogeny).

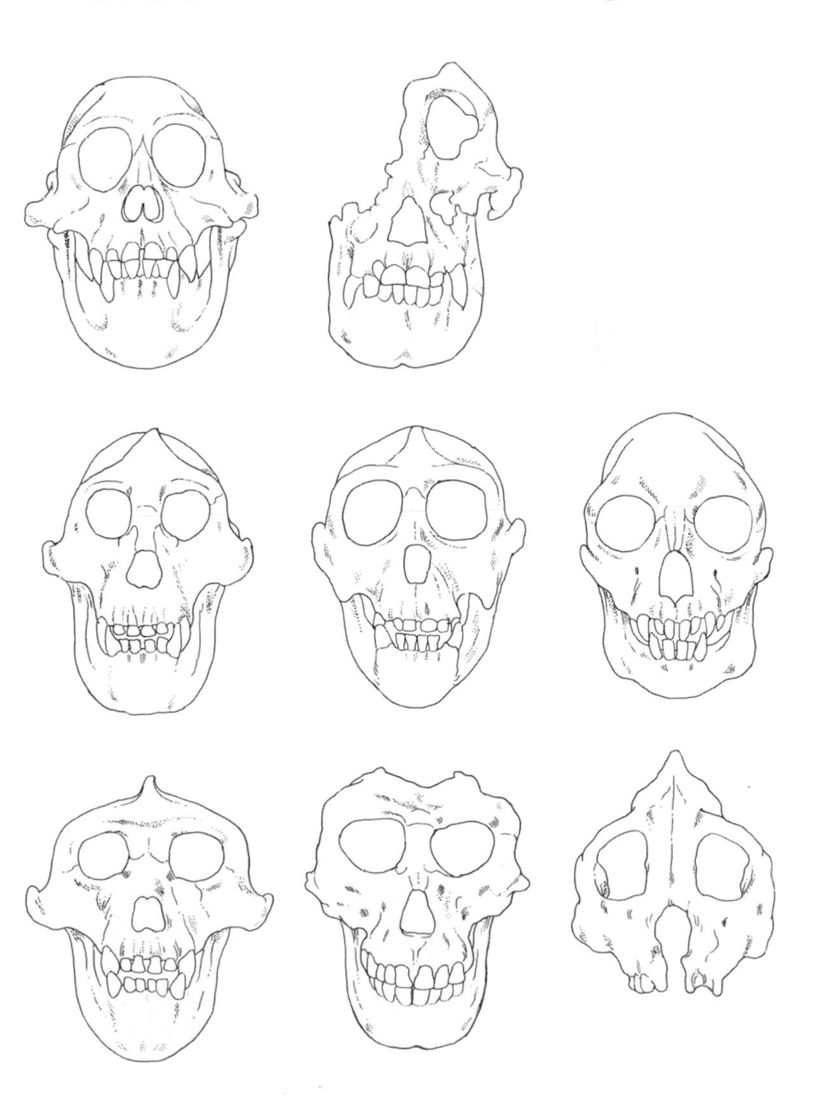

Variation in the form of fossil ape skulls: the top row shows the similarities shared by the orang utan (left) and Sivapithecus *(see next section); the middle row shows the buttressed skulls of* Dryopithecus *(left) and* Oreopithecus *(middle) compared with the much more gracile skull of* Proconsul *(right); and the bottom row compares the skulls of* Graecopithecus *(left) with* Ankarapithecus *(middle), with the much earlier skull of* Propliopithecus *on the right.*

Orang utan Ancestors

Early in the 20th century, extensive fossil collections in the Siwalik Hills in northern India produced two fossil apes. Work had been going on for much of the 19th century without success, but after these initial finds, numerous specimens began to appear, and this abundance has continued in recent decades through the work of the Harvard expeditions to Pakistan in the Siwalik deposits that continued across the border. Fossils know no political boundaries, and in fact the Siwalik Hills are the foothills of the Himalayas, extending all along the southern boundaries of these much greater mountains.

In the 1930s, an American palaeontologist, G. Edward Lewis, found new specimens that he considered were very like those of recent humans. These fossils appeared to have rounded lower jaws, small canine teeth and a non-projecting face, all then thought to be uniquely human characters. Thus was born *Ramapithecus*. For several decades *Ramapithecus* was accepted as the earliest human ancestor, which put the origin of humans back to around 12 million years ago. Three events in the 1960s and 1970s were to show this to be false. One was the application of molecular biology to the study of relationships between species, which showed that the African apes are more closely related to humans than is the orang utan, and their separation was more recent than the time of *Ramapithecus*. The second event was the demonstration that the African specimens that had been identified as *Ramapithecus* did not have the three characters mentioned above as characterizing *Ramapithecus*. The third and most important event was the discovery of new and better fossils that showed that first of all *Ramapithecus* should be grouped with the fossil ape *Sivapithecus*, and secondly that the combined *Sivapithecus* sample was actually related to the orang utan, not to humans.

Links with the orang utan

The new specimens of *Sivapithecus* showed it to be remarkably orang-like. The skull has the same

Reconstruction of the environment of Sivapithecus indicus. *Postcranial bones of this species indicate a partly terrestrial adaptation, and the environment was a mixture of subtropical monsoon forest and more open areas.*

dished shape, concave when viewed from the side; the distance between the eyes is very narrow, unlike almost all other apes (it is also narrow in *Ankarapithecus*, which led to its being identified for a time as *Sivapithecus* also); the cheek region of the face is broad and faces towards the front rather than partly to the side as in most apes (it faces the front also in *Dryopithecus*, which has led some anthropologists to link this fossil also with *Sivapithecus* and the orang utan); and the detailed morphology of the connection between the floor of the nose and the mouth, what is termed the naso-alveolar region, is just like that of the orang utan and different from all other apes, both fossil and recent.

(Above left) The maxilla of Sivapithecus sivalensis, *a small species of this genus that was formerly called* Ramapithecus.

(Above) An orang utan shows off its agility by hanging from one arm whilst using both feet and its other hand to steady a durian fruit.

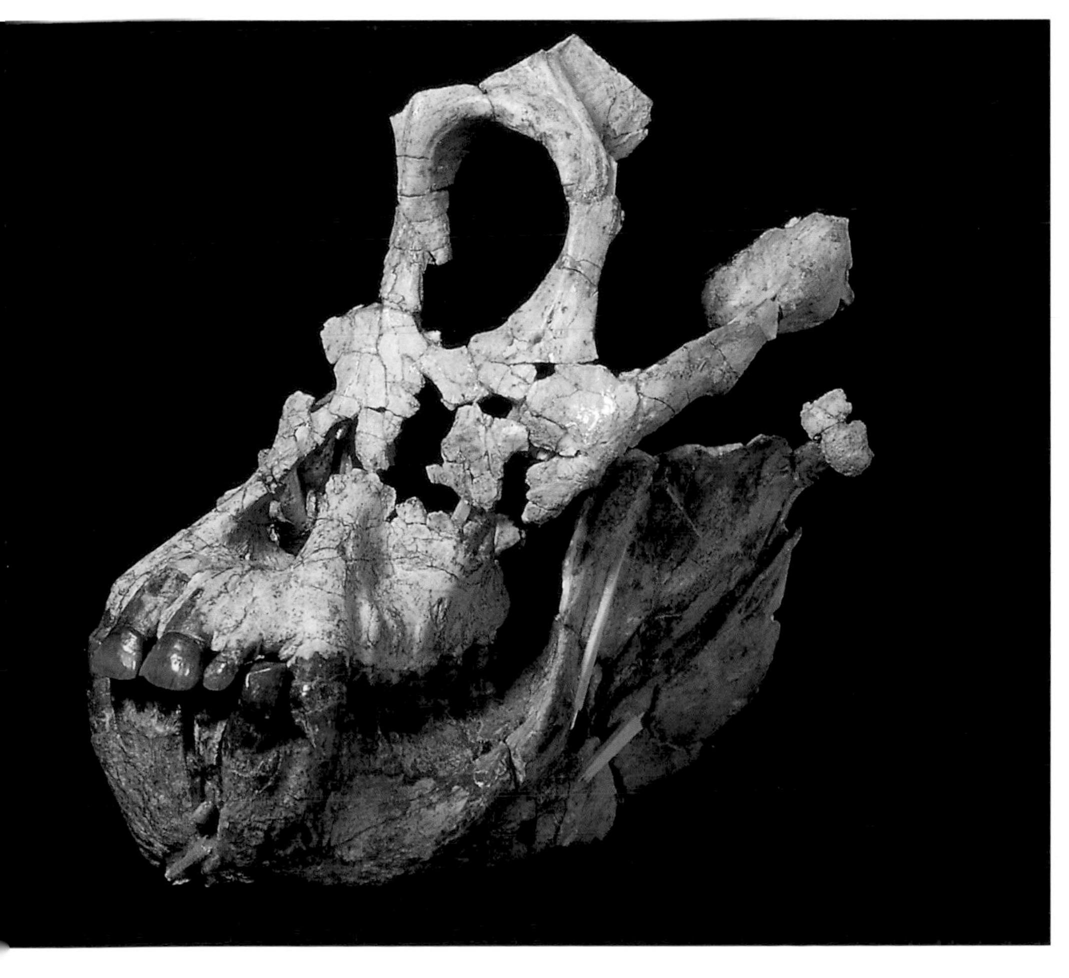

(Left) The skull of Sivapithecus indicus *showing its similarities with the orang utan: the concave or dished face, the narrow distance between the eyes, and the expanded zygomatic (cheek) region.*

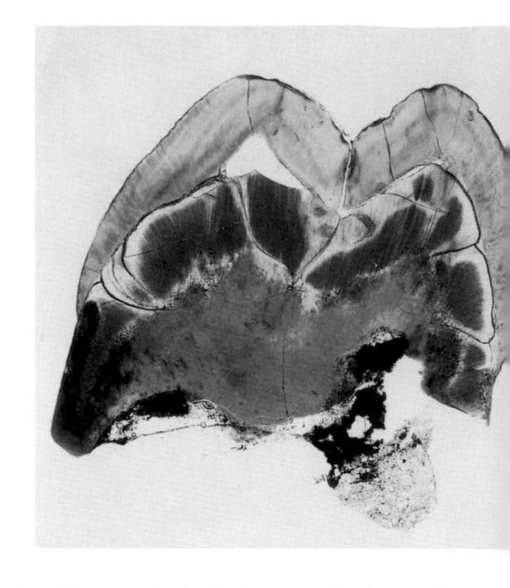

(Left) Four mandibles of Sivapithecus *species showing the size range from* S. indicus, *upper left, to* S. sivalensis *(formerly* Ramapithecus*), right.*

(Right) Cross section of a tooth of Sivapithecus sivalensis, *showing the thick enamel.*

(Below) Comparison of Sivapithecus indicus *skull (middle) with a chimpanzee on the left and an orang utan on the right. Note again the dished face and the lack of brow ridges shared by the fossil ape and the orang utan.*

In many other respects, *Sivapithecus* is just like other fossil apes living at the same time. Its teeth are little different from those of *Kenyapithecus*, *Griphopithecus* or *Ankarapithecus*, and indeed all these fossils have at one time or another been included in the genus *Sivapithecus* on the basis of this similarity. The molars of *Sivapithecus* had thick enamel and were relatively large, as in these other genera, and like them its lower jaw was very robust. Its skull was lightly built, not heavily buttressed as in living great apes. None of these characters serve to distinguish it from other fossil apes. Its arm bones indicate an above branch adaptation although it was almost certainly partly ground-

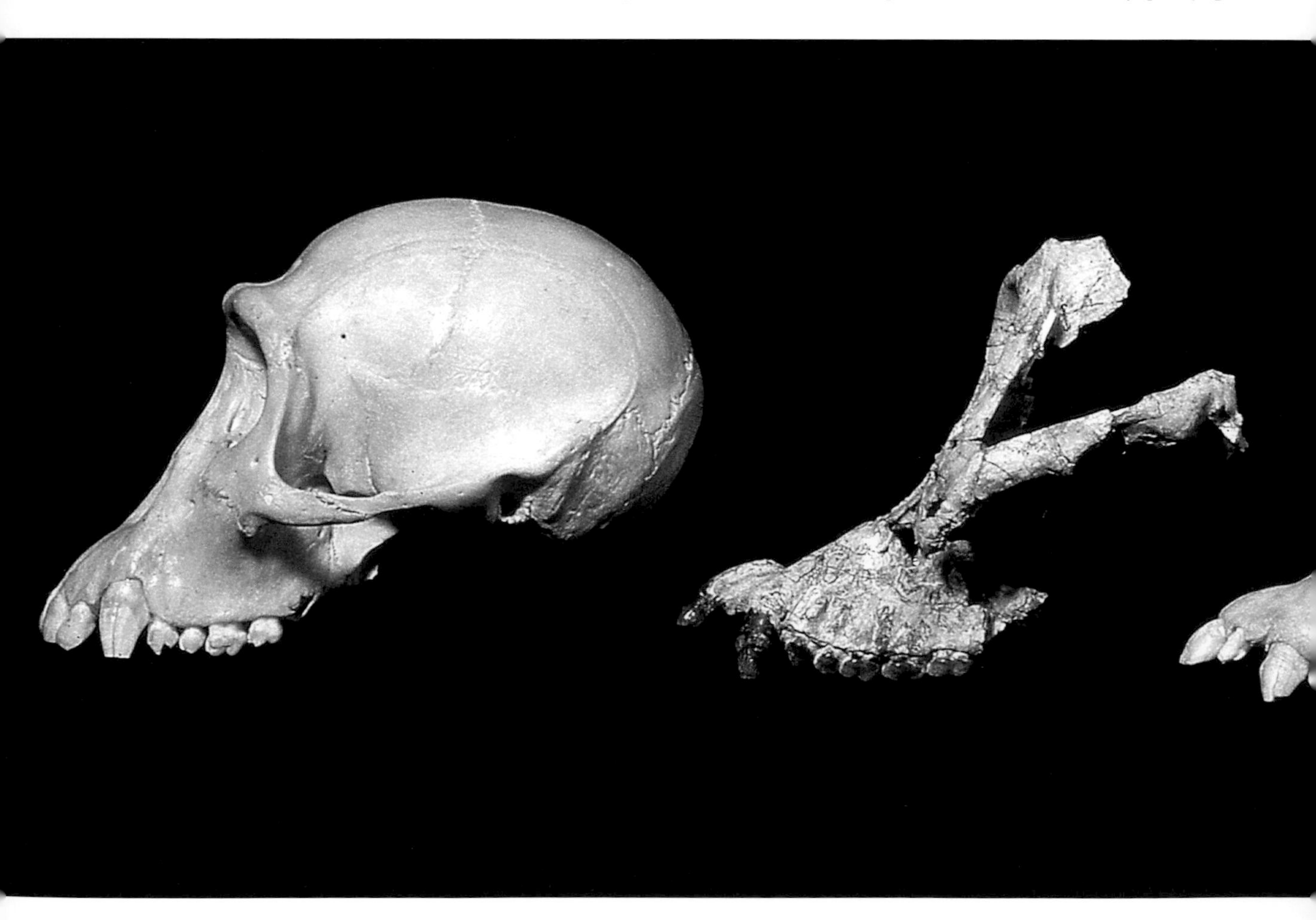

living. All these characters differ from the suite of characters shared by the living apes and presumed to have been present in their common ancestor.

It is important to be sure of the relationship of *Sivapithecus* with the orang utan, for this is the only evidence that we have that links any fossil ape with any living species. It gives us important information on the dating, for instance, of the evolutionary event when the line leading to the orang utan split off from that leading to African apes and humans. The earliest fossil known for *Sivapithecus* is just over 12 million years old, and this tells us that this split must have occurred by then, and the separation of the line leading to humans from the African apes must therefore have been later in time. *Sivapithecus* survived in Pakistan until between 7 and 8 million years ago, a similar point in time to a new supposed orang utan relative, *Khoratpithecus piriyai* ('Ape from Khorat', Thailand, named in honour of the finder, Piriya Vachajitpan). This is similar in both age and morphology to *Lufengpithecus lufengensis* ('Ape from Lufeng') from southwest China. These fossil apes have little similarity to the orang utan, and although they are at the right place at the right time, the nature of their relationship, if any, is unclear.

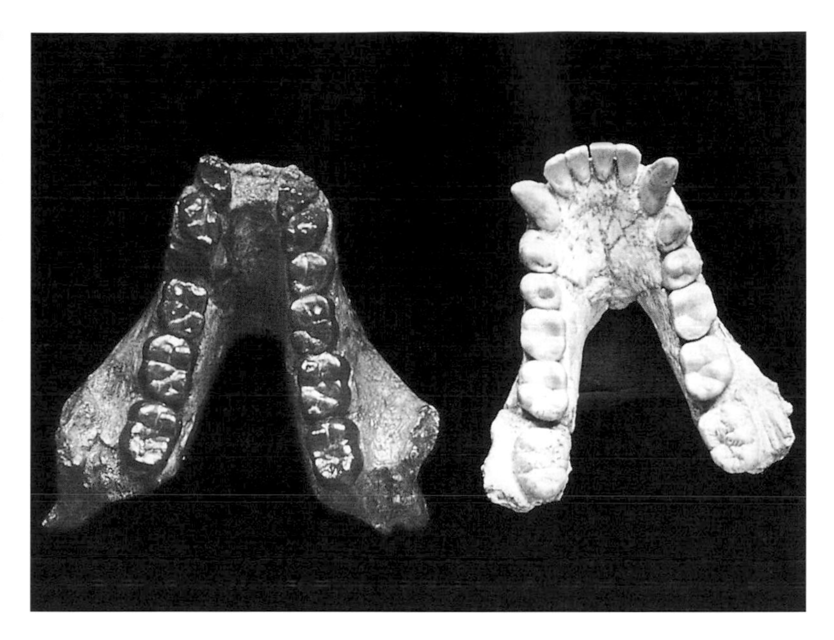

The mandible of Gigantopithecus bilaspurensis *(left), which was contemporaneus with* Sivapithecus *and is probably ancestral to later species of* Gigantopithecus. *It is here compared with a mandible of* Graecopithecus *(see next section).*

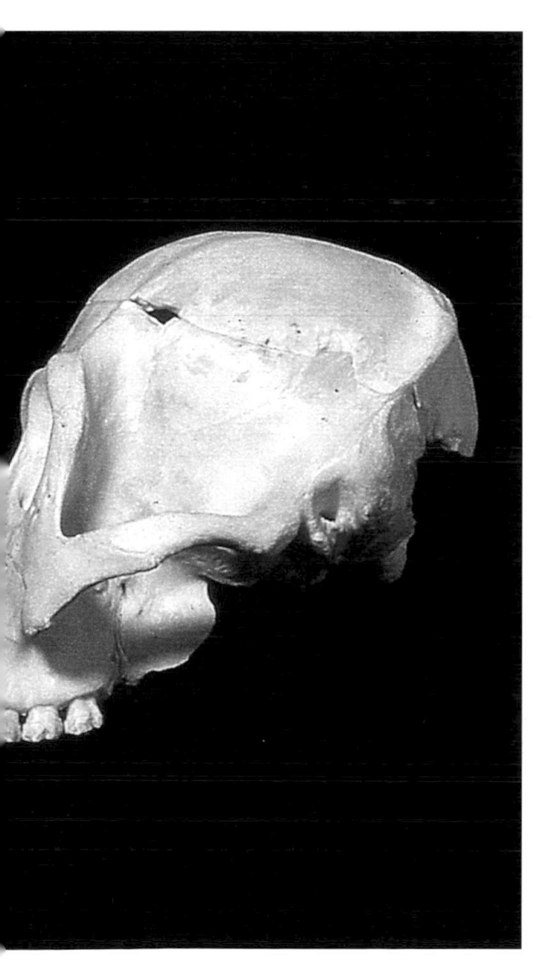

Ground living versus life in the trees

The environment occupied by *Sivapithecus* was not the typical forest habitat to which living apes are adapted. It was a mixed environment composed of subtropical forests, open woodlands and perhaps even some non-wooded country as well. This is similar to the probable environments of *Kenyapithecus* in Africa and *Griphopithecus* in Turkey, although the later Turkish ape *Ankarapithecus* was probably adapted for more open conditions. *Sivapithecus* lived in this habitat in much the same way as the other thick-enamelled apes, eating fruit and spending part of its time on the ground, for its limb bones show that it was partly adapted for ground-living. Herein lies a problem, for the living orang utan is committed to life in the trees, and it is puzzling to have a possible ancestor living on the ground and missing the adaptations for suspensory locomotion in trees that is common to all the living apes. There is little or no evidence of adaptations for suspension in *Sivapithecus*, and this has caused some anthropologists to doubt the orang utan affinities of *Sivapithecus*. As yet there is no resolution to this disagreement.

Brief mention should be made here of the giant fossil ape *Gigantopithecus* ('Giant ape'), which probably was descended from *Sivapithecus*. One species of *Gigantopithecus* indeed is known from the same deposits in India as *Sivapithecus*; the later fossils came from further east in China and Southeast Asia. *Gigantopithecus* carried tooth enlargement and jaw robusticity to its extreme, and while its body may have been little bigger than living male gorillas (although there is no direct evidence of this), its teeth were much larger. Its large size and its survival into the late Pleistocene makes *Gigantopithecus* a possible candidate for tribal memories of the Yeti.

The Ancestry of the Living Apes

Two groups of fossil ape are of particular relevance to later hominid evolution. These are *Graecopithecus* ('Greek ape', sometimes known as *Ouranopithecus*) and *Dryopithecus* ('Oak ape'). They fall into the two distinct groups of fossil ape mentioned earlier: the thick-enamelled apes which occupied seasonal habitats with a semi-terrestrial lifestyle, and thin-enamelled apes living in subtropical forests and with below-branch suspensory locomotion. *Graecopithecus* represents the former group and is known from several mandibles and maxillae (opposite) and a fairly complete skull (p. 112). It has a number of characters that link it with the African ape and human lineage, especially the relatively vertical face, the morphology of the nose and the canal linking it with the mouth. Only two phalanges are known from the postcranial skeleton, not enough to determine the positional behaviour of *Graecopithecus*, but their morphology suggests a strong terrestrial element in its behaviour. Its relationship with other fossil apes is unknown at present, although it has been linked with both *Ankarapithecus* and *Dryopithecus* in the past. Its similarities with extant hominines may indicate that it was close to their ancestry.

The enigmatic *Dryopithecus*

When it was first found last century at St Gaudens in France, *Dryopithecus* was known only from a few jaws and a small number of isolated teeth. In the middle part of this century, new discoveries have

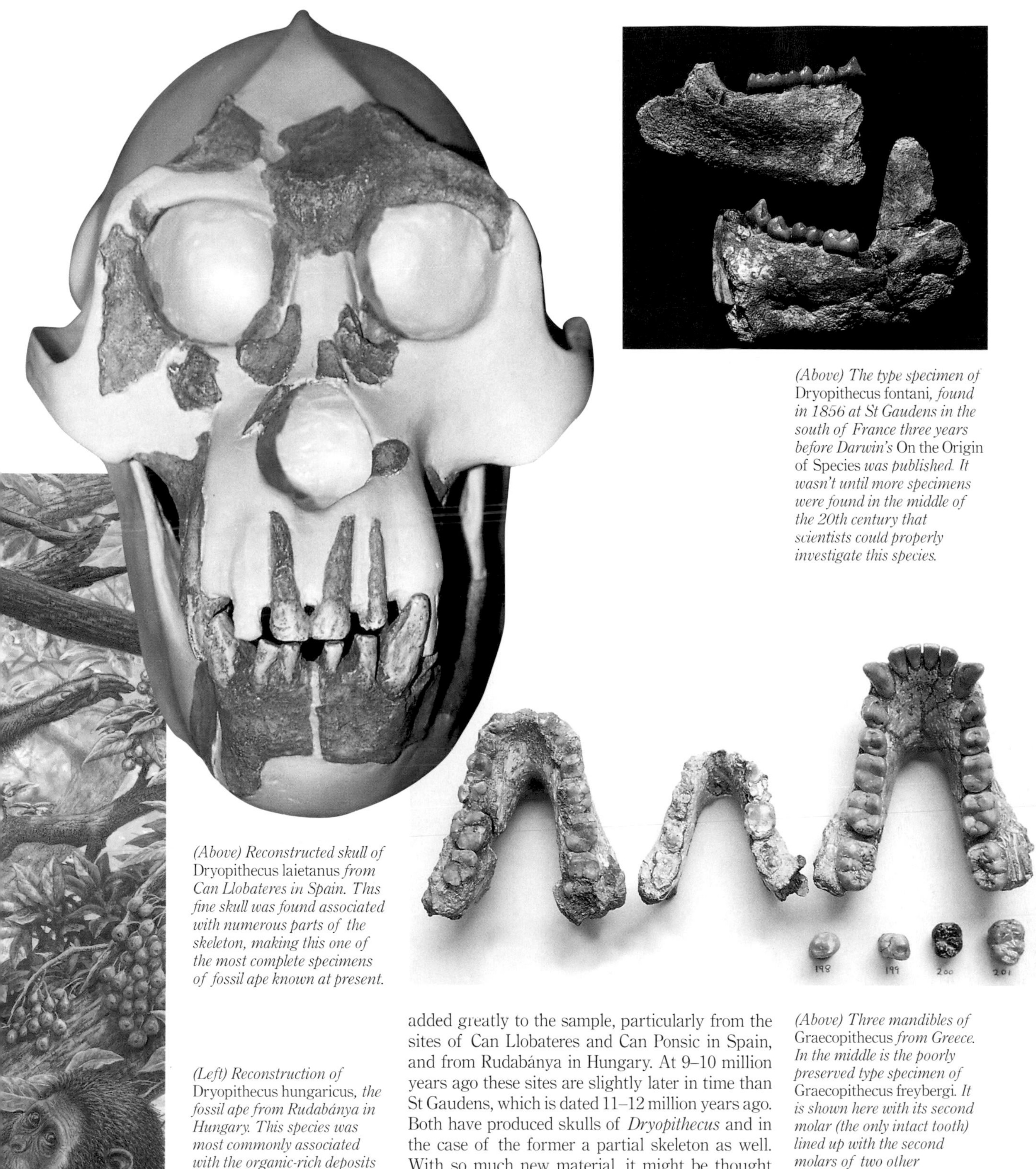

(Above) The type specimen of Dryopithecus fontani, *found in 1856 at St Gaudens in the south of France three years before Darwin's* On the Origin of Species *was published. It wasn't until more specimens were found in the middle of the 20th century that scientists could properly investigate this species.*

(Above) Reconstructed skull of Dryopithecus laietanus *from Can Llobateres in Spain. This fine skull was found associated with numerous parts of the skeleton, making this one of the most complete specimens of fossil ape known at present.*

(Left) Reconstruction of Dryopithecus hungaricus, *the fossil ape from Rudabánya in Hungary. This species was most commonly associated with the organic-rich deposits formed when the Pannonian Lake had receded (see page 67), leaving shallow valleys with swamp forest in which it is probable that the fossil ape lived.*

added greatly to the sample, particularly from the sites of Can Llobateres and Can Ponsic in Spain, and from Rudabánya in Hungary. At 9–10 million years ago these sites are slightly later in time than St Gaudens, which is dated 11–12 million years ago. Both have produced skulls of *Dryopithecus* and in the case of the former a partial skeleton as well. With so much new material, it might be thought that *Dryopithecus* would be both the best known fossil ape and the one with the clearest evolutionary affinities. The former is certainly true, but there is actually more disagreement over what its evolutionary position is than there is for any other fossil ape.

(Above) Three mandibles of Graecopithecus *from Greece. In the middle is the poorly preserved type specimen of* Graecopithecus freybergi. *It is shown here with its second molar (the only intact tooth) lined up with the second molars of two other mandibles of the same species (but which are named* Ouranopithecus macedoniensis *by some scientists).*

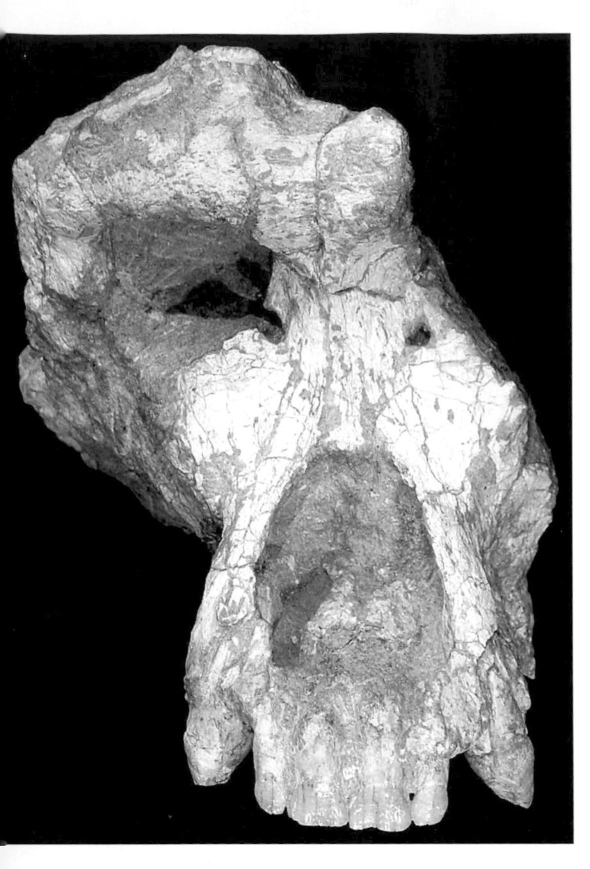

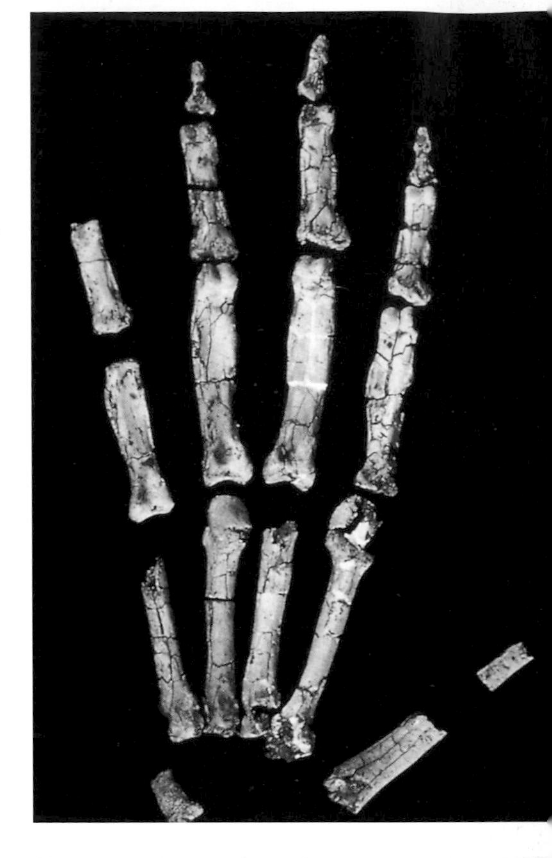

(Right) The nearly complete hand of Dryopithecus laietanus *found associated with the skull on the previous page. The proportions of the hand are remarkable, with very elongated finger bones compared with the metacarpals. There is no parallel with this among living primates, but it would seem to indicate a powerful grip in the fossil species.*

(Above) The skull of Graecopithecus. *The upper part of the skull has been distorted by crushing, and there is considerable bone loss over the surface of the skull. The lower face is more vertical than in most living and fossil apes, and the nose and premaxilla are similar to the condition in hominines, which may indicate a relationship with the living African apes and humans.*

The skulls are lightly built, as in other fossil apes, and there are fairly well-developed brow ridges, as in *Ankarapithecus*. This feature is different from the *Sivapithecus* skull, and other differences are that the nose is narrow and the distance between the eyes is broad. Like *Sivapithecus*, however, the cheek region is flat and is directed forwards, as it is also in *Ankarapithecus*. The anatomy of the floor of the nose and mouth, a highly diagnostic region where *Sivapithecus* closely resembles the orang utan, has the primitive ape condition in *Dryopithecus* and differs little from *Proconsul* and *Griphopithecus*. Its teeth are primitively small and lack the thickened enamel of the other fossil apes from the same period and in fact differ from *Proconsul* teeth only in minor details such as loss of the additional ledge running along the insides of the upper teeth (the cingulum) that is present in *Proconsul*.

A life in trees

It is this combination of characters that has produced the confusion over the relationships of *Dryopithecus*. In some characters it is like the orang utan, in others it is like the African apes, while in others still it is like neither and is simply primitive. However, in one respect it is not primitive at all and differs from all other fossil apes bar one. Its limb bones are clearly adapted for suspension of the body in trees, just like the living apes, and it is this characteristic that is so conspicuously lacking in

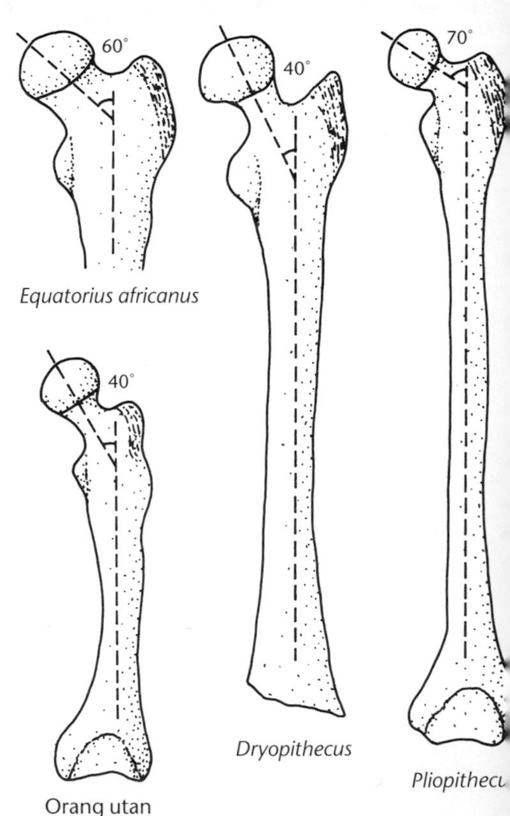

(Right) One of the femora of Dryopithecus laietanus *compared with an orang utan and two other fossils,* Pliopithecus vindobonensis *from Neudorf and* Equatorius africanus *from Maboko Island. The vertically oriented head of the femur of* D. laietanus *is similar to that of the orang utan, indicating high mobility of the lower limbs.*

the other fossil apes. In particular, its legs were highly mobile in a manner very like the orang utan, and it had extremely large and powerful hands, with the fingers greatly elongated and having powerful muscles for grasping. These characters point to life in trees with little or no access to the ground. This is consistent with what is known of the habitat it lived in, which was dense subtropical forest.

On the basis of these characters of the limb bones, it may be that *Dryopithecus* is the fossil ape most closely related to the living great apes, which is the conclusion of some workers, but it remains a problem that these same characters are not present in *Sivapithecus*, for which there is good evidence of relationship with one of the living great apes, the orang utan. If the common ancestor of the great apes had these characters for suspensory activity, as in *Dryopithecus*, and if *Sivapithecus* is on the line leading to the orang utan, it should also have these characters. But it does not.

Oreopithecus, an island species

One fossil ape that may share some suspensory characters with *Dryopithecus* is *Oreopithecus* ('Swamp ape'). This ape is known only from Italy and Sardinia from younger late Miocene deposits, but it is well known from a nearly complete skeleton and a number of jaws and teeth. It had long arms and mobile legs, adaptations for life in the trees as in *Dryopithecus*, but it has recently been claimed that it also had adaptations for living on the ground. Furthermore, it is claimed that when on the ground it walked on two legs like humans, not on all-fours like apes. *Oreopithecus* is also strange in a unique specialization of its teeth, which are unlike those of any other fossil ape, and it may be that these peculiarities are due to the environment in which it lived, which was an island cut off from the European mainland. This was part of Italy at a time when it was encircled by sea, and the area of the fossil sites was covered in dense subtropical swamp forests. It may be that the different adaptations of the teeth came about because of isolating effects of being cut off on an island, where also the lack of predators is known to produce some odd effects. Were it not for these, *Oreopithecus* might appear rather more similar to *Dryopithecus* and could said to be closely related to it.

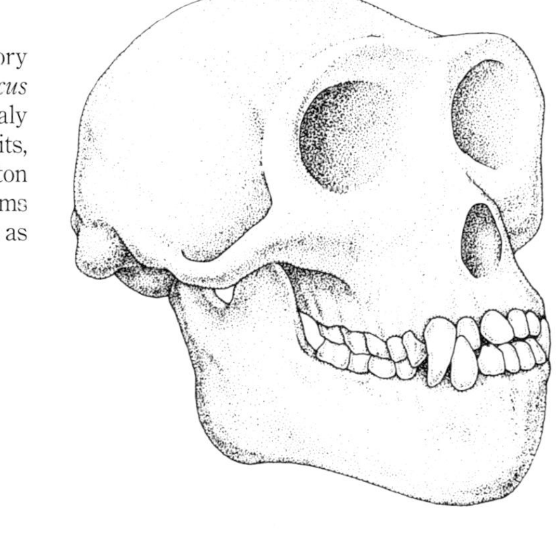

(Left) The reconstructed skull of Oreopithecus bambolii *from Italy. The face is relatively short, unlike the long faces of the living apes, and this gave rise to early ideas that* Oreopithecus *was a human ancestor. The skull is now known to be similar to that of other fossil apes, however, and nothing to do with human ancestry.*

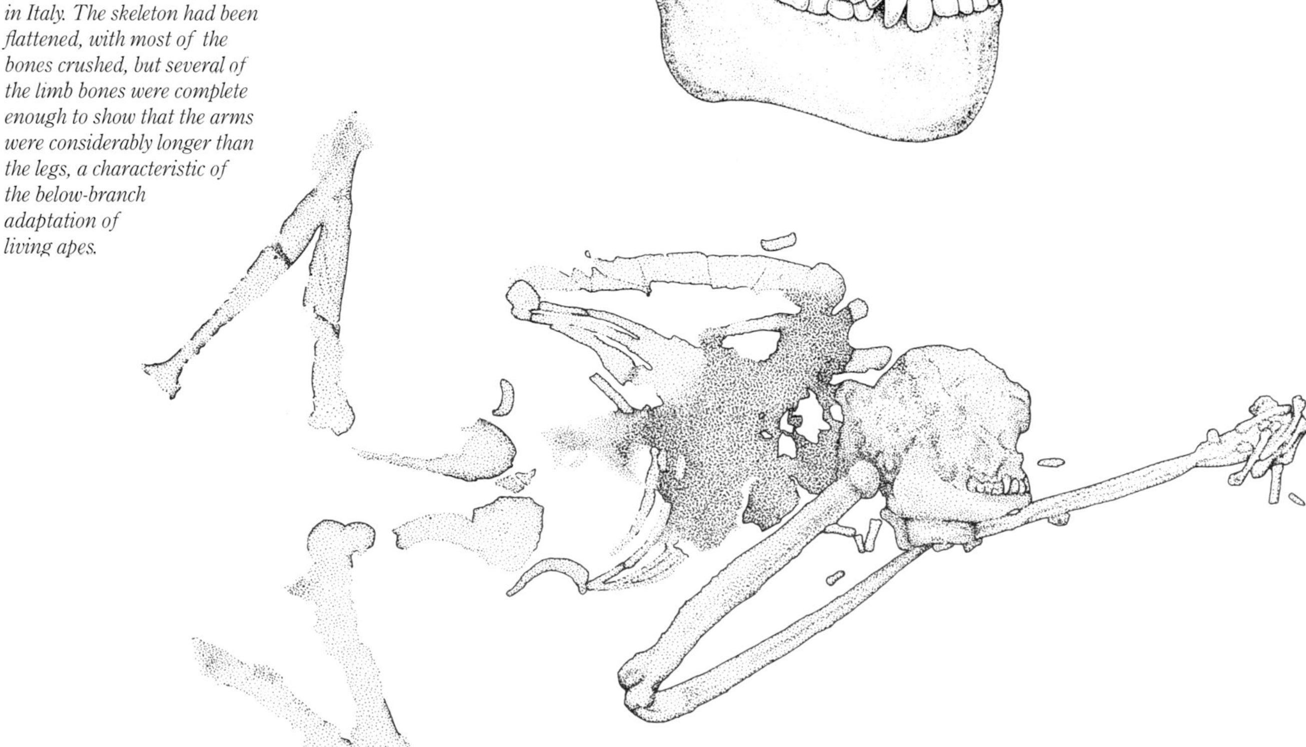

(Below) The skeleton of Oreopithecus *as it was recovered from coal deposits in Italy. The skeleton had been flattened, with most of the bones crushed, but several of the limb bones were complete enough to show that the arms were considerably longer than the legs, a characteristic of the below-branch adaptation of living apes.*

Late Miocene Apes and Early Human Ancestors

So far we have been examining the range of fossil apes to discover if any of them might be singled out as a possible human ancestor. It is very apparent that this is not possible, for there is nothing to link any one of them with human ancestry. What we can do, however, is to outline the range of adaptations of the late Miocene fossil apes in the period 10–6 million years ago, and since it is probable that early humans emerged from some part of this range, it will give a good idea of what to expect in such an early human ancestor. We will do this briefly and then proceed to describe all the likely candidates for human ancestry amongst the late Miocene and Pliocene hominines.

Two groups of fossil ape have been distinguished above. One has robust jaws, enlarged molar teeth with thick enamel, and some buttressing of the face to accommodate chewing stresses caused by the large teeth and a hard fruit diet. They lived in seasonal woodland to open forest environments and were adapted to some extent to ground living. The other group inhabited wetter, less seasonal forests and lived in trees employing a form of locomotion that involved some degree of suspension from overhead branches. Their jaws were more lightly built and their teeth not enlarged, so that their diet must have been soft fruits. The latter group is known only from Europe and Asia; the former group is known from both these continents and is also known from Africa. This may be significant, for all the early evidence for human evolution during the period 6–2 million years ago is African.

Distinguishing human ancestors

The key question to ask in distinguishing the earliest human ancestor from the apes is what are the characters that set them apart from the apes? Various characters have been proposed, such as enlarged brains, reduced sexual dimorphism (differences between males and females), upright bipedal (two-legged) walking, enamel thickness and enlarged molar and premolar teeth. The first two can be discounted immediately, for increase in brain size occurred late in human evolution, after 2 million years ago, and sexual dimorphism remained high for about as long. Thick enamel on the teeth is common to most hominines, but as we have seen, it was also widespread in fossil apes. The same is true of enlarged molars in one group of fossil apes. The only character we are left with to distinguish human ancestors is bipedalism, but we have to ask, is this enough? Could not some fossil apes with no connection with human ancestry have experimented with bipedal walking? Does this make them human ancestors? Several groups of fossil ape have been seen to have been partly ground living, a likely precondition to the development of bipedalism. We will now examine the evidence with these considerations in mind.

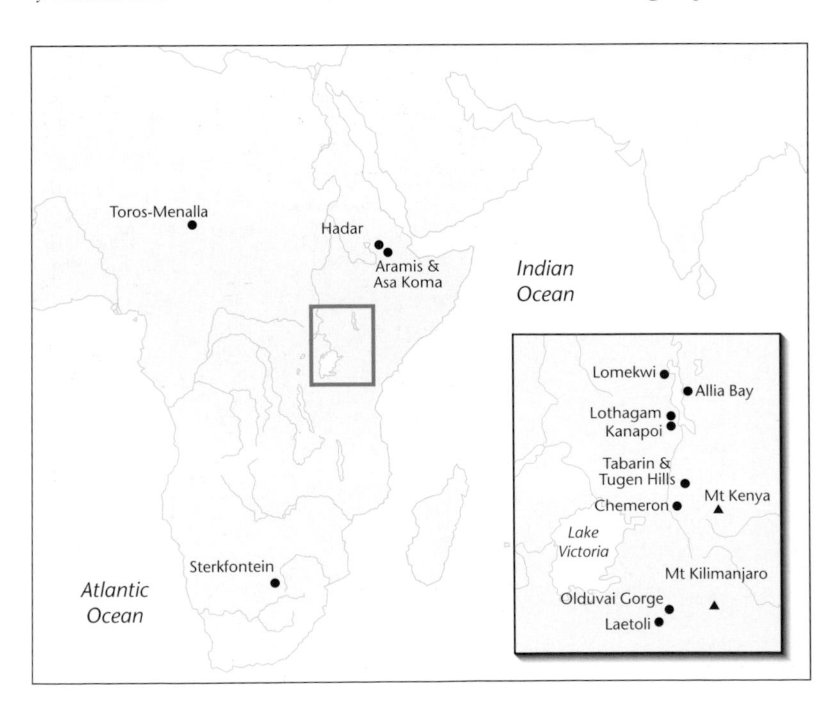

(Below) Map showing localities in Africa of the late Miocene and Pliocene hominins.

(Right) The field crew screening the sands of the desert around Lake Chad, looking for more fragments of hominin bone.

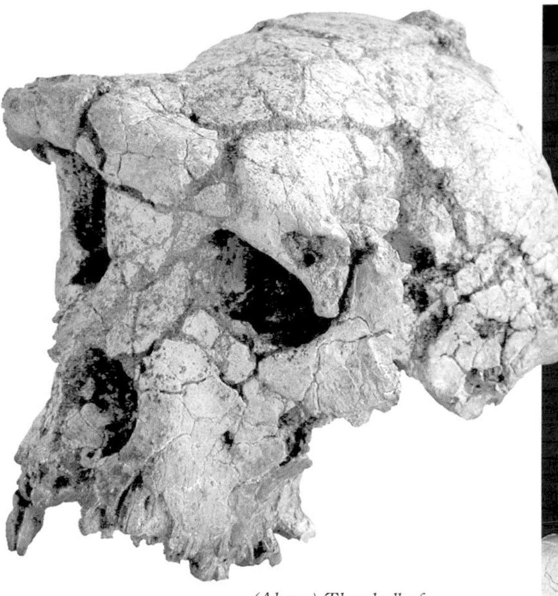

(Above) The skull of Sahelanthropus tchadensis *from Toros-Menalla, Chad. There is no indication at present whether this species was able to walk bipedally.*

(Above) Michel Brunet holding the Chad skull (with a chimpanzee skull for comparison), the reward of many years of persistent searching for fossils in this inhospitable region.

The earliest hominines?

One of the earliest possible hominine fossils (dated 7–6 million years ago) has been discovered recently, not from the Rift Valley of East Africa, but from the site of Toros-Menalla near Lake Chad. This is named *Sahelanthropus tchadensis* ('Human fossil from the Sahel, Chad'). The remarkably complete skull was small, with small teeth and a short face, but the upper face has strong brow ridges and the teeth have thick enamel. This is a combination not seen in any fossil ape, which generally have long faces lacking prominent brow ridges and larger molars and canines. It is also a combination not seen in later hominines, as we shall see.

Comparable in age is another recently discovered fossil, *Orrorin tugenensis* ('Original man from the

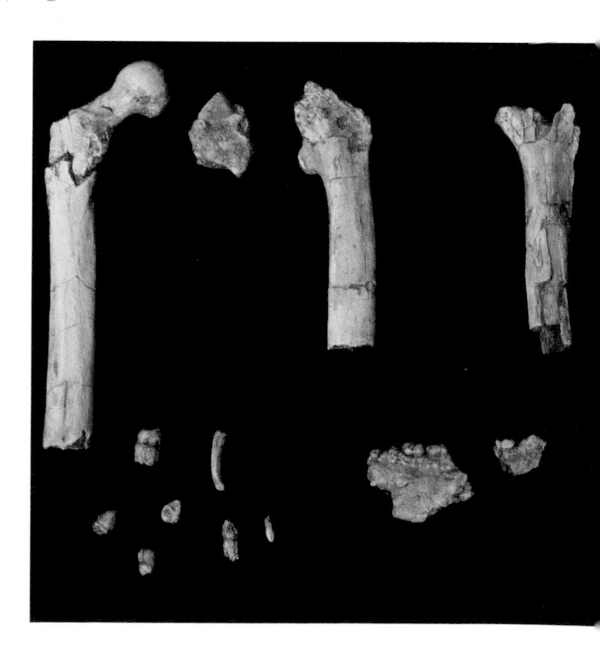

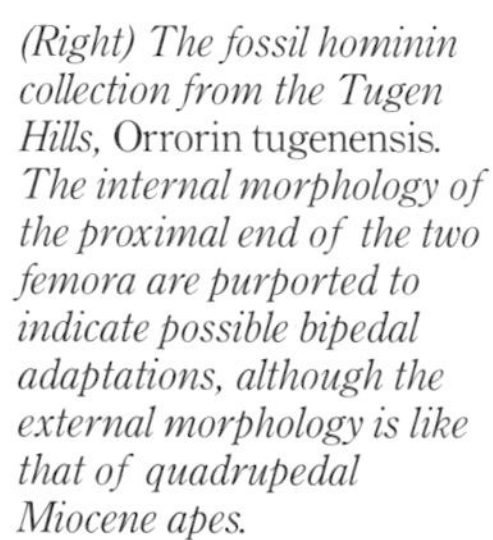

(Right) The fossil hominin collection from the Tugen Hills, Orrorin tugenensis. *The internal morphology of the proximal end of the two femora are purported to indicate possible bipedal adaptations, although the external morphology is like that of quadrupedal Miocene apes.*

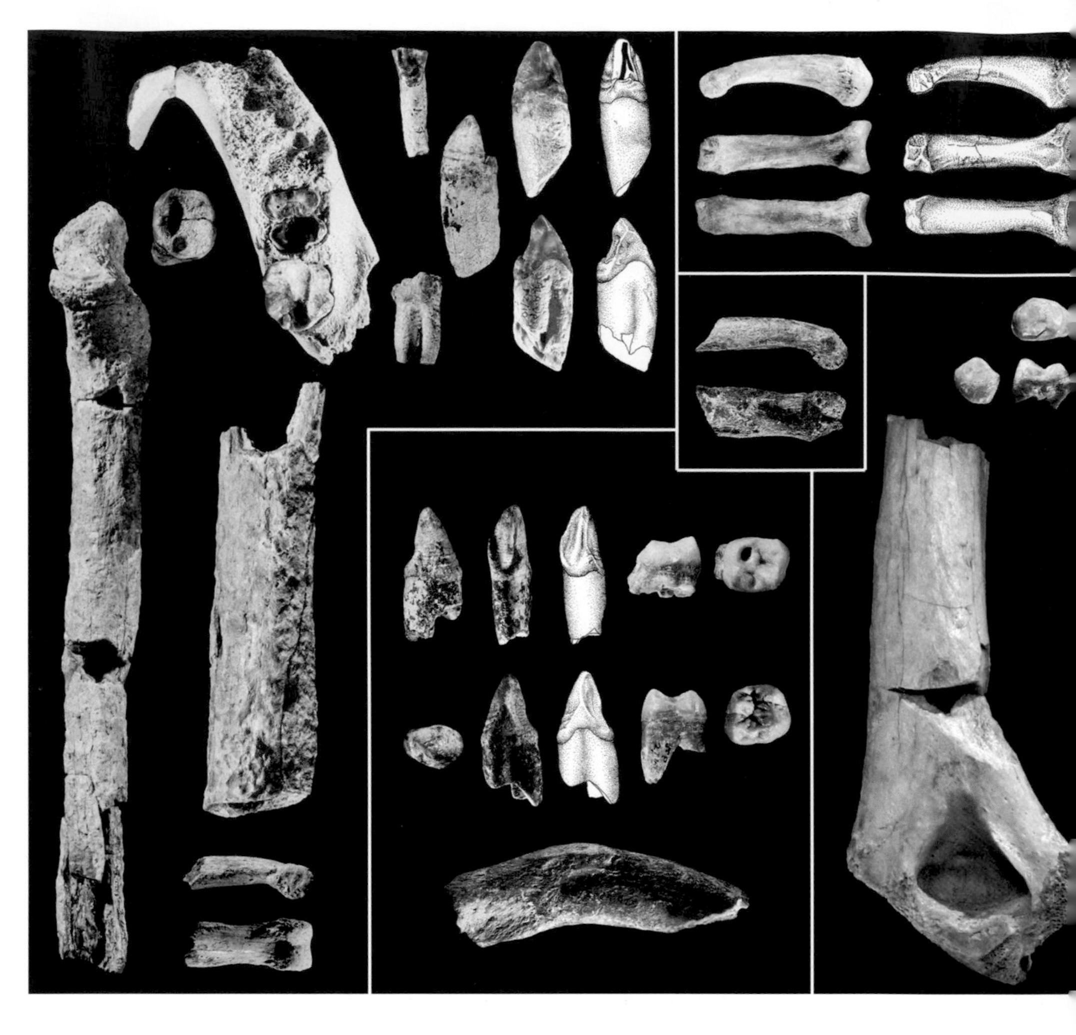

The collection of fossils representing what is now identified as a separate species of Ardipithecus, A. kadabba. *There is no clear indication whether this species could walk bipedally or not.*

Tugen Hills') found in northern Kenya. This is dated to between 6 and 5.8 million years ago and is represented by a few teeth and limb bones, which show some interesting differences from *Sahelanthropus*. The teeth have thick enamel, as in the Chad hominine and in later hominines (and one of the groups of fossil apes), but the canine is large and pointed like those of apes, both conditions the opposite to that seen in *Sahelanthropus* and later hominines. The femur (thigh bone) has a larger head than in later Pliocene hominines, which has been claimed as evidence of bipedalism, although in this character it is like that of similar sized fossil apes as well as extant chimpanzees.

Another new genus and species was described a few years ago from a site at Aramis, in Ethiopia. This is *Ardipithecus ramidus* ('Ground ape at the root'), but in 2004 new remains of *Ardipithecus* were assigned to the species *A. kadabba*, formerly a subspecies of *A. ramidus*. It comes from deposits in Ethiopia (Asa Koma locality 3) dated to 5.8–5.2 million years ago, and it is distinguished mainly on the basis of a higher crowned upper canine with functional honing against the lower third premolar. In this character it is similar to *Orrorin tugenensis*, with which it also shared relatively small molars, although it is not clear if this is a primitive retention or a shared derived character. Similarly, the enamel thickness is also ambiguous, for the last common ancestor of hominins could either have been large toothed with thick enamel or small toothed with thin enamel (see above). Some aspects of the teeth later occurring in *A. ramidus*, which is dated to 4.4 million years ago, are similar to *A. kadabba*, especially the relatively small molars and premolars, but it lacked the evidence of a honing canine/premolar complex. None of these early hominins show the enlarged molars and premolars of later australopithecines (see below), and in particular, one of the milk teeth that is enlarged in all australopithecines is small and ape-like in the Aramis fossil. This is not much to go on, and since in most other respects the Aramis fossil was remarkably ape-like, it has been distinguished from later fossil humans at the generic level.

These three taxa, all more than 5 million years old, have been assigned to three different genera, even though each is based on incomplete material that does not permit comparison. Where common parts occur, such as canines and molars, they are similar to each other, indicating that they could in fact all be the same thing, but evidence is lacking as to what it may be related to, ape or human. Retention of primitive characters, such as the honing third premolar, does not in itself indicate ape affinities, but neither does the scanty evidence of possible bipedality indicate human affinities. The jury is still out on this one.

The australopithecines

At Kanapoi in Kenya some more complete jaws and a partial skull of a young individual have been recently described. The lower jaws in particular are remarkably like those of thick-enamelled late Miocene apes from Europe and Africa, but the limb bones, which include part of the elbow joint and part of the lower leg, show that the Kanapoi fossil walked bipedally. The site has been dated to between 4.2 and 4.1 million years ago, and on the basis of differences in anatomy, a new species of *Australopithecus* ('Southern ape') has been recognized named *A. anamensis*. The australopithecines were amongst the first fossil hominins described from Africa, the earliest one being *Australopithecus africanus* ('Southern ape from Africa') described in 1925. While similar to Miocene apes in their teeth and jaws, this fossil provides the earliest definite evidence of upright bipedal walking, but even so the evidence from these early sites is sparse and hard to interpret.

The flat-faced skull of Kenyanthropus platyops. *Apart from the flattened face, this hominin has many similarities with* A. afarensis, *and it is not clear at this stage whether it is really distinct from this species.*

Another skull is known from a site called Lomekwi on the western shores of Lake Turkana in Kenya. This is *Kenyanthropus platyops* ('flat-faced skull from Kenya'), and it was found with many other tooth and skull fragments. It dates back to about 3.5 million years ago, and it bears a remarkable similarity to a much later hominin fossil, *Homo rudolfensis*, as seen particularly in the skull from East Turkana, numbered 1470. It is not clear at this stage if this similarity has any phylogenetic significance.

By far the best evidence for Pliocene hominins comes from the period between 4 and 3 million years ago. One of the most complete fossils yet discovered is still in the process of excavation at the time of writing. This is a fossil hominin known as Little Foot, which was discovered almost by chance and against considerable odds. In 1994, the anthropologist Ron Clarke found a number of hominin foot bones that had been overlooked in previously collected material from Sterkfontein. These bones displayed a mix of ape and human characteristics, indicating that the species could both walk on the ground and climb trees. Three years later, he spotted a section of bone protruding from the breccia which fitted one of the foot bones which had been collected 65 years previously. Little Foot is a near-complete skeleton of an (as yet unnamed) early hominin with short arms and short fingers, rather like modern humans. It was clearly bipedal, but on its foot the big toe is separate from the other toes which would have enabled it to grasp branches to climb trees.

There were also dramatic discoveries from the same time period of a hominin called *Australopithecus afarensis* ('Southern ape from Afar') from Tanzania and Ethiopia. The earliest discovery from Ethiopia was a knee joint from Hadar, with just the ends of the two bones of the upper and lower leg, but because of the particular structure of the knee in bipedal humans this was enough to show that this early find was from a bipedal human ancestor. In the five years following this first discovery, many additional fossils were recovered from Hadar, including the famous skeleton nicknamed Lucy and a collection of 13 individuals which has been suggested to be the remains of a social group killed by some catastrophic event. Most of the fossils from Hadar are dated between 2.8 and 3.3 million years ago, although there are a few fragments older than 4 million years. At this time the area was a lake with marshy areas and rivers providing a lush environment very different from the arid and hostile environment today.

Slightly earlier in time is a site in Tanzania called Laetoli. A collection in the 1930s had yielded what was actually the earliest australopithecine fossil to be found in East Africa, and many more have been found since. The deposits at Laetoli are between 3.5 and 4 million years old and are situated close to Olduvai Gorge, and the environment at that time was probably similar to that of today, seasonal woodlands and dense riverine woodland and forest. A unique find at Laetoli were footprints preserved in ashfalls from nearby volcanic eruptions, both now attributed to *A. afarensis*. Olduvai and Laetoli are close to the western rim of the Rift Valley, which has a row of volcanoes along its length, and one of these volcanoes was emitting ash eruptions at this time. Several of these ash falls coincided with light rains, and because the ash is rich in salts it hardened quickly when wet, thus preserving the footprints of many of the animals living in the area, including a row of footprints of *A. afarensis*. These show beyond any doubt that *A. afarensis* walked upright on two feet, although the absence of any adult limb bones from Laetoli means that we cannot be sure of the exact nature of the bipedal adaptation that was present at that time (see pp. 187–189).

Other specimens attributed to *Australopithecus afarensis* include a lower jaw from Lothagam in northern Kenya dated to the latest Miocene between 5 and 5.5 million years ago. The specimen was originally identified as *Australopithecus africanus*, but with the discovery of *A. afarensis*, it could be attributed to this species on the basis of its age . Its australopithecine characters are its robust jaw and enlarged teeth, both, as we have seen, characters of one of the late Miocene groups of fossil ape. Another lower jaw found years later at the site of Tabarin, dated to 4.1 million years ago, is identified to *A. afarensis* for the same reasons. Similarly, a set of foot bones from the site of Sterkfontein in South Africa is the same age as *Australopithecus afarensis* in eastern Africa at 3.5–3.3 million years ago. This fossil appears to have an ape-like divergent big toe for grasping branches, and more complete remains are currently being excavated from Sterkfontein, and these should provide valuable additional information.

This completes our review of Pliocene human ancestors between 6 and 3 million years ago. They are known mainly from eastern Africa, but clearly they also lived in central and southern Africa and were a widespread and successful group. In the next section we will describe the later stages of human evolution, including the diversification of the australopithecines and the origin of the genus *Homo*, but first, in order to demarcate the time and place of origin of the human lineage, it is necessary to go into some detail in describing the anatomy of some of these Pliocene fossils.

Early australopithecine anatomy

The shoulder joint: The shoulder joint is poorly known in the fossil record. This is because the bones of the shoulder are relatively weak, and the great development of muscles around the shoulder make it an attractive body part for meat-eaters to consume. As a result, the shoulder is almost the first part of the body to be destroyed after death, and only in relatively complete skeletons, which have not been killed or eaten by meat-eaters, is the shoulder preserved as a fossil. One specimen is known in the Pliocene, the shoulder of *Australopithecus afarensis* in the skeleton Lucy from Hadar, Ethiopia. This has a mixture of ape and human features: the latter is seen in the less rounded shape of the articular

Reconstruction of the habitat and way of life of Australopithecus afarensis. *Their preferred habitat was probably forest and woodland, where they foraged for food on the ground and up trees, much as chimpanzees do today, but they almost certainly also ventured out into more open areas as well, where they would have walked bipedally.*

surface of the upper arm where it articulates with the shoulder, indicating less mobility of the arm at the shoulder; and the former is seen in the direction the shoulder joint points, upwards as in the apes rather than sideways as in humans, indicating that the arm was frequently directed upwards, as when hanging from branches.

The elbow joint: The Kanapoi specimens of *Australopithecus anamensis* include part of the elbow joint, the lower end of the upper arm (humerus). This bone lacks the specialized ridging present on apes which is related to their knuckle-

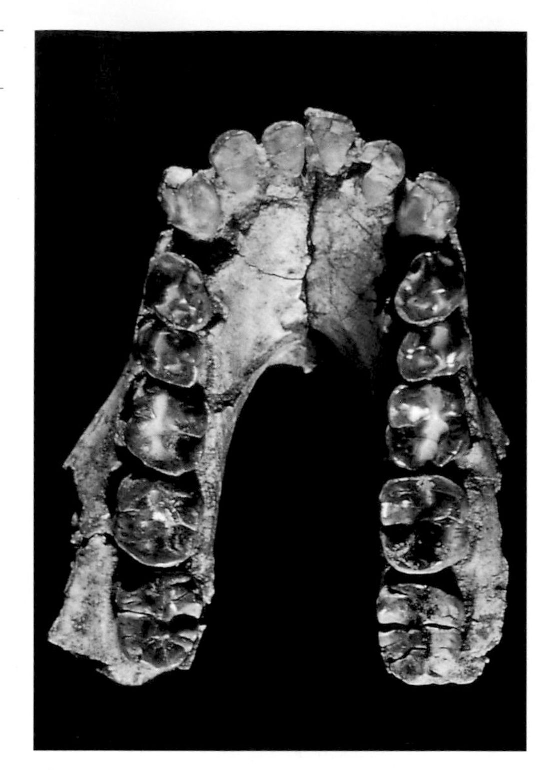

The mandible of Australopithecus anamensis *from Kanapoi. The jaws and teeth of this species are remarkably similar to those of fossil apes, with parallel tooth rows and large, thick-enamelled teeth, but their association with limb bones showed that this species walked bipedally.*

walking form of locomotion. There is no evidence, therefore, that this Pliocene species was descended from a knuckle-walking ancestor such as the chimpanzee. In fact this fossil elbow is so much like modern human elbows, it was initially classified in the genus *Homo*. The very early date for the Kanapoi deposits therefore came as something of a surprise, for at just over 4 million years old the Kanopoi specimen is much older than any other known record of *Homo*. Probably mainly as a result of its greater age, this bone is now grouped with the new fossils from Kanapoi as *A. anamensis*.

The hand: The bones of the hand are another body part not often preserved as fossils. Many of the hand bones are preserved, however, in the skeleton of *A. afarensis* (Lucy), and they show few adaptations similar to modern humans. The proportions of the thumb in relation to the other fingers is the main similarity with the human hand, but in most other respects Lucy's hand resembles a generalized ape hand, although without the specialized knuckle-walking features of chimpanzees and gorillas. The fingers were elongated and curved, indicating a hand adapted for powerful grasping, as in apes, although not as much as in the orang utan. Lucy's hand was not adapted for precision gripping, although because of the relatively greater length of the thumb (it will be remembered that apes have extremely short thumbs in relation to the rest of their hands), her hands would have been more versatile than ape hands. The overall picture, therefore, is that the hand of *A. afarensis* was still primarily adapted for powerful grasping, and since the evidence of the shoulder joint in the same skeleton indicates that the arm was habitually directed upwards over the head, the evidence for a way of life involving frequent suspension from branches of trees is very compelling.

The hip bone – evidence for walking upright: If the arms provide evidence for the climbing abilities of the early australopithecines, it is the evidence from the leg bones that indicates they could also walk bipedally. One of the most diagnostic bones in this respect is the hip, for humans differ greatly from apes in this bone because of their unique form of locomotion. The hip bone in apes is high and narrow and faces forwards; the hip bone in humans is wrapped around the sides of the body to provide support when standing upright, so that it is much wider than ape hip bones and it is also much shorter. Once again it is the skeleton of Lucy, *A. afarensis*, that provides the evidence at this early stage of human evolution, and Lucy's hip is like that of humans in being wide and short. In fact the upper part of the hip, the iliac blade, is even wider than in modern humans, and Lucy's iliac blade does not wrap around the sides of the body but is sited towards the back and faces forwards as in apes. Since it is the forward extension of the hip that provides the attachment for the muscles that enable us to keep our balance when standing upright, it seems likely that its absence in australopithecines meant that they would have had difficulty keeping upright while standing still. It is thought that in order to keep upright they would have had to keep moving in what might have been little more than a shuffle.

The area where the two sides of the hip bone connect with the back bone, called the sacrum, reveals another human-like feature of australopithecines. In apes, this articulation is narrow and is positioned in front of the leg articulation (the acetabulum), so that the full weight of the upper body passes across the front of the hip. This is as we would expect in animals like apes that habitually move on all four legs. In humans, however, the articulation with the back bone is wide and is behind the leg joint, so that the upper body is balanced when standing upright. The australopithecines had the human condition, except that as was the case with the ilium, the sacrum was even wider than in humans. While this shows that *A. afarensis* could stand upright, it still would have had difficulty balancing because of the greater sideways broadening of the whole of the hip.

The thigh bone: In *A. afarensis*, the neck of the femur (thigh bone) is also elongated, a feature probably related to the extra wide hip bones. The neck of the femur is the part of the upper leg bone

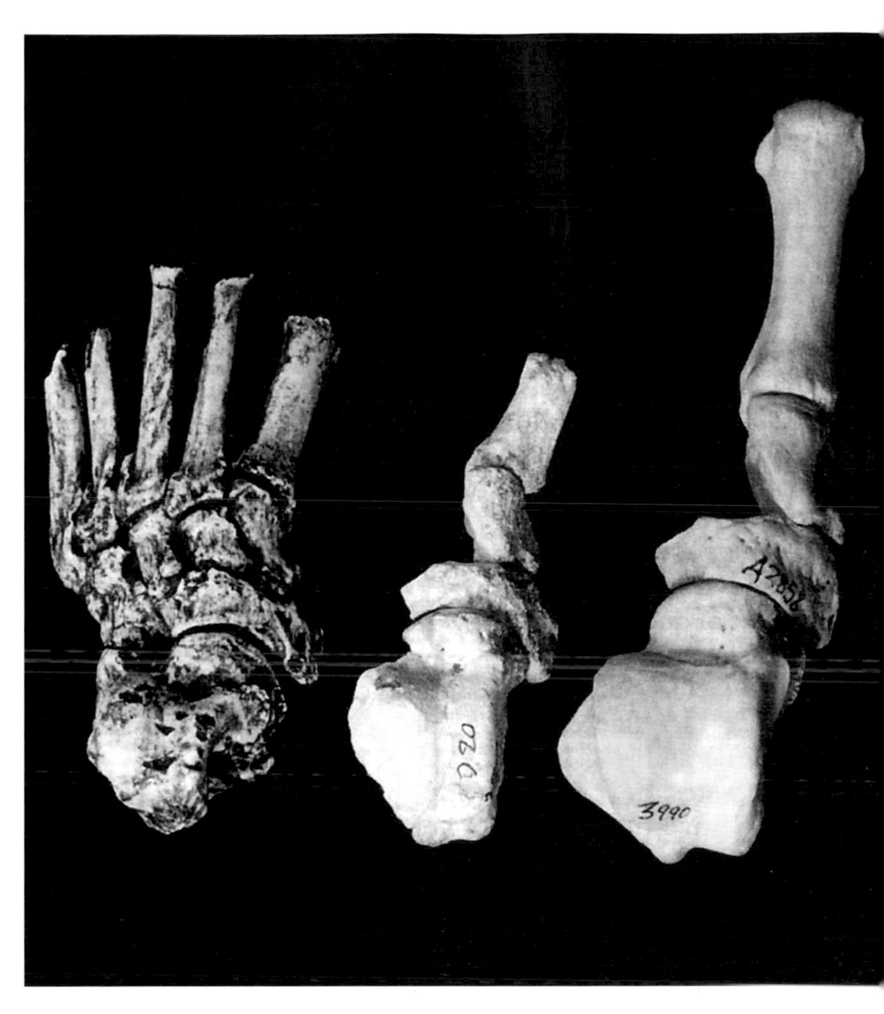

The apparently divergent big toe of the Sterkfontein foot (middle) compared with the straighter toes of a modern human foot (right) and the foot of Homo habilis *from Olduvai Gorge.*

that connects the main shaft of the upper leg to the articulation with the hip. Apes and humans have rather similar proportions for the upper part of the femur, and the later australopithecines differ from both in the great length of the neck and the small diameter of the head that forms the joint with the hip. *A. afarensis* is intermediate between them, but the earlier species *Orrorin tugenensis* has a shorter femoral neck and larger head like those of Miocene apes. It looks very much as if it is the later australopithecines, *A. africanus* and *Paranthropus robustus*, that are specialized, for it also seems to be the case that the greater the width of the hip, the longer the neck of the femur. These later species are similar, however, in having relatively short legs, just like the great apes.

The knee joint: The knee joint is another significant joint surface in bipedal locomotion. The knee was also the first fossil found that showed that *A. afarensis* was unequivocally bipedal. For a human to stand with feet and knees together, the upper leg has to angle in strongly because the tops of the legs are widely separated at the hip. The upper leg therefore makes a sharp angle with the lower leg at the knee joint, and this is quite different from the ape knee. Apes walk and stand with feet and knees apart, so that upper and lower legs are in a straight line, but *A. afarensis* has the human condition in its knee angle. In most other respects, however, its knee retains the ape morphology, so that it is basically an ape-like joint, with primary adaptations for tree-living, but modified in this one way for bipedal walking. It is interesting, however, that these adaptations for tree-dwelling in *A. afarensis* are most marked in the small specimens, while the larger specimens are more human-like. Does this mean there was different behaviour in males and females? Or might there be more than one species represented by *A. afarensis*?

The foot: The foot of *A. afarensis* also has a mixture of ape and human characters. The angle of the foot with respect to the lower leg in *A. afarensis* is perpendicular, as it would be in bipedal humans, and there is some development of the arches of the foot. On the other hand, the toes are long and curved, like the fingers of the hand, and there is some indication that the big toe was divergent from the other toes and so could be used for grasping. The bones of the big toe are incomplete and damaged in *A. afarensis*, and so this last conclusion is uncertain, but there is a similar aged foot from the South African site of Sterkfontein, nicknamed Little Foot, that appears to show the same thing.

Evidence for both tree-living and bipedal walking: This evidence strongly suggests that *A. afarensis* retained striking adaptations for tree living, particularly the grasping foot and mobility at the knee and hip joints, while at the same time it had clear adaptations for bipedal walking in all three areas, namely hip, knee and foot. None of these adaptations, however, are precisely the same as are found in living humans, and most of them are exactly the adaptations that would be expected in any fossil ape that was already partly terrestrial and was living in seasonally open woodland/forest environments. In other words, it is not certain that evidence of bipedal adaptations are by themselves conclusive proof that these Pliocene australopithecines were ancestral to humans. It can be questioned further if the combinations of bipedal adaptations in the early australopithecines is homologous (genetically the same descended from a single common ancestor) with those of later hominins, for apparently similar functions can evolve convergently through different evolutionary trajectories.

The jaws and skull: So far we have not mentioned the skull and jaws, although these are by far the most common body parts found as fossils. The jaws of *Ardipithecus* are lightly built and the teeth are

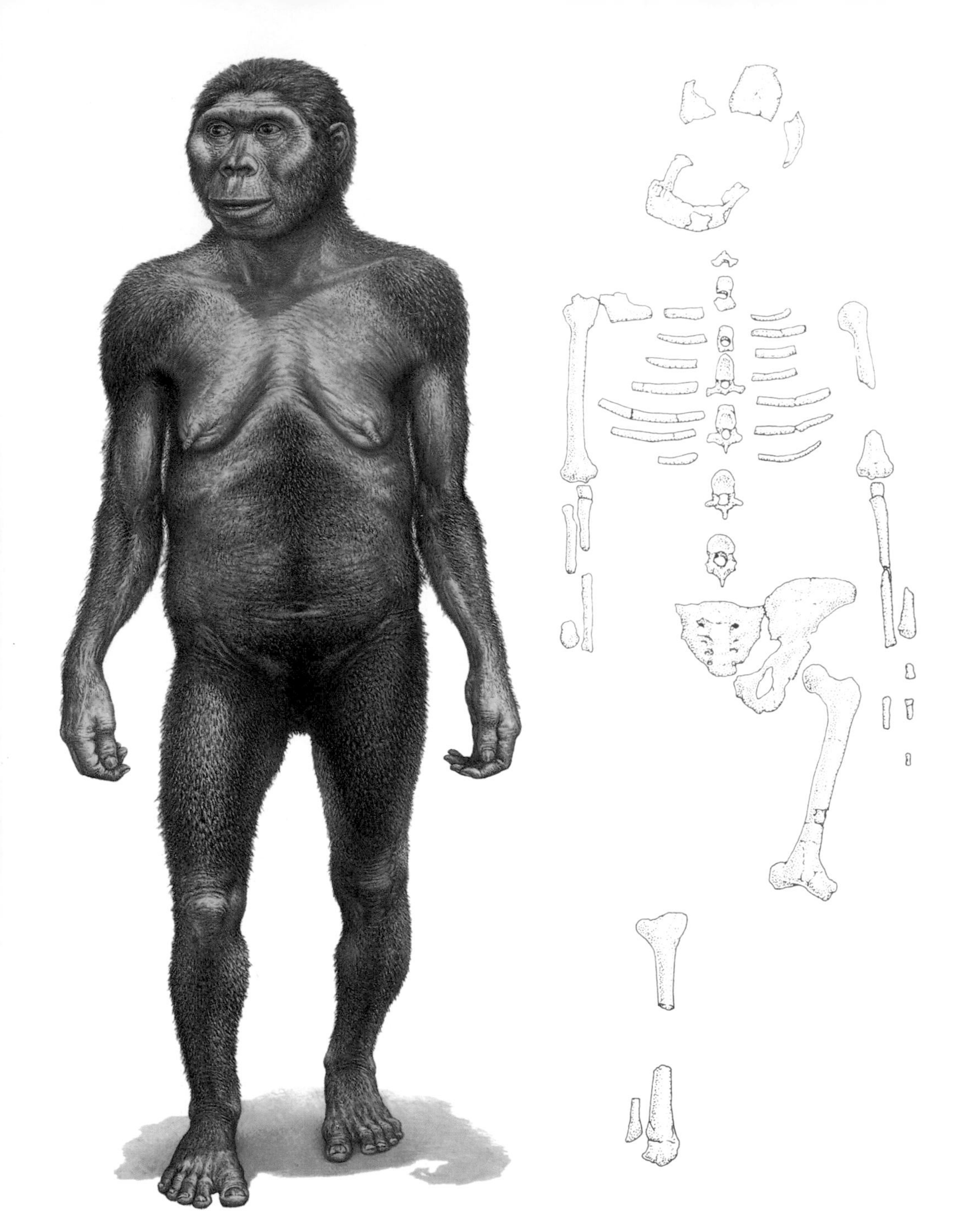

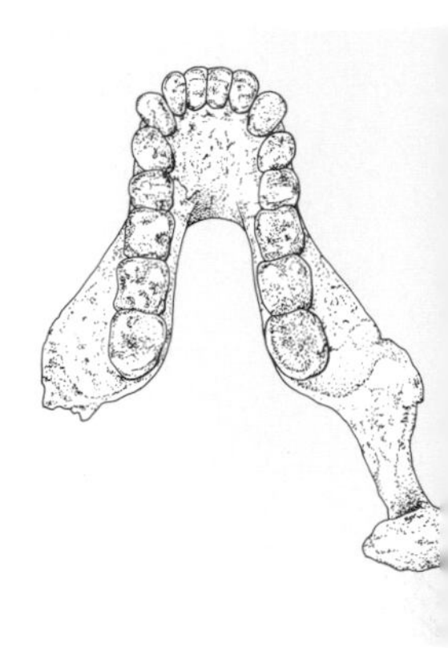

(Left) The partial skeleton of Australopithecus afarensis *from Hadar in Ethiopia, nicknamed Lucy by Don Johanson, her discoverer, with a reconstruction of Lucy shown on the left. The morphology is clearly that of a bipedal walker, but the relatively short legs make it unlikely that Lucy walked like humans do today.*

not enlarged, unlike the late Miocene fossil apes but similar to living chimpanzees. The teeth of the other early hominins such as *Sahelanthropus* and *Orrorin* are similarly small, but it is difficult to say how much so. Tooth size should be seen as relative to body size, but these early hominins are too fragmentary to provide accurate estimates of body size. The early species of *Australopithecus*, such as *A. anamensis*, are quite different, with robust jaws and large teeth that are remarkably ape-like. The jaws and teeth of *A. afarensis* are similar, and its skull is small, no larger relative to body size than those of apes, and the teeth are 1.7 times the size expected for their body size. The enlargement of the teeth continued into the later australopithecines, with tooth sizes up to two and a half times larger than expected. The later australopithecines developed a complex system of buttressing of the skull to absorb the stresses set up by their huge teeth during chewing food, but these were apparently not present in *A. afarensis* (and are not known for other Pliocene species).

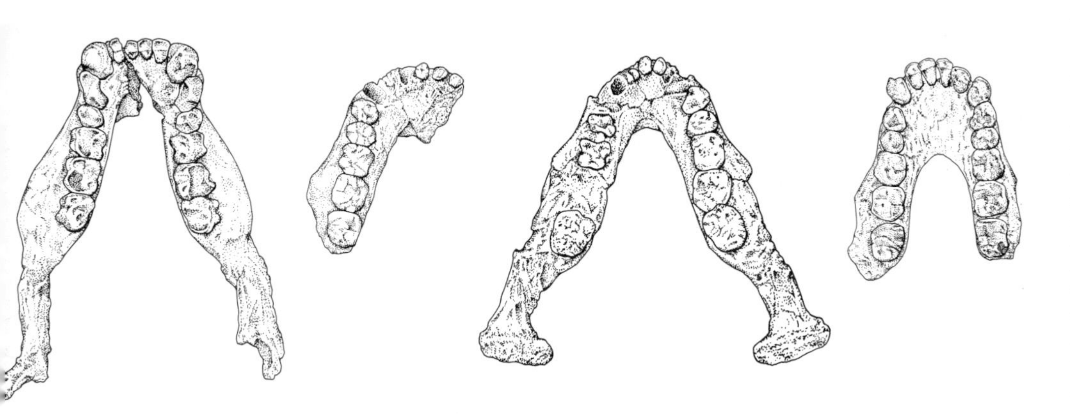

(Left) The shape and structure of the lower jaw appears to have changed little from Miocene apes to Pliocene hominins. From the left are two fossil ape jaws, Ankarapithecus meteai *and* Sivapithecus indicus*, both with robust jaws and straight tooth rows. To the right are three australopithecine jaws,* Australopithecus afarensis*, 'Lucy' (also* A. afarensis*) and* Australopithecus anamensis*, which conservatively retain similarly robust jaws and straight tooth rows.*

In a curious parallel with the late Miocene fossil apes, there appear to be two groups of Pliocene human ancestors. One has small teeth and lightly built jaws; the other massive jaws with large and thick-enamelled teeth. The latter group, all species of *Australopithecus*, combined a mixture of bipedal adaptations with a few ape-like adaptations for suspensory life in the trees. Nothing is known of the locomotor abilities of the former group, *Ardipithecus ramidus*, although the claim has been made that it too walked upright, but this requires verification since the claim was based on fragments of skull. We are left, therefore, with more questions than answers: was *Ardipithecus* ancestral to *Australopithecus*? Was *A. anamensis* ancestral to *A. afarensis*? Or were either of them ancestral to later species of *Australopithecus*? There is no evidence this early for any of them having a stone tool culture, but might they have made tools of perishable materials as does the chimpanzee today? Were indeed any of them ancestral to either the later australopithecines or to the genus *Homo*? We do not have the evidence to answer these questions at present, but new field-work is underway in East and South Africa to collect additional fossils. It is to be hoped that this will answer some of these tantalizing questions.

What's in a Name?

In this section, the Pliocene hominin *Australopithecus afarensis* has featured prominently, but there is much controversy over the naming of this fossil hominin. This is an esoteric issue but one of some importance, for the naming of fossils has implications for their evolutionary relationships. Is *A.afarensis* a member of the genus *Australopithecus*? or is it ancestral to *Australopithecus* and *Paranthropus*? (in which case it must belong to a different genus); is it the same as other species such as *A. anamensis*? How is it related to later members of the Hominini such as *Homo*, *Paranthropus* or *Australopithecus*? These are issues that are all affected by the naming of this fossil species.

The genus *Australopithecus* was named by Raymond Dart in 1925 and the species name *afarensis* by Don Johanson and others in 1978. The type specimen of *A. afarensis* (that is, the specimen on which the name is based) is a specimen from Laetoli, in Tanzania, shown here, but there is a prior named species from this site which is almost certainly the same taxon. This is *Praeanthropus africanus* named by the German anthropologist Hans Weinert in 1950. It was recognized that the new fossils were the same as Weinert's 1950 species, but because there was already a species with the name of *africanus* in the genus *Australopithecus*, named by Dart in 1925, a new species name had to be coined, hence *afarensis*.

There is some doubt, however, that the species *afarensis* properly belongs in *Australopithecus*, and if it were to be removed the genus name *Praeanthopus* would become available again. The species name *africanus* given by Weinert in 1950 should also have been available, because its naming predated the 1978 date of *afarensis*. However, in a 1999 ruling by the International Commission for Zoological Nomenclature (Opinion 1941), the species name *africanus Weinert 1950* was suppressed, so that the correct name as matters stand at present should be *Praeanthropus afarensis*.

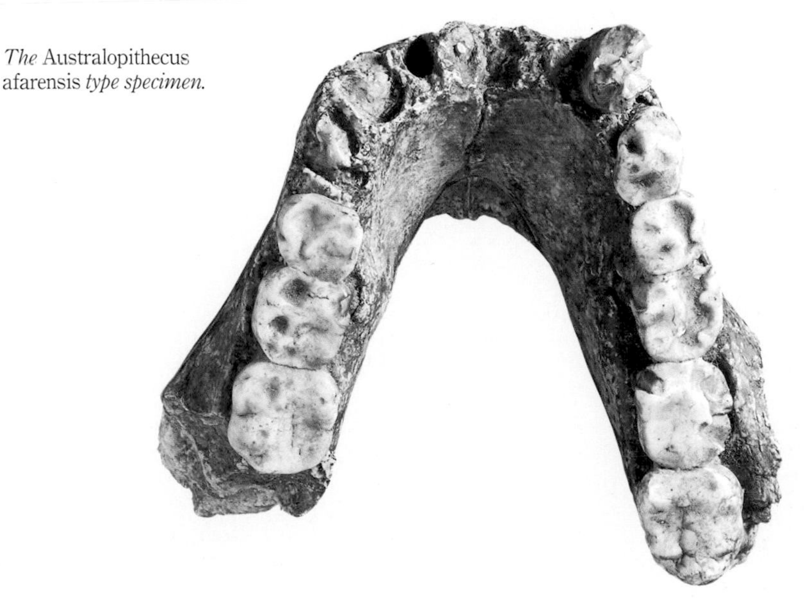

The Australopithecus afarensis *type specimen.*

Australopithecus africanus

(Above) This skull of an immature primate, found at Taung in South Africa, was passed to Raymond Dart in 1924, and he published it as representing a new species called Australopithecus africanus. *Many of the world's most distinguished scientists were not convinced of Dart's claims that* Australopithecus *was a close relative of humans, considering instead that it was an extinct ape more closely related to the gorilla and chimpanzee. Subsequently Dart was proved to be largely correct.*

At the beginning of this century, no fossil evidence had been discovered to support Charles Darwin's educated guess that Africa was the ancestral homeland of humans. Early human fossils had been recovered from Europe and Java, but there were no significant finds from Africa. Things began to change with the discovery of a fossilized human cranium in a mine near Broken Hill, in what was then Northern Rhodesia (now Kabwe, Zambia), in 1921. Then, at the end of 1924, a fossilized skull was found in a limestone quarry at Taung, in South Africa. It was studied by a newly established Professor of Anatomy, named Raymond Dart, and in 1925 he published a paper in the scientific journal *Nature*, making some remarkable claims about the fossil. It consisted of the face and front of the skull of an immature individual, since its jaws contained a combination of milk and permanent teeth. With the skull was a nearly complete replica of the inside of the braincase, which had been created by limestone sediment filling the brain cavity.

Dart argued that the fossil showed a combination of ape-like and human features, and that its teeth and brain shape were particularly human. He thought that its brain must have been large and advanced in structure, and that it probably held its head erect on its spine, meaning that the creature walked upright, as we do. Dart named it *Australopithecus africanus* ('Southern ape from Africa'), and declared that it was probably closely related to humans, and might even represent a human ancestor. Moreover, he believed that these australopithecines were carnivorous, and that they took their prey back to the limestone caves in which they lived.

Dart's claims were treated with great scepticism by the scientific establishment, particularly in England. This was partly because of judgments about Dart's youth and relative inexperience and partly because of the immature nature of the fossil itself, since young apes may look more human than adult apes. Additionally, despite Darwin's opinion, some scientists were committed to Asia rather than Africa as the homeland of humans. Others felt that 'Piltdown Man' (now known to be a fake) demonstrated the paramount role of Europe, and a quite different ancestor than *Australopithecus africanus* in early human evolution. Finally, the Taung fossil was thought to be only about 500,000 years old, and was therefore viewed as being too late to be a genuine human ancestor. Instead, it was considered to represent a peculiar kind of ape, paralleling humans in some ways.

Dart is largely vindicated

Dart hoped for more fossil material to vindicate his claims, but it was to be another ten years before the South African limestone sites yielded up further important finds. Then, excavations at a limestone cave called Sterkfontein produced an adult australopithecine skull. At first it was named a different species, but in time it was recognized as belonging to the same species as Dart's child. Although Taung never produced further hominin fossils, Sterkfontein contained hundreds of them, and in the last few years, an older part of this cave system has yielded up a virtually complete australopithecine skeleton dated at over 3 million years. Sterkfontein was soon supplemented by another site called Makapansgat where the majority of the finds were of jaws and teeth. Such rich additional finds have shown that Dart was correct in some of his claims about *Australopithecus*. These early hominins did indeed show human features in their jaws and teeth – for example, their front teeth, including the canines, were relatively small compared with those of apes. Hip and leg bones show that these australopithecines were small-bodied but regularly walked upright, sharing this unique method of locomotion

(Right) The site of Sterkfontein in South Africa has been made a World Heritage Site. It is one of the richest sites in Africa for many fossil species dating between 2 and 3 million years old, including Australopithecus. *Although Dart believed that australopithecines lived in the caves, it is more likely that the fossils worked their way in through natural processes or the activities of carnivores.*

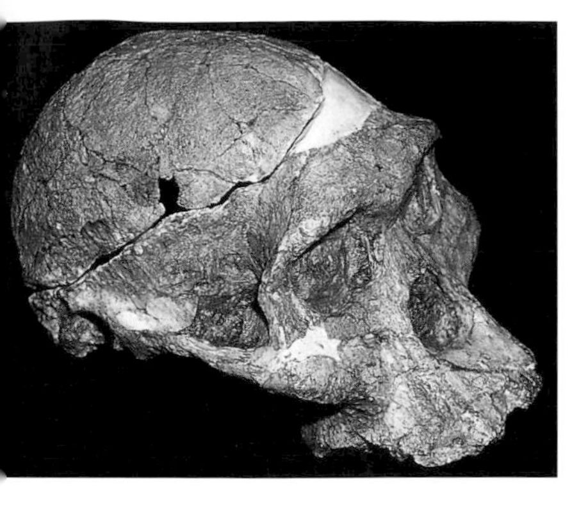

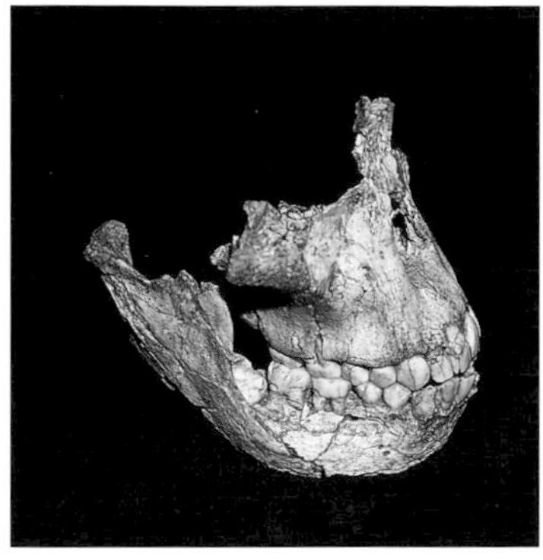

with humans. However, in other respects, Dart went far beyond the evidence in his claims. The brain cavities of the australopithecines were actually still only ape-sized, although expert opinion is divided over whether the shape of their brains was already showing a more human organization. And in overall body shape, they were still ape-like, with short legs, and suggestions in their limb, hand and foot bones that they still spent time in the trees.

(Above) Two of the best-preserved australopithecine fossils from Sterkfontein. The skull of 'Mrs Ples' (originally named 'Plesianthropus transvaalensis'*) is shown on the left, and the jaws of find number Sts53, containing one of the best-preserved upper and lower dentitions of the species, on the right.*

Hunters or hunted?

It also seems likely that they were mainly vegetarian, and that they were not active hunters. Instead, evidence suggests that the australopithecines were hunted, not hunters. Their bones probably accumulated in the caves not because they were living there, but because their body parts fell into the caves as the prey of true carnivores, such as leopards and hyenas. As far as we know they were not tool-makers and did not make fire. In their troop structures, patterns of growth and lifestyle they were probably much more like apes than humans. However, by comparing the animal fossils found alongside them with those from other dated sites, it is now known that they were much more ancient than formerly believed – not about 500,000 years old, but closer to 3 million. Thus their combination of ape-like and human features seems much less puzzling now than it was for Dart and his contemporaries. But their place in human evolution is still uncertain some 80 years after Dart's bold pronouncements. The wheel has turned almost full circle and many experts now echo the doubts raised in 1925 that these enigmatic creatures may only represent an extinct side-branch in the human evolutionary story.

(Above) Dart believed that the australopithecines were carnivores and active hunters, who took the remains of prey back to their cave homes. This reconstruction shows them emerging from their caves to hunt game, using tools of stone and wood.

(Below) Sterkfontein has produced several partial skeletons of Australopithecus. *It has recently been suggested that these remains, numbered Sts 14, could even represent the body of 'Mrs Ples'.*

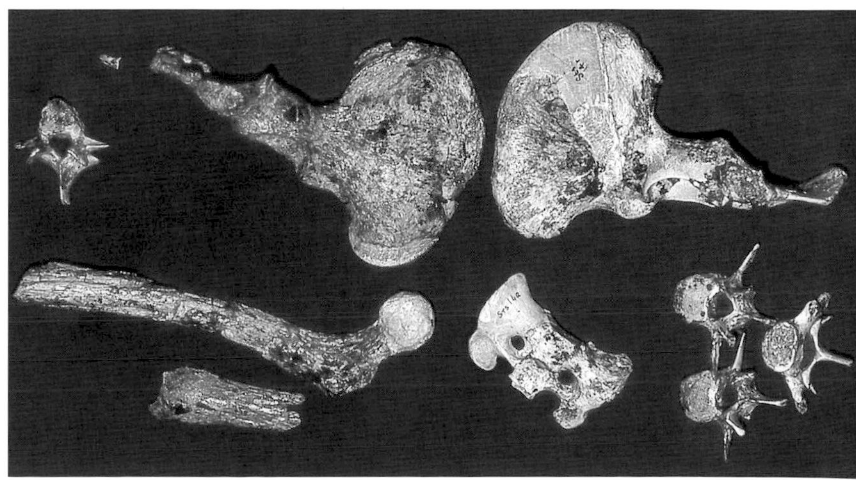

Robust Australopithecines

(Below) This map shows many of the main African localities of early hominin fossil discoveries. The sites can be divided into those associated with the sedimentary basins of the African Rift Valley, ranging from Ethiopia in the North to Malawi in the South, and those found in the limestone caves of South Africa, such as Swartkrans, shown undergoing excavation (right).

Bahr el Ghazal
Middle Awash
Bouri
Omo
Koobi Fora
West Turkana
Chesowanja
Peninj
Olduvai Gorge
Matema
Broken Hill
Makapansgat
Drimolen
Sterkfontein
Kromdraai
Swartkrans
Taung
0 500 miles
0 500 kilometres

As the search for more of Dart's australopithecines extended to new sites in South Africa, evidence began to accumulate that there was more than one type of this creature. The sites of Swartkrans and Kromdrai contained fossils which were clearly related to those which he had described, but the skulls seemed much more heavily built, and the back teeth were even larger than those of *A. africanus*. In fact, in some cases, there was a large crest of bone running along the top of the skull, similar to those found on large gorillas and chimpanzees. This crest developed to provide the maximum surface area for large jaw muscles. While some researchers felt that the variation might merely reflect differences in sex – in other words, the smaller types represented by Dart's child were females, while the more massive individuals were males – other scientists argued that another species of australopithecine was present. Their view eventually prevailed, and the new species was called *Australopithecus robustus*: hence these more heavily built forms became known as the 'robust', as distinct from the 'gracile' (meaning slender), australopithecines. As with the gracile forms, bones of the rest of the skeleton suggested that these

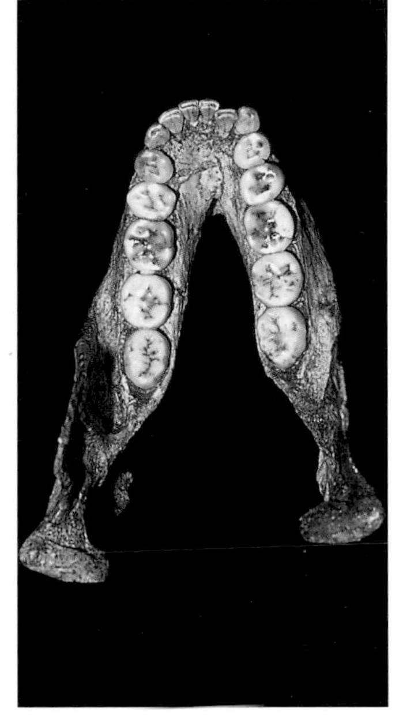

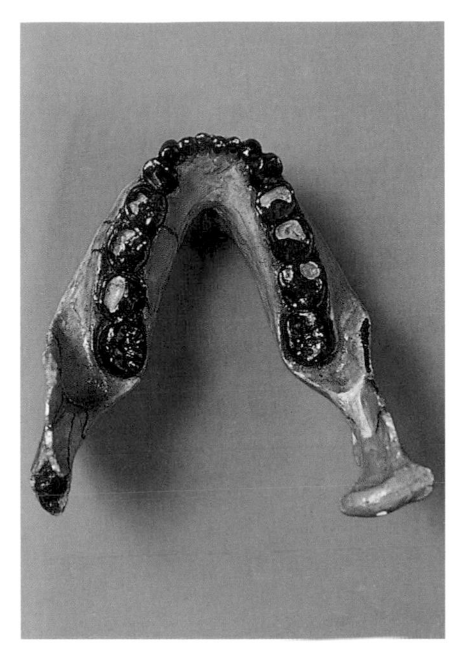

creatures also regularly walked upright. Their body size was certainly larger than that of the graciles, and they also had somewhat larger brains. While the smaller species had a brain size of about 400 ml, similar to that of chimpanzees, the brain size of the more robust form was comparable with that of the gorilla – about 500 ml. In the past few years, several new sites containing robust australopithecines have been discovered in South Africa, including Drimolen, which has produced a nearly complete skull of this species.

Because the two species were found in separate caves, it was not possible to tell whether they were contemporaneous or not. However, dating evidence now suggests that *A. africanus* lived about 3–2.5 million years ago, while the robust species probably lived between about 2 and 1.3 million years ago. This has led to suggestions that the gracile form evolved into the robust, perhaps because the increasing dryness of the environment encouraged adaptation to a diet richer in nuts, seeds, roots and tubers. Although Dart favoured the idea that the australopithecines of South Africa were carnivorous hunters, evidence suggests that they

(Above) This comparison shows lower jaws of robust australopithecines from South and East Africa with characteristic thick bone, small front teeth and massive back teeth. The mandible on the left (slightly crushed) is from Swartkrans, assigned to Paranthropus robustus, *that on the right from Peninj in Tanzania, assigned to* Paranthropus boisei.

(Opposite) This nearly complete skull from Swartkrans (find number SK 48) found in 1952 was one of those that finally established the existence of two forms of australopithecines in the cave deposits of South Africa. These were Dart's original Australopithecus africanus *('gracile australopithecines'), and the second were the robust australopithecines, often now assigned to a separate genus called* Paranthropus. *SK 48 shows the crested skull, large flat face and large back teeth characteristic of this group.*

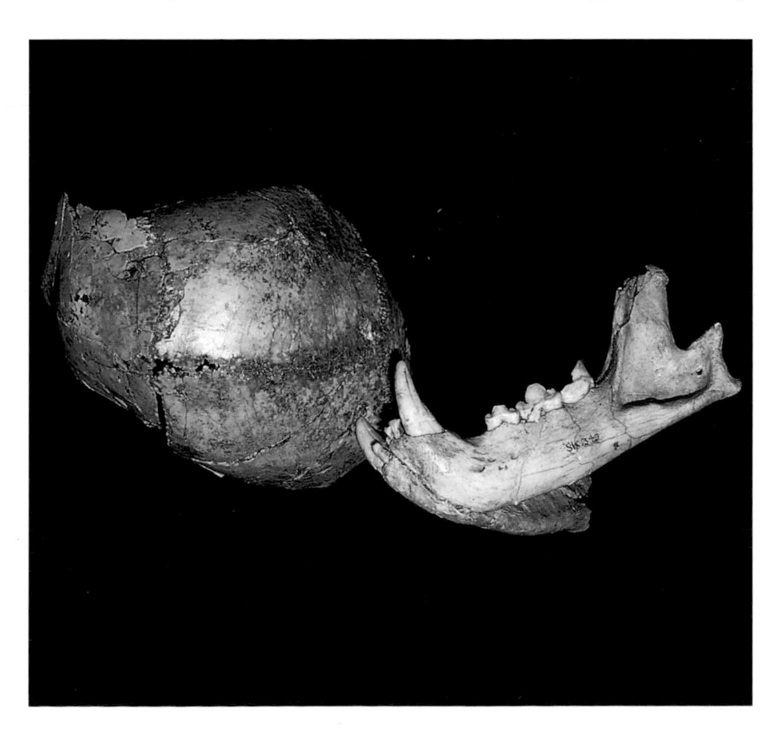

(Left) This picture shows the skull of a young robust australopithecine (SK 54) from Swartkrans, which has depressions at the back reminiscent in size and spacing of punctures made by the canine teeth of a leopard. That comparison is validated here using a fossil lower jaw of a leopard from the same site.

(Above) The cave site of Swartkrans, near Sterkfontein, started to produce important fossils in 1948, when Robert Broom found part of a human-like jawbone, followed soon afterwards by skull parts, now known to represent robust australopithecines. It was excavated more systematically from 1965 by C.K. ('Bob') Brain, who studied the whole process by which fossil remains had accumulated in the cave.

were in fact the hunted, the prey of predatory carnivores. One young robust australopithecine even has the puncture marks of leopard canines in the back of its skull. However, recent research on the chemical composition of their teeth suggests that the diet of the robust australopithecines probably did include meat, or insects such as termites, the latter perhaps collected using probes made of bone.

Robust australopithecines in East Africa

As excavations developed along the Rift Valley of East Africa, fossil evidence of australopithecines began to be found there, too. Mary Leakey discovered a skull in some of the oldest deposits at Olduvai Gorge, Tanzania. Originally called '*Zinjanthropus boisei*', it became the type specimen of a new species of robust australopithecine which was even more robust than the South African form, with thicker jaws and bigger molars. This form was also subsequently discovered in Ethiopia, at sites like Omo, and Kenya, at sites like Koobi Fora. More recently, many scientists have accepted that the differences between the gracile and robust forms are even more fundamental, and have favoured a greater distinction than the species level. They have resurrected an old name for the robust forms of South Africa – *Paranthropus* ('Alongside man'), and so the species of robust australopithecines hence become *Paranthropus robustus*, and *Paranthropus boisei*. The dating of the South and East African robusts suggests they are approximately contemporary, between about 2 and 1.3 million years ago, although it would appear that the East African form was more specialized in its teeth, reflecting an even greater concentration on chewing its foods. It is assumed by many experts that the two species evolved from a more ancient common ancestor, and this may have been discovered in Kenya, Ethiopia and Malawi, since a more primitive robust australopithecine called *Paranthropus aethiopicus*, has been found there. It has a somewhat smaller brain and much more projecting face than the later and possibly descendant species, but the characteristic crest on the top of the skull is already well developed.

The relationship between the robust australopithecines and early humans has been much discussed. The two forms clearly overlapped at the

(Right) Louis Leakey pictured with the skull of Olduvai Hominid 5 (known to the Leakeys as 'Dear Boy', but also nicknamed 'Nutcracker Man') in London in 1959. At this stage, he still believed that 'Zinjanthropus' boisei *was the real manufacturer of the oldest stone tools excavated at the site, and was a human ancestor. Soon afterwards, the discovery of* Homo habilis *fossils at Olduvai forced a rethink.*

time of both *Homo habilis and Homo erectus* in Africa, and then the robusts became extinct. It has been suggested that once *Homo erectus* became an accomplished hunter, the robusts might have become prey, and were eventually hunted to extinction by their cousins. This is certainly possible, but it is equally possible that environmental change, or dietary competition from successful radiations of pigs or monkeys may have been responsible. Whatever their fate, the robust australopithecines represented one of the most remarkable evolutionary experiments in the hominid family.

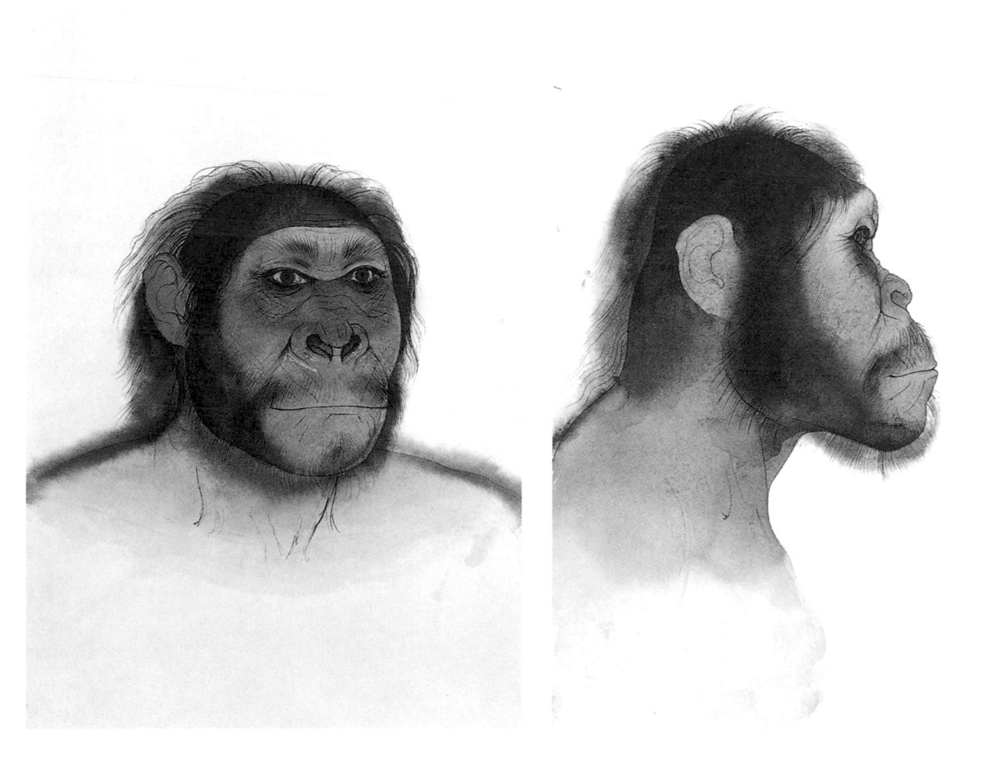

(Left) Reconstructions of the face of Paranthropus boisei, *based on the fossil material from Olduvai and Peninj. Although ear and hair form is conjectural, the very flat nose and face and the massive jaws are undoubtedly accurate.*

The Origins of Humans

Defining what we mean by 'human' is not a straightforward task. For example, in behaviour, humans are characterized by great complexity compared with other animals, and one of our special features is our great reliance on toolmaking and tool-using. Other animals utilize objects, or may modify them to a specific design (e.g. a sea otter breaking mollusc shells with a rock, or a bird using twigs to build a nest), but most of this is simple or even instinctive behaviour. Humans used to be defined as 'Man the Toolmaker', and the appearance of stone tools in the prehistoric record was viewed as the key evidence for the evolutionary appearance of humanity. However, it was then discovered that chimpanzees, our closest living relatives, sometimes modified grass stems in order to 'fish' for termites in their termite nests, and many anthropologists have since accepted that the difference between human and ape toolmaking is a quantitative rather than qualitative one.

(Below) Observation of wild chimpanzees shows that they use tools for such purposes as threatening each other, catching termites, and cracking nuts. These behaviours vary from group to group, suggesting that chimpanzees pass on such behaviour socially as 'culture'.

(Below right) The chimpanzee vocal tract is not designed to produce the variety of sounds that a human can make.

Language as a defining characteristic

Other behavioural evidence has also been used in attempts to recognize the first humans. Language is a complex system of communication, and it has been argued that this is one of our defining features. While animals may communicate with each other through gestures or sounds, these are not usually arranged in complex sequences, and can only deal with immediate events. Many attempts were made in the past to teach apes to talk, but these were unsuccessful, and we now know that they could not have succeeded because the throat in apes is differently structured, and simply cannot produce our range of sounds. However, it was later discovered that chimpanzees could learn to use symbols or key boards to 'talk' to humans and to other chimpanzees. They could learn a vocabulary of up to 200 'words', and could arrange these in simple sequences to, say, ask for food, or describe objects. Nevertheless, no ape has been able to master a larger vocabulary, deal with abstract concepts, or talk about the distant past or future, which humans can do.

So human language is certainly unique, and determining its time of origin would be vital in recognizing a key development in human history. The outside surface of the human brain does show some structural features that are important in the production of speech, and if these are damaged,

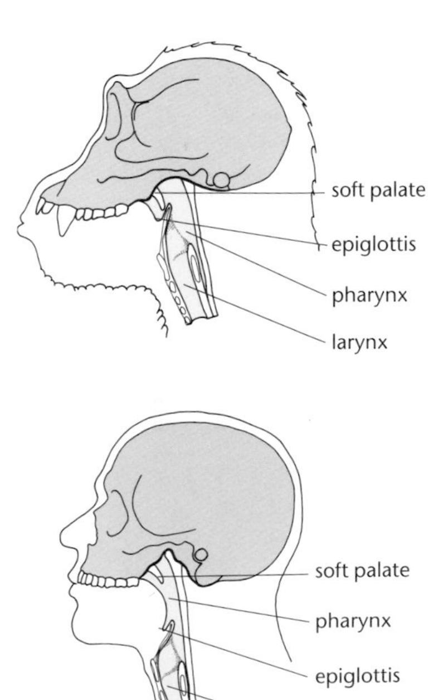

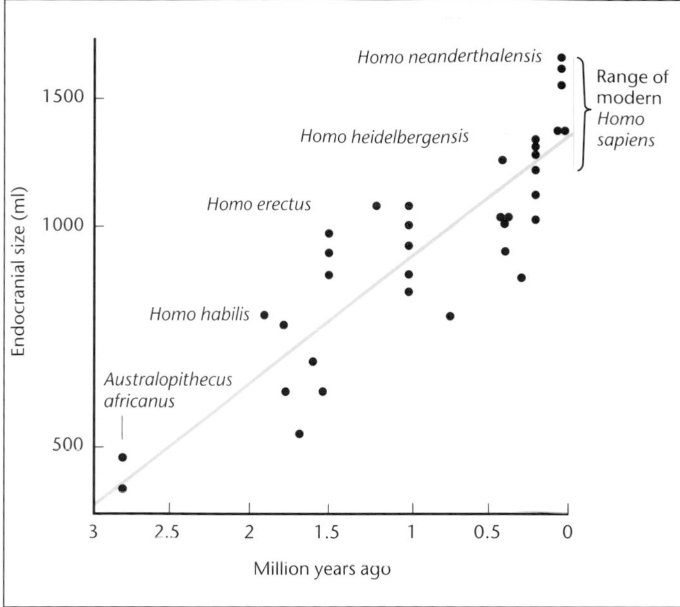

speech may be deficient or lost completely. However, trying to recognize equivalent surface features on endocasts (replicas taken from the brain cavities of fossil skulls) is difficult, and of debatable value. It is probably more realistic to try to assess the presence of language from the degree of behavioural complexity which we can infer from the archaeological record, and on that basis, many scientists now believe that language was not present early in human evolution.

Bodily features as defining characteristics

Because behaviour does not 'fossilize' directly, many scientists have preferred to recognize humans from bodily features, which do get preserved. Walking upright is a unique human adaptation, but we now know it developed early in human evolution, when many other human features were still to evolve, so most experts do not use this feature to recognize true humans. Our long-legged body shape is also distinctive, but this seems to have originated later in time – perhaps 1.8 million years ago. A large brain is another special human feature, and the size of the brain cavity can be measured quite accurately in fossil skulls. Brain size does also have to be related to body size since, other things being equal, a larger animal will generally have a larger brain. Thus gorillas usually have larger brain sizes than chimpanzees, although there is no evidence that this gives them a greater intelligence – in fact, the reverse could be true in this case.

Relative brain size does seem to be a better indicator, and on this basis it appears that the earliest members of our evolutionary line were hardly larger-brained than present-day apes. It was not until about 2 million years ago that relatively larger brains first evolved in our lineage, and for many scientists, this marks the true origin of humanity. Yet other characteristics have been proposed to mark the appearance of real humans – for example, the relative size and projection of the face, the evolution of a prominent nose or the development of the extended growth period and delayed maturity which characterizes humans today. In reality, many of these features evolved gradually, and at different rates, and it could not be expected that they evolved suddenly as a 'package'. Thus recognizing the first 'humans' is likely to remain a matter of great controversy, as it was for most of the last century.

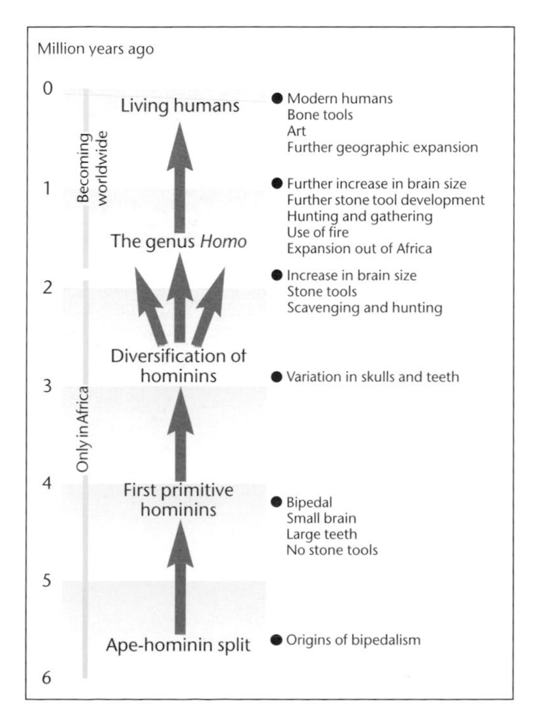

(Above left) Although great apes cannot imitate human sounds, they show a surprising ability to use words when these are coded in the form of objects, computer symbols or hand signs. This shows that their brains are structured (potentially) to create and use quite an extensive vocabulary.

(Above right) Brain size has more or less trebled during the evolution of the human lineage over the last 3 million years, although there was also an increase in overall body size over that time, meaning that the proportionate increase was actually somewhat less.

(Left) The course of human evolution and the appearance of human characteristics over the course of the last 6 million years.

Early Homo

Reconstruction of Homo habilis *at a kill site. Partial skeletons of large and medium-sized mammals surrounded by stone tools have been excavated at sites in Tanzania, Kenya and Ethiopia dating from nearly 2 million years ago. The animal bones are often marked with cuts suggesting butchery, and breakage suggesting the extraction of marrow. It is not always clear, however, whether the carcasses were the result of human hunting, or human scavenging from the kills of other carnivores. Sometimes, however, the cut marks do seem to precede evidence of scavenging by animals such as hyaenas, suggesting that people did have primary access to the dead animal.*

As we saw on p. 68, crude stone tools had been found in ancient deposits at Olduvai Gorge in Africa long before any signs of their manufacturer had been discovered. But in 1960, after many years of searching, the Leakeys finally found fossils which seemed to represent the first toolmakers at Olduvai. Over the next couple of years enough material was collected to show that the hominin concerned was definitely distinct from the robust australopithecine that had been found there in 1959. The assemblage of fossils included upper and lower jaw fragments, skull parts, hand and foot bones, and leg bones. The teeth in the jaws were differently proportioned from those of australopithecines, since they were relatively larger at the front and smaller at the back, and thus more like those of humans. The skull bones were thin, did not have any bony crests on them, and were large enough to suggest that brain size was significantly bigger than that of any australopithecine. Louis Leakey decided that the material was sufficiently human to be classed within our own genus *Homo*, as a new species called *habilis*, meaning 'handy', because of its assumed toolmaking ability. He and his collaborators believed that this species represented the most ancient and primitive of all humans.

Fossil finds from Koobi Fora

Homo habilis was not well received by the scientific community. Some scientists felt that the material was not complete enough for definite opinions, others felt that it merely represented a new kind of australopithecine, while yet others felt that it consisted of a mixture of australopithecine and genuine early human fossils. Nevertheless, *Homo habilis* gradually gained scientific credibility, and new fossils were found at Olduvai and elsewhere. Louis's son, Richard, initiated a new research project in northern Kenya, at Koobi Fora, on the eastern side of Lake Turkana (formerly Lake Rudolf). He was soon rewarded with finds of stone tools like those found in the earliest layers at Olduvai Gorge, as well as the remains of robust australopithecines, dated at nearly 2 million years old. So there was immediate speculation about whether *Homo habilis* would also be found there. The discovery of the fossil skull known by its catalogue number KNM-ER 1470 seemed to fulfil these expectations, since it was clearly large

(Left) There has long been debate about variation within the species Homo habilis. *This comparison shows a small-brained hominin (number KNM-ER 1813 from the Kenya National Museum site originally called East Rudolf) on the right and a larger brained hominin (number KNM-ER 1470) on the left. Are these skulls merely small (female?) and large (male?) variants of a single species, or do they represent different species – perhaps* Homo habilis *(1813) and* Homo rudolfensis *(1470)?*

(Right) Malawi lies at the southern end of the African Rift Valley and it was suspected that early hominins must have lived there. This was confirmed in 1993, when a thick lower jaw was discovered at Uraha by a team led by the German palaeontologist Friedemann Schrenk. Find number UR 501 (pictured here), dated at about 2.4 million years, bears a close resemblance to a mandible found in Kenya (KNM-ER 1802), and both have been assigned to Homo rudolfensis.

brained (estimated volume about 750 ml, compared with australopithecine figures of about 400–500 ml), and lacked any sign of the cresting on the top of the skull which large robust australopithecines showed. However, it also had a large and very flat face and, although its teeth had been lost, their sockets indicated that they must have been rather large.

One species or two?

Matters became even more complicated when further finds of possible early humans were found at Koobi Fora. These included skulls with rather human faces and teeth, but small braincases, no larger than those of australopithecines, and even an example of an apparent *habilis* with a crest on the top of the skull, like that of an ape or a robust australopithecine. As a result, a number of scientists have suggested that more than one species is represented in the fossils assigned to *Homo habilis*. They have suggested that the larger form represented by the 1470 skull should be called *Homo rudolfensis*, while the smaller species would retain the name *habilis*. Because of this increased complexity, it is no longer clear which, if any, of these early forms of *Homo* might be ancestral to later humans, and which were the toolmakers. And there is one further species that some experts believe could be a human ancestor, an australopithecine species called *A. garhi* from about 2.5 million years ago. Found at Bouri in Ethiopia, the material consists of skull and jaw parts and separately, arm and leg bones with rather human proportion, and cut-marked animal bones implying both stone tool use and meat-eating at this early date.

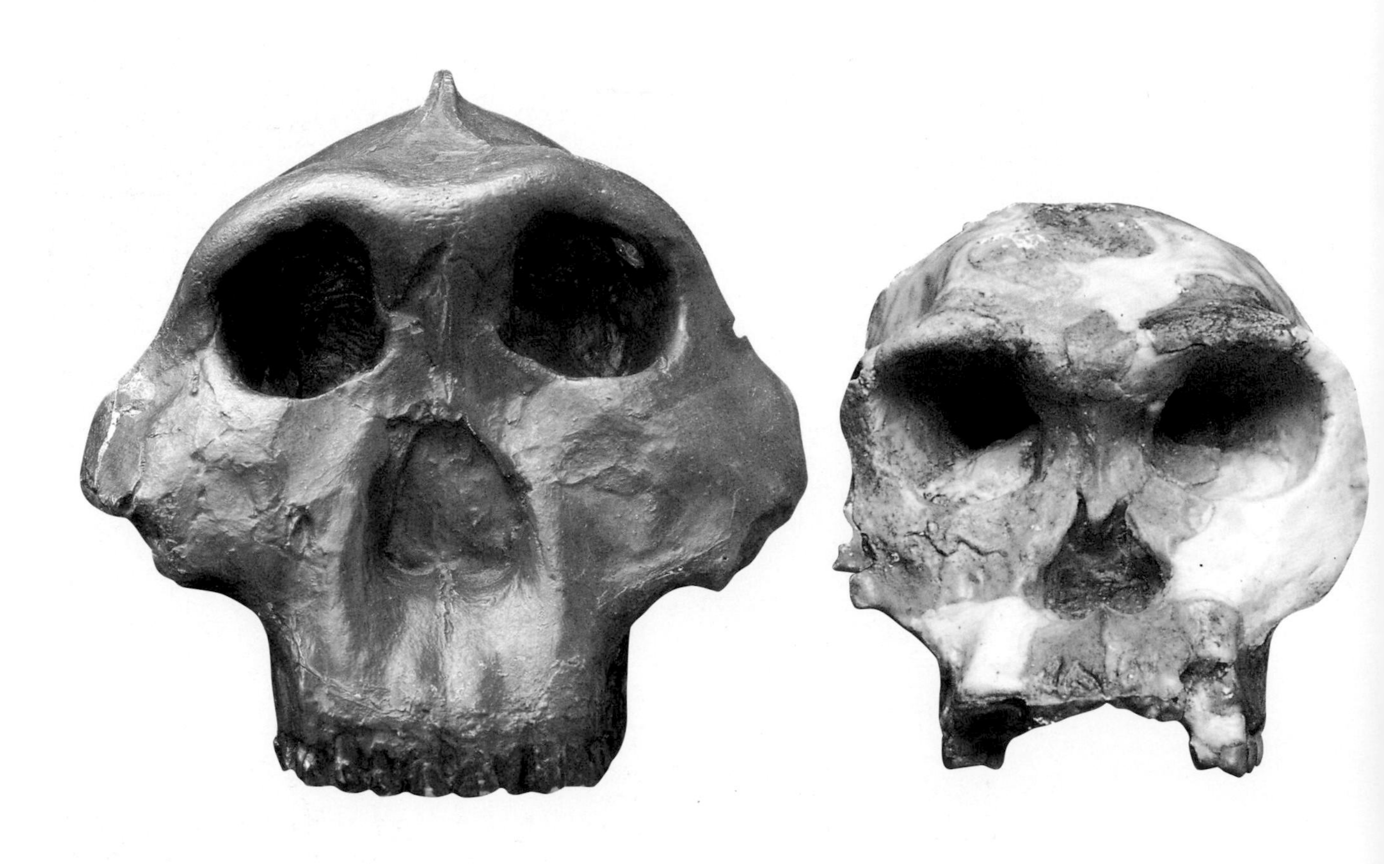

Despite the suggestive evidence from Bouri, it is also unclear what kind of skeletons these early human-like forms had. Some rather large leg bones were found in levels also containing *Homo rudolfensis* fossils, suggesting human-like body size and proportions for this species, but a recent discovery at Olduvai paints rather a different picture for *Homo habilis*. Olduvai Hominid 62 consists of an upper jaw which has human proportions, but the associated skeleton is tiny, and its proportions are decidedly non-human. The arms are relatively long compared with the leg bones, and this implies that the body proportions of this creature were ape-like, perhaps even more so than that of Lucy, the famous australopithecine skeleton from Ethiopia, which is over a million years older. The human status of the small *habilis* species therefore remains doubtful. Equally, the large face, back teeth, and thick jaws of *rudolfensis* do not look particularly human, and some scientists have made an evolutionary link with the 3.5-million-year-old form called *Kenyanthropus platyops*, implying these might represent an entirely separate and extinct hominin lineage. Thus, despite their disparate human-like features in teeth, face or brain size, the human status of the varied '*habilis*' group of fossils remains a matter of dispute.

(Above and right) Homo habilis *also shows variation in the lower jaw. This comparison shows a mandible from Kenya (KNM-ER 1802), left, and the original mandible of* Homo habilis, *Olduvai Hominid 7. The former is now often assigned to* Homo rudolfensis, *while the latter remains as* Homo habilis.

(Opposite) The skull, right, numbered Olduvai Hominid 24 (OH 24), was discovered by Peter Nzube in 1968 at Olduvai in Tanzania. Squashed nearly flat when it was discovered, it was nicknamed (in what may now seem to be an act of political incorrectness) 'Twiggy', after the famous model of the time, who was apparently rather flat-chested. More recently it has generally been assigned to Homo habilis. *Here it is compared with the much larger skull of* Paranthropus Boisei – *Olduvai Hominid 5.*

(Right) For some workers, true humans only appeared with the species Homo ergaster *(for some, early* Homo erectus*) – exemplified by skull KNM-ER 3733 shown here. They argue that it was only with these forms that human features in the skull, jaws, teeth and body shape first appeared. In this viewpoint, species such as* Homo habilis *and* Homo rudolfensis *are really more closely allied to the australopithecines.*

Homo erectus

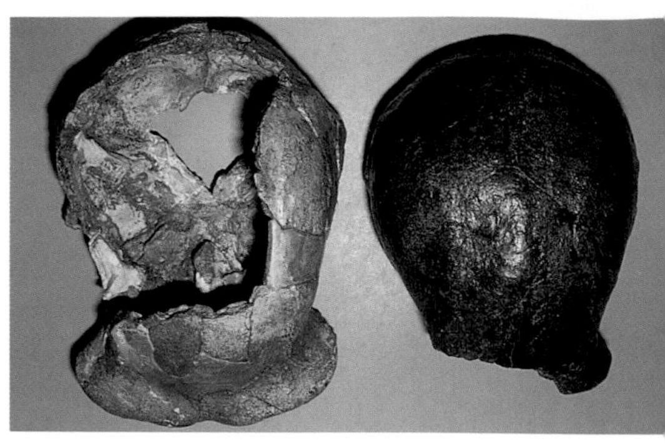

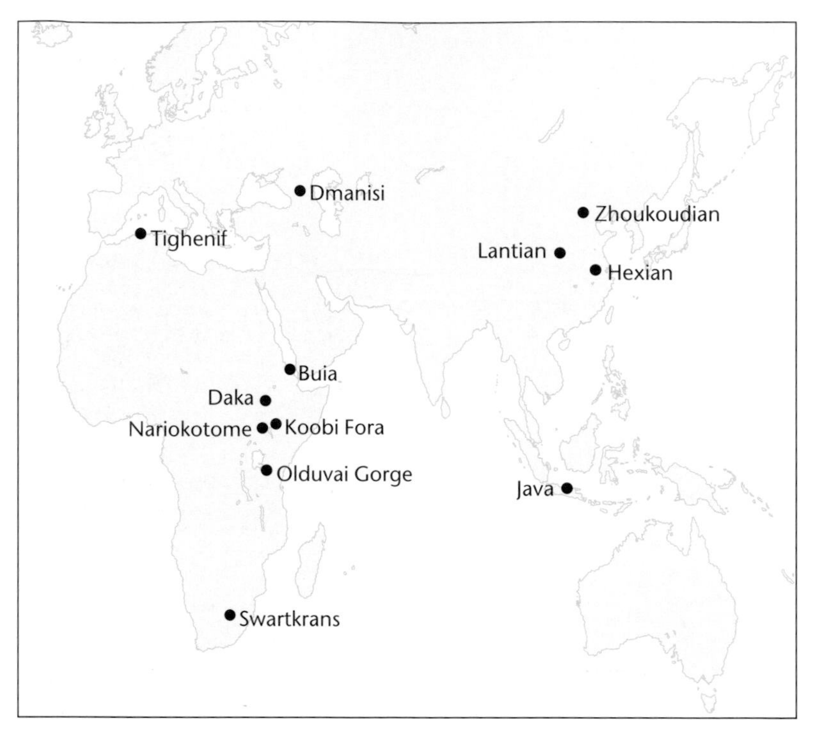

(Above) Map showing sites of Homo erectus *finds.*

(Above right) This comparison is a view from the top of Homo erectus *skulls from the African and Australasian regions. On the right is Sangiran 2 from Java, Indonesia, and on the left, the partial skull Olduvai Hominid 9 from Tanzania. Differences in size and robusticity probably reflect variation in time and space, and also a difference in sex (the Javanese skull is younger in age and may be that of a woman, the Tanzanian skull is probably male).*

During the 19th century, a German biologist called Ernst Haeckel developed a series of hypothetical stages in human evolution, based on his belief that the gibbon represented the closest living approximation to our ancient ape ancestor. Haeckel believed that Darwin was wrong, and Asia, rather than Africa, was our ancestral homeland. He named one of the hypothetical pre-human stages *'Pithecanthropus alalus'* ('Ape man without speech'), and argued that it would have lived in Southeast Asia. In 1889, a young Dutch doctor called Eugene Dubois decided to prove Haeckel right by finding actual evidence of this 'Pithecanthropus'. He obtained a posting as an army doctor to the island of Java in the Dutch East Indies, and astonishingly within two years, he had found fossil evidence of a primitive stage in human evolution. His discoveries included a thick, long and low skull cap, with a large brow ridge, and a very human-looking thigh bone. Accordingly, following Haeckel, he named his find

'*Pithecanthropus*', but gave it a different species name because of the upright posture he inferred from the femur – '*erectus*'. We now know this species as *Homo erectus* ('Erect man'), because it is generally recognized that it is indeed human.

Java and China

By 1940, many more remains of this species had been found in Java, and in new regions such as China. There, the site of Zhoukoudian, near Beijing, produced numerous erectus fossils which were initially assigned to '*Sinanthropus pekinensis*' ('Chinese man of Peking'), but which were later amalgamated with the fossils of 'Java Man' as also representing *Homo erectus*. The characteristics of this species were now clear. The skull was relatively

A Homo Erectus *skull from Java (right) compared with an early African skull (left) often assigned to* H. ergaster *and the more recent Petralona skull (centre), often assigned to* H. heidelbergensis.

(Below) A reconstruction of Zhoukoudian showing a group fashioning tools of bamboo and quartzite, and cooking meat on a fire. In the background dangerous hyaenas are being driven away – it seems likely that humans and hyaenas regularly competed both for shelter and food. Recent research has cast doubt on the extent of human fire use at Zhoukoudian, however.

(Left) Dragon Bone Hill, the location of the caves of Zhoukoudian, near Beijing, China. Fossils, believed to be the remains of dragons, were collected from here and ground up for traditional medicines.

elongated, but low, with a flat forehead. It was very broad across its base, and poorly expanded higher up. There was a strong bar of bone above the eye-sockets (the supraorbital torus), and another running across the back of the head (the occipital torus). The braincase was quite voluminous compared with those of earlier hominins, reaching a capacity of about 1,000 ml, 75 percent of average values today, but its walls were thick, and often reinforced by extra bone (keels). The bones of the rest of the skeleton were heavily built, with thick walls, suggesting that erectus had a lifestyle which placed heavy demands on the skeleton. The teeth were large, but human, although they were set in thick-boned jaws which lacked a chin at the front.

Dubois had no real idea of the antiquity of the Javanese fossils, but we now know that the earliest ones date from about 1.5 million years, while the youngest may be surprisingly recent, perhaps less than 100,000 years old. Chinese *erectus* fossils span a similar time range, from over 1 million, perhaps down to about 250,000 years. Over this time span, the species showed a moderate increase in average brain size (as estimated from the inside of the skull), perhaps correlated with an increase in body size, but few other significant changes. Nevertheless, some scientists believe that *Homo erectus* in Asia was gradually evolving into modern humans – this view is a fundamental tenet of multiregional evolution (see pp. 140–143).

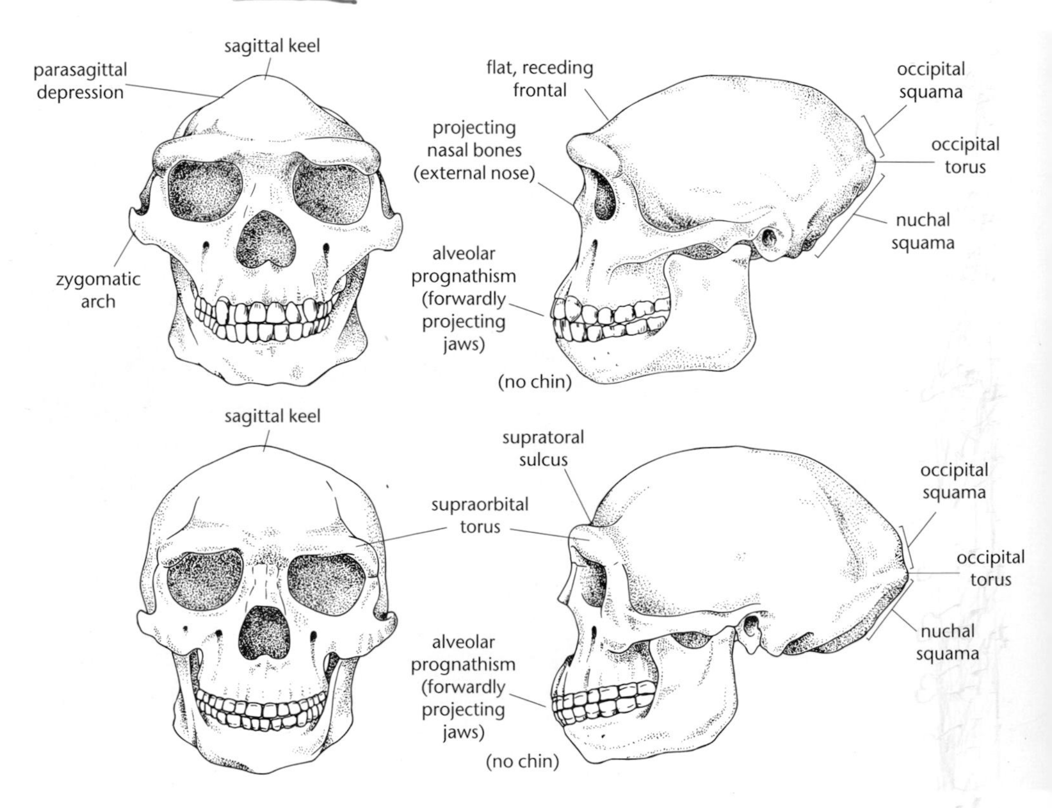

(Right) This comparison shows reconstructions of Homo erectus *skulls from China and Java supervised by the German palaeoanthropologist Franz Weidenreich. The upper reconstruction is based on more ancient fossils from Sangiran, Java, probably male, and the lower on younger material from Zhoukoudian, China, probably female. The Javanese fossils are generally more robust, with thicker and stronger skulls, the later Chinese individuals more lightly built and somewhat larger brained. Weidenreich believed that the Javanese skulls could be linked to the ancestry of native Australians, while the Zhoukoudian fossils may have represented ancestors for recent oriental peoples.*

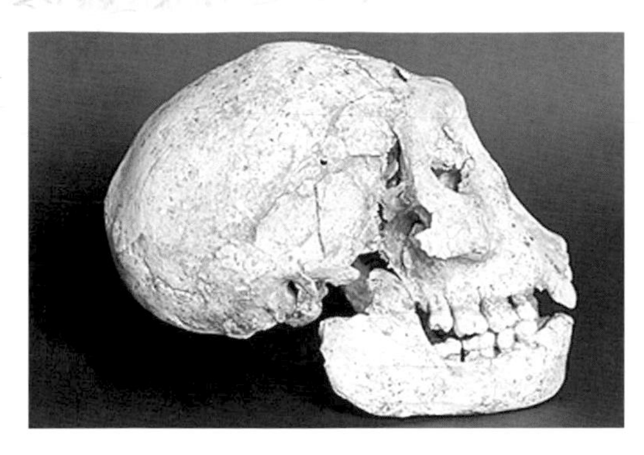

The origins of *Homo erectus*

The origins of this species are uncertain. There are arguably very early examples of *erectus* outside Africa in the Caucasus (the Georgian site of Dmanisi), where several skulls, jaws and parts of skeletons have been discovered and dated to about 1.8 million years ago. Some workers regard these as an early form of *erectus*, while others see links to more primitive African species, even including *H. habilis*. Recently it has even been suggested that a new species, *'Homo georgicus'*, is represented. However, there is considerable variation in the Dmanisi collection of human fossils, so it is possible that more than one species of early human is present. Java itself has volcanic rocks supposedly associated with early *erectus* fossils as ancient as 1.8 million years, but many workers are cautious about the precise association and age. Most experts believe that *erectus* originated in Africa perhaps 1.9 million years ago. A skull cap assigned to this species was found by the Leakeys in Olduvai Gorge, and dated to about 1.2 million years ago, but even earlier finds have been made near Lake Turkana, in northern Kenya. Skulls similar to those of *erectus* have been excavated from Koobi Fora, and dated at about 1.8–1.7 million years old, and most spectacularly, a nearly complete skeleton of a boy was found at Nariokotome on the western side of the lake.

The Nariokotome boy

His skeleton is dated at about 1.5 million years, and represents the most complete ancient human skeleton yet discovered. The boy died at about 11 years of age, to judge from modern human developmental standards, but studies of microscopic growth lines in his teeth suggest that he had been growing very fast and was closer to 8 at death. He must have been well over 1.5 m (5 ft) tall by then. Estimates of his probable adult height are close to 1.83 m (6 ft), and his physique was comparable with that of present-day African populations in the region today – he was tall, long-legged and slender, with narrow hips and shoulders. This body shape is an ideal one for humans in hot, dry conditions, since it maximizes the body's surface area, and thus helps to prevent overheating. He showed some peculiarities in his spine which some people think were caused by disease or injury, and his ribcage was differently shaped from our own. The head already showed characteristic *erectus* features in the lack of a chin, the big, flat, but jutting face and broad nose, the developing but strong brows, and the brain capacity of about 900 ml (compared with a modern male average of about 1,350 ml). However, some experts argue that the early African forms of the species are distinct enough to be classified as a different species, *Homo ergaster* ('Working man'), because they are less extreme in their features than the later Asian *erectus* people. In this scenario, *ergaster* evolved in Africa from something like *Homo habilis*, and then spread out rapidly to tropical and subtropical Asia, where the descendant *erectus* species developed its more robust features.

(Above left) This small human-like skull is numbered D2700 and was discovered at Dmanisi in Georgia in 2001. Its estimated age is close to 1.8 million years and it shows how primitive these earliest known Asian humans were. Features such as the brain size of only about 600 ml have led to suggestions that the Dmanisi material represents an earlier stage of human evolution than H. ergaster *and* H. erectus, *perhaps closer to the African species* H. habilis *or* H. rudolfensis. *Yet another suggestion is that the material represents a new species,* 'H. georgicus'.

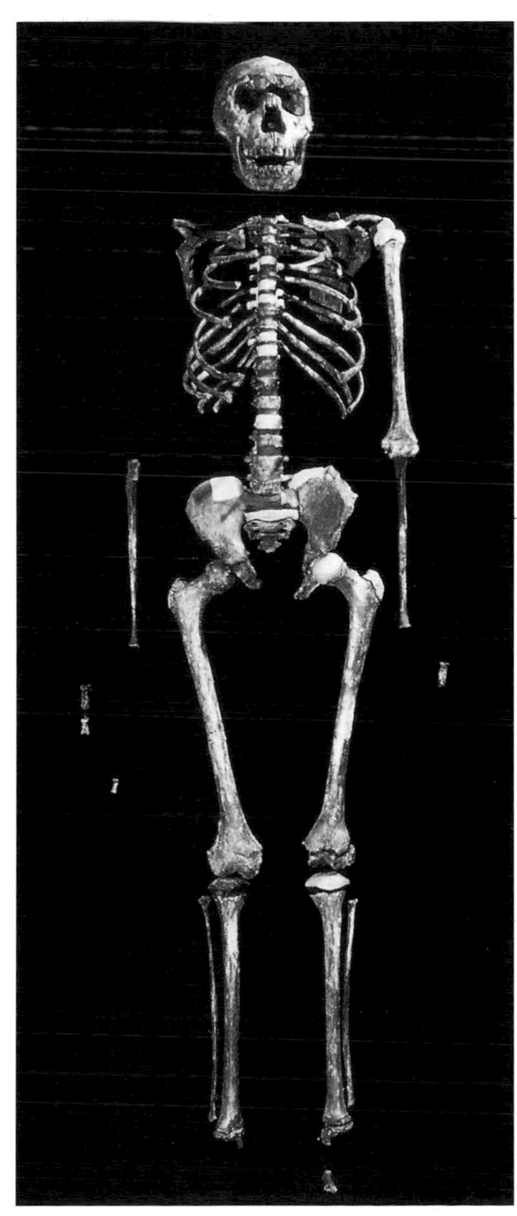

(Right) This skeleton numbered KNM-WT 15000 was discovered at Nariokotome, West Turkana, Kenya in 1984. It was dated at about 1.5 million years and has been assigned to H. ergaster *or early* H. erectus. *While the skeleton does have some unusual features, it is essentially human in form below the neck, with a tall, linear body shape well-adapted to life in hot, dry conditions. Above the neck, the skull and jaw already show the characteristic features of* H. ergaster/H. erectus, *even though this youth may only have been 8 years old at death.*

Models of Recent Human Evolution

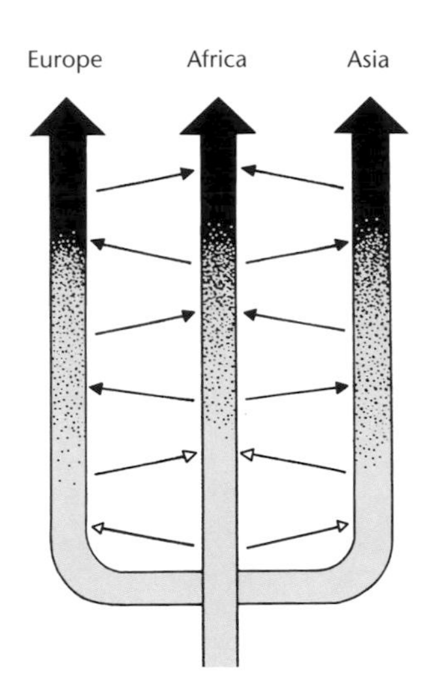

Discussing the origin of our species involves an understanding of the evolution of the special characters which living humans share, for example, a more lightly built skeleton compared to other human species, a higher and rounder braincase, a retracted face, smaller brow ridges, and a more prominent chin. But we also need to examine the evolution of the characters which distinguish different geographic populations today, the regional or 'racial' characteristics, such as the more projecting nose of many Europeans, or the flatter face of most orientals. There are two diametrically opposed views on how our species *Homo sapiens* evolved from its assumed ancestor, *Homo erectus*, with many intermediate views between these extremes. The two extreme views differ quite radically over where and when we developed our special 'modern' features, and when we began to evolve our regional differences.

The Multiregional Model

Supporters of one extreme view, the Multiregional Model (now very much a minority view), believe that *Homo erectus* gave rise to *Homo sapiens* across its whole range, which, about 1 million years ago, included Africa, China, Indonesia, and, perhaps Europe. According to this view, when *Homo erectus* dispersed around the Old World over a million years

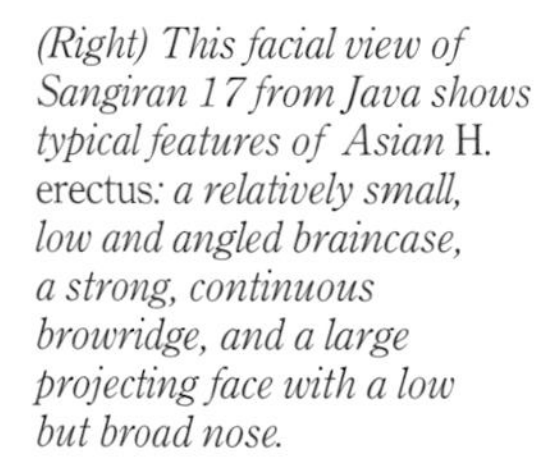

(Right) This facial view of Sangiran 17 from Java shows typical features of Asian H. erectus*: a relatively small, low and angled braincase, a strong, continuous browridge, and a large projecting face with a low but broad nose.*

(Opposite) This facial comparison of a modern human skull (left) with a Neanderthal from France (La Ferrassie) shows some of their shared advanced features compared with Homo erectus*: a higher, rounder skull, containing a larger brain, and a narrower, more lightly built face. However, the skulls contrast in browridge form (still large in the Neanderthal) and in the shape of the middle of the face: the Neanderthal has inflated cheekbones and a large, projecting nose.*

ago, it gradually began to develop both the modern features, and the regional differences that lie at the root of modern 'racial' variation. Particular features in a given region developed early on, and persist in the local descendant populations of today. For example, 400,000-year-old Chinese *Homo erectus* specimens had the same flat faces, with prominent cheekbones, as modern oriental populations. Indonesian *Homo erectus* of 700,000 years ago had robustly built cheekbones and faces that jutted out from the braincase, characteristics found in modern Australian Aborigines. In Europe, another line of evolution gave rise to the Neanderthals, who, according to this view, were the ancestors of modern Europeans. Features of continuity in this European lineage include prominent noses and midfaces. Recent supporters of multiregional evolution such as Milford Wolpoff and Alan Thorne have emphasized the importance of gene flow (interbreeding) between the regional lines, which prevented them from diverging and speciating, and allowed new traits to spread from one population to another across the inhabited world. In fact, they regard the continuity in time and space between the various forms of *Homo erectus* and their regional descendants to be so complete that they should all be regarded as representing only one species: *Homo sapiens*.

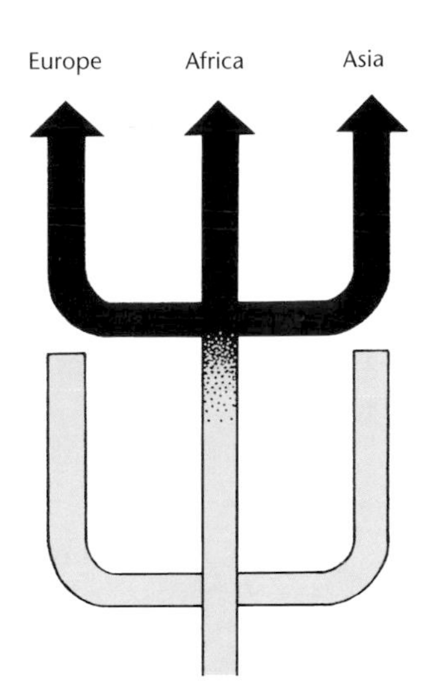

Models of Recent Human Evolution

(Left and opposite) These diagrams illustrate two contrasting models of recent human evolution. Opposite, Multiregional Evolution, essentially the evolution of only one species – Homo sapiens *– over a 2-million-year period; and on the left the Out of Africa model, with the recent evolution and spread of* Homo sapiens *from Africa, and the accompanying replacement of more ancient lineages outside of Africa.*

The 'Out of Africa' Model

The opposing view is that *Homo sapiens* had a restricted origin in time and space. Modern proponents of this idea, such as Gunter Bräuer and Chris Stringer, focus on Africa as the most important region. Some argue that the later stages of human evolution, like the earlier ones, were characterized by evolutionary splits, and the coexistence of

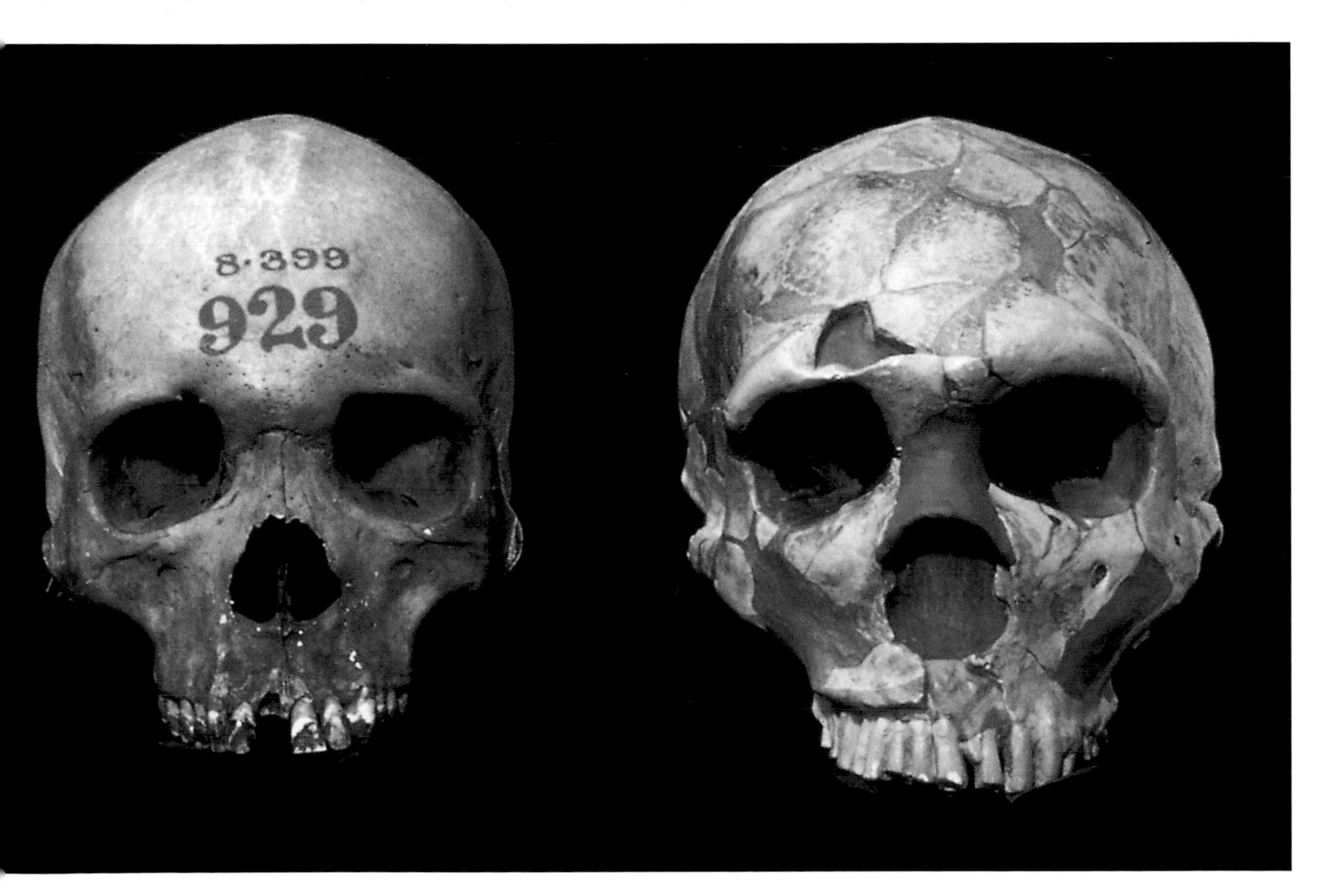

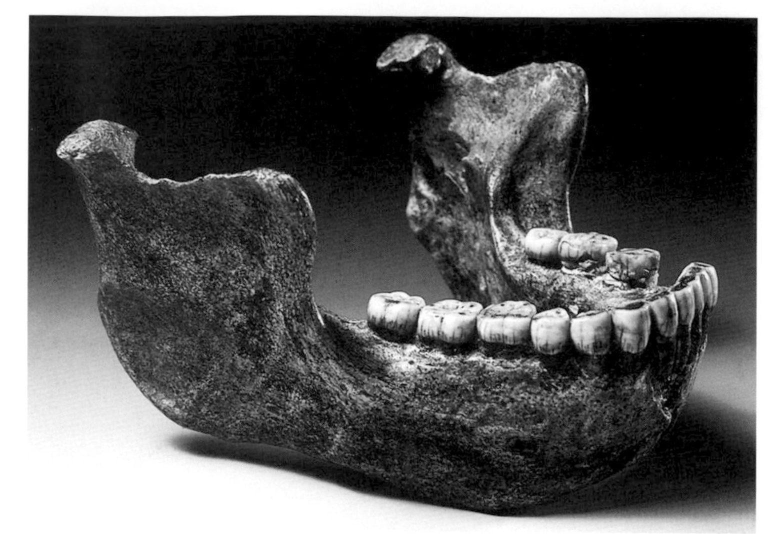

The Mauer jaw was discovered in a sand quarry near Heidelberg, Germany, in 1907. One year later it was named as representing a new human species, Homo heidelbergensis. *For many years this massive and chinless jawbone was instead regarded as a European example of* Homo erectus, *but recently its original name has increasingly come back into use for the European human populations of about 500,000 years ago.*

(Below) Here two famous French fossils are compared to show the characteristic features of their respective populations. They are the 'Old Men' of (left) La Chapelle-aux-Saints (a late Neanderthal), and of Cro-Magnon (an early modern human). Under the Multiregional and Assimilation models, these populations could have exchanged genes quite extensively, compared with the replacement of the Neanderthals implicit in the 'Out of Africa' model.

separate species. They recognize an intermediate species between *Homo erectus* and *Homo sapiens*, called *Homo heidelbergensis*. On this view, by about 600,000 years ago some *erectus* populations in Africa and Europe had changed sufficiently in skull form to be recognized as a new species, *Homo heidelbergensis*, named after a 500,000-year-old jawbone found at Mauer, near Heidelberg, in Germany. Members of this species had a less projecting face, more prominent nose, and a more expanded braincase than *erectus* fossils. *Homo heidelbergensis* is known from Africa, Europe and possibly China, between about 600,000 and 300,000 years ago.

From the 'Out of Africa' viewpoint, after about 400,000 years, heidelbergensis apparently gave rise to two descendant species, *Homo sapiens* and *Homo neanderthalensis*, the former evolving in Africa, and the latter in Europe and western Asia. About 100,000 years ago, the African stock of early modern humans started to spread from the continent into adjoining regions and eventually reached Australia, Europe, and the Americas (probably by 60,000, 40,000, and 15,000 years ago respectively). Regional (racial)

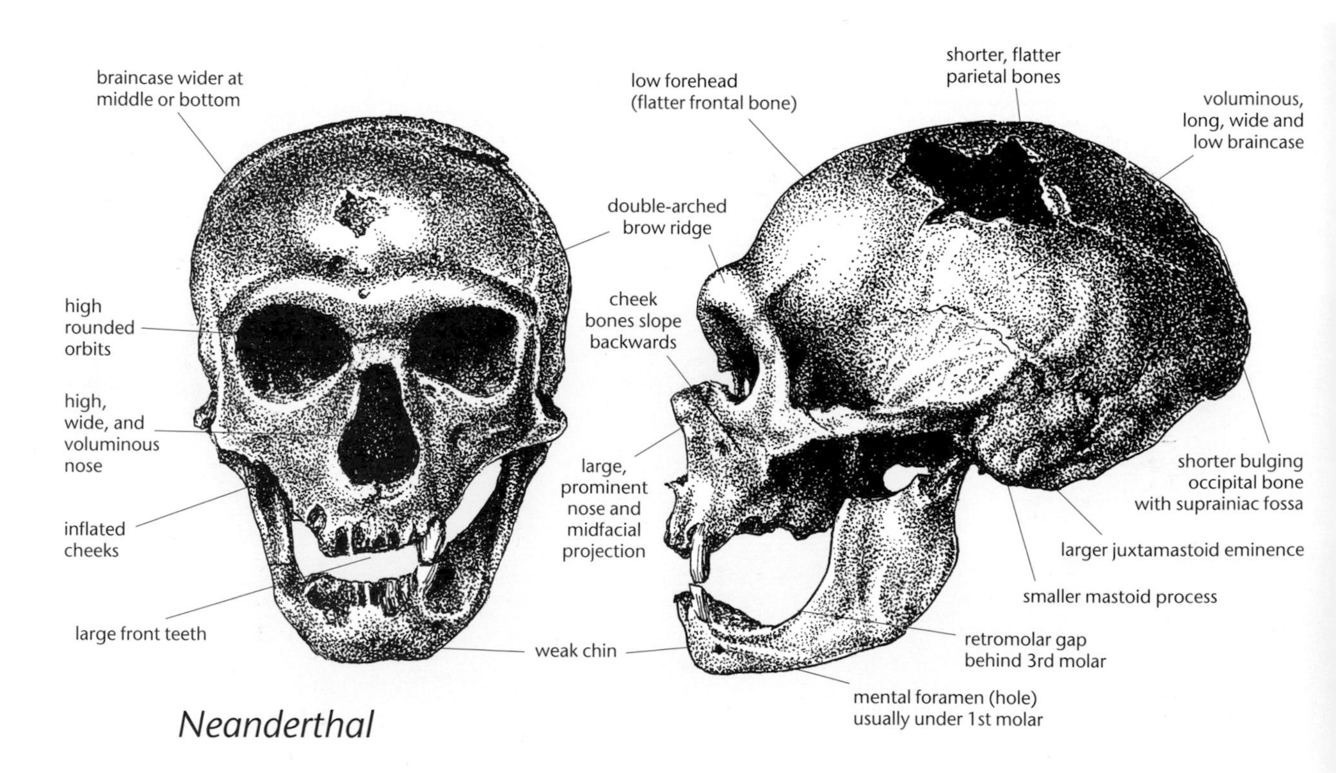

variation only developed during and after the dispersal, so that there was no continuity of regional features between *Homo erectus* and present inhabitants in the same regions. Like the Multiregional Model, this view accepts that fossils assigned to *Homo erectus* evolved into new forms of human in inhabited regions outside Africa, but argues that these non-African lines became extinct without evolving into modern humans. Some, such as the Neanderthals, were replaced by the spread of modern humans into their regions. Bräuer has argued that the replacement may not have been absolute, so that the populations dispersing from Africa interbred, to a greater or lesser extent, with resident archaic people – for example, in Europe with the Neanderthals, or in Java with the descendants of *Homo erectus*.

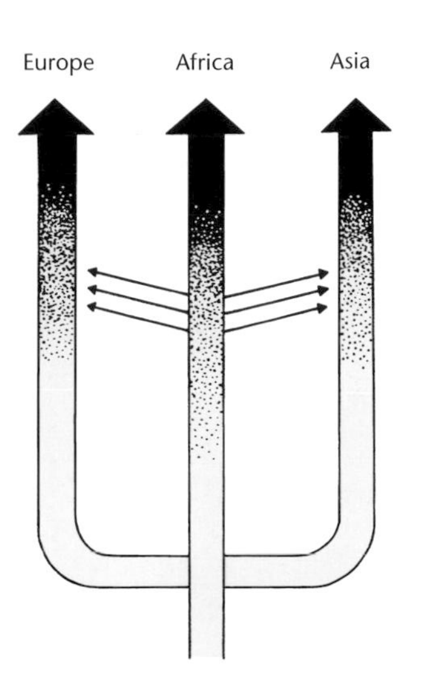

Models of Recent Human Evolution

There are other models of recent human evolution apart from the extremes of Multi-regionalism and 'Out of Africa'. This diagram represents the Assimilation model of the American palaeoanthropologist Fred Smith, and depicts a more gradual spread of modern human features from Africa, accompanied by intermixture with local human lineages in Europe and Asia.

Intermediate views

As mentioned above, there are also intermediate views about the origins of our species. Some workers, such as Fred Smith and Erik Trinkaus, believe that, while Africa was predominant in the story of human evolution, modern features spread more gradually from there, through interbreeding between adjacent populations. So, for example, populations in northern Africa exchanged genes with those in the Middle East, and they, in turn interbred with people in Asia Minor. From there, gene flow could have occurred with Neanderthal populations in Europe. Thus new genes spreading from Africa mixed with those of the native Neanderthals, catalysing an evolutionary transition to modern humans without large-scale invasions or replacements.

Yet other views have continuity in some regions outside Africa, but not all. Thus the Neanderthals could have become completely extinct, but the descendants of archaic people in China or Indonesia could have evolved into modern people in those regions. Testing and distinguishing intermediate models which involve gene flow or assimilation outside of Africa is more difficult than comparisons between the extremes of classic Multiregionalism and a pure 'Out of Africa' scheme.

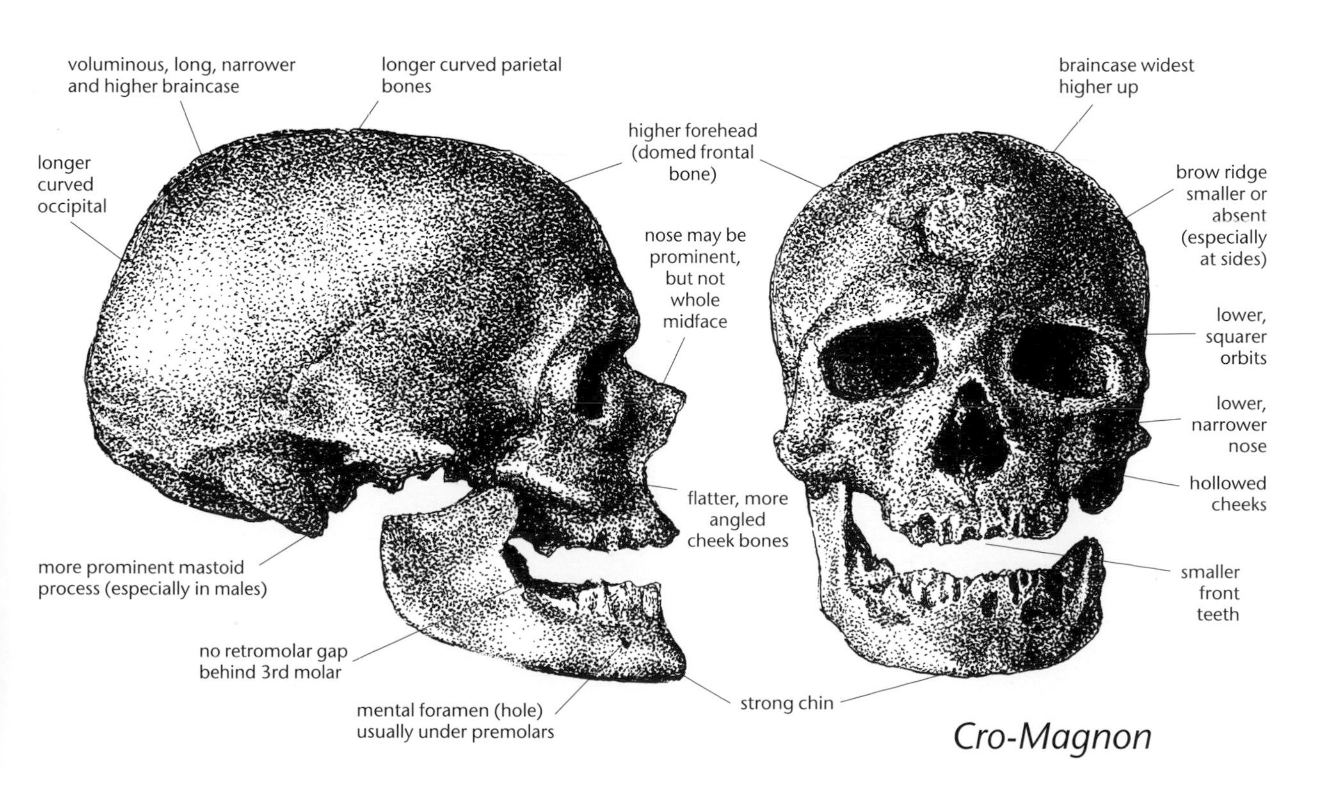

Cro-Magnon

The Early Occupation of Europe: Gran Dolina

The time of the arrival of the first people in Europe is still uncertain. As we have seen, *H. erectus* (or *H. ergaster*) had left Africa by about 1.8 million years ago, but initially this species must have remained in tropical or subtropical environments. The move into Europe and northern Asia would have been more challenging, since nearer the poles day length varies more, seasons are more strongly expressed, and in particular winters are longer and more severe, and growing periods for vegetation are shorter. All of these factors would have posed problems for what was still basically a tropical ape, and it may not have been until humans were proficient hunters or scavengers that they could finally survive the northern winters when plant resources would have been less reliable or available. The use of fire would have been a great aid to adaptation, but evidence of its early employment (for example at Zhoukoudian) has recently been questioned. Nevertheless, by about 1 million years ago there is reason to believe that people were at least on the fringes of Europe.

(Below) The Ceprano skull from Italy may be about 800,000 years old. It resembles both H. erectus *and* H. heidelbergensis, *but has been named as a new species* H. cepranensis. *However, another possibility is that it represents an adult version of the (perhaps) similarly aged partial child's skull from Gran Dolina assigned to* H. antecessor.

Africa to Europe: three possible routes?

Three ancient routes into Europe have been postulated: the most obvious one is via the Levant – that is the corridor connecting Northeast Africa and western Asia, where countries like Israel and Lebanon are today. Two other possible routes could have led more directly from North Africa into southern Europe via routes which are under the Mediterranean now. The more westerly one would have been from what is now Morocco into what is now Spain or Gibraltar, while a central one might have led from what is now Tunisia to Italy via Sicily. These last two routes might have been easier if the sea level was affected by the periodic growth of the ice caps, thus lowering the Mediterranean and exposing more land. However, it is not thought that there were ever continuous land bridges between Africa and Europe in these two regions, so some kind of raft would have been required by any early human pioneers contemplating such a journey. Since there is only disputed evidence of such an early use of sea-going craft (from the island of Java to neighbouring Flores, about 800,000 years ago), most experts believe that the eastern route into Europe via western Asia was the most viable. And this region does have evidence of human occupation from at least a million years ago. In Israel, the archaeological site of Ubeidiya is dated about 1.54 million years old, and much further north, as we have seen, the site of Dmanisi in Georgia has well-preserved human fossils of even greater antiquity.

Early evidence from Europe

Comparably early evidence from Europe is either absent or at best, highly contentious. It has been claimed that artifacts from sites in France and southern Spain date from over 1 million years ago, but there are numerous unresolved problems surrounding these sites. However, there is now strong evidence that humans were indeed in southern Europe by about 800,000 years ago. From an open site in central Italy called Ceprano, a skull-cap which resembles those of both *H. erectus* and *H.*

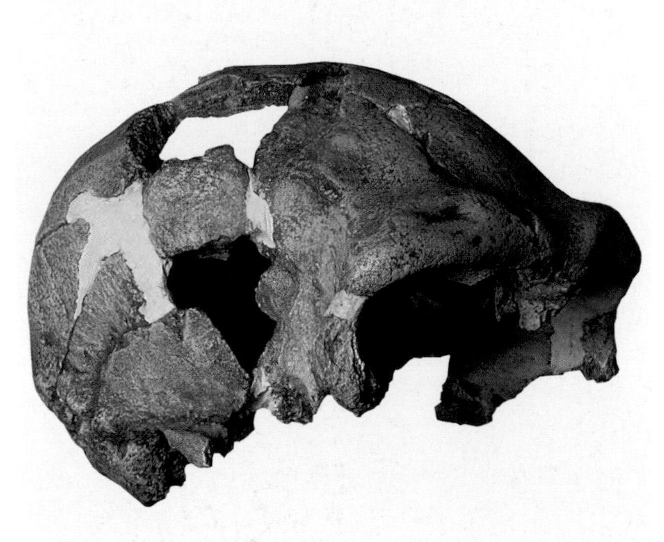

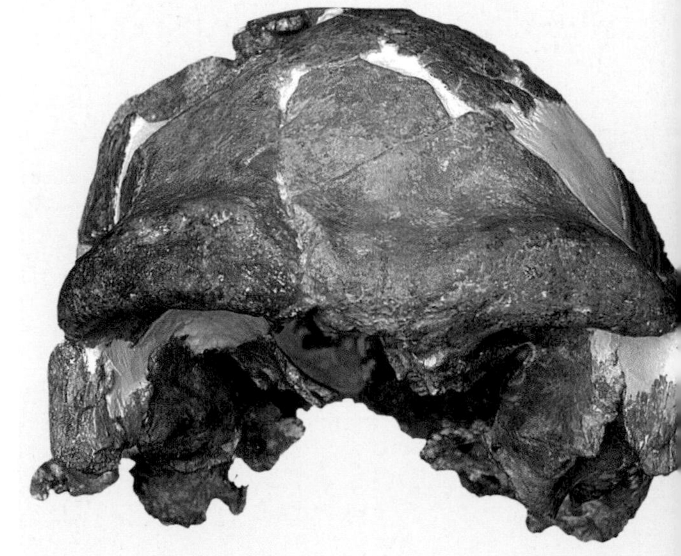

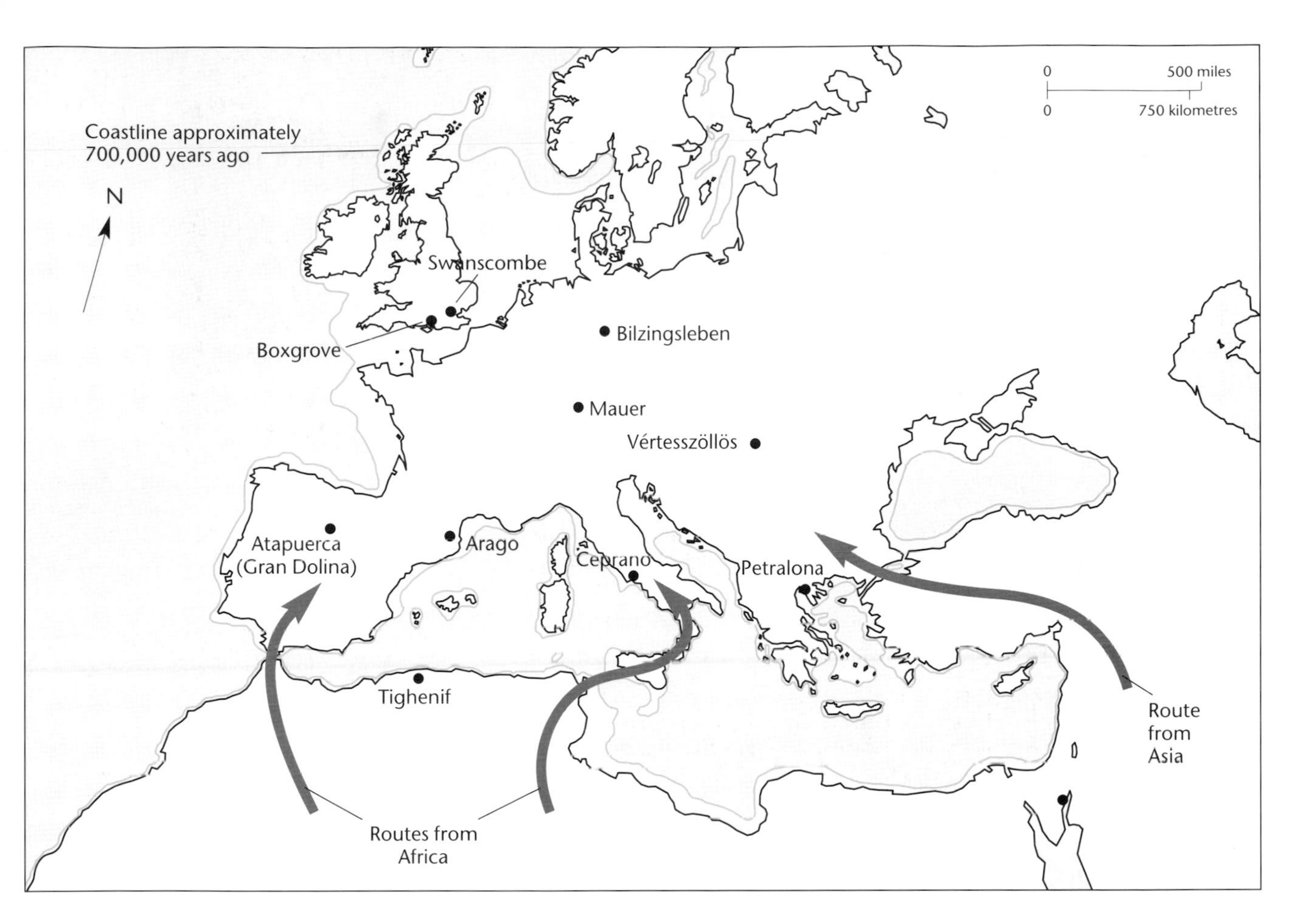

heidelbergensis (pp. 148–151), has been discovered, and has been dated at about 800,000 years. From a similar age, a site in northern Spain called Atapuerca has produced the broken bones of several individuals. The Sierra de Atapuerca is a limestone hill which has been sectioned by a railway cutting. This cutting has exposed a number of old cave chambers or openings filled with earthy

(Above) This map shows some of the early sites of human occupation in the Mediterranean region and Europe. Although some workers have proposed that animals and humans could have entered Europe from Africa across the Strait of Gibraltar or an ancient landbridge linking Africa and Sicily, the most likely route for human dispersal is via western Asia.

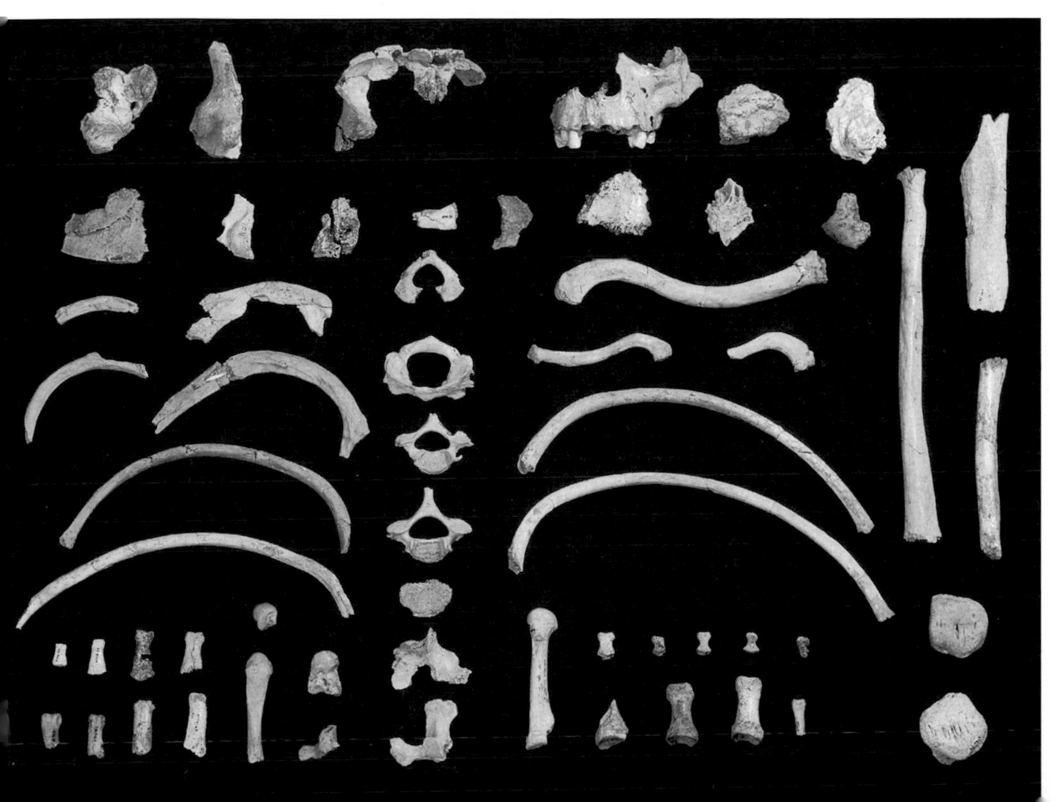

(Left) The collection of human bones from level TD6 at Gran Dolina date from about 800,000 years ago. They are from several adult and immature individuals, and some show cut marks indicative of cannibalism. They have been assigned to the species H. antecessor, *but resemblances to approximately contemporary material from Tighenif in Algeria, originally named* 'Atlanthropus mauritanicus', *have also been noted.*

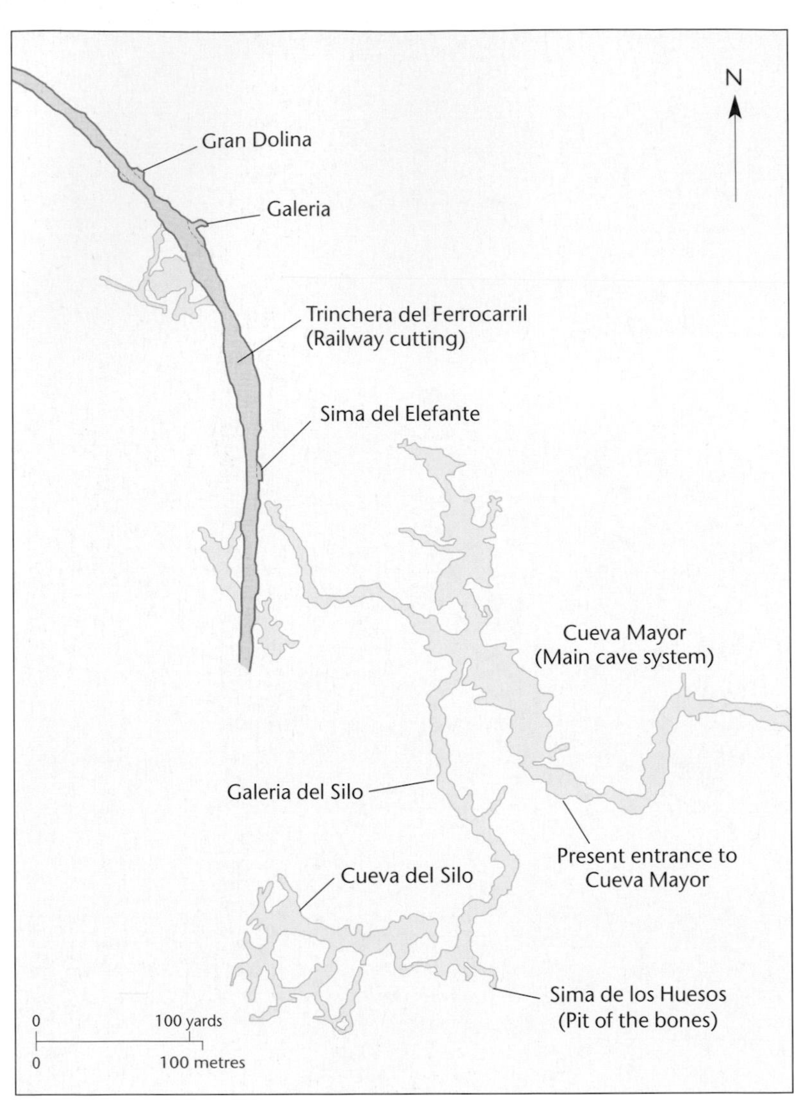

(Left) A plan of the rich sites of the Sierra de Atapuerca, near Burgos, in northern Spain. Some of the oldest sites containing evidence of human activity, such as Gran Dolina, were exposed in a old railway cutting ('Trinchera del Ferrocarril'), while the younger and extraordinarily rich site of Sima de los Huesos (Pit of the Bones) lies deep within a cave system ('Cueva Mayor').

sediments, and one of these called Gran Dolina has yielded the ancient human remains. They lie immediately under a level which records the last time that the Earth's magnetic poles underwent a major switch in orientation – a switch which has been dated at about 780,000 years ago.

The bones are from several individuals and are mostly those of children. They include the bones of a forehead, a face, a lower jaw, teeth, arm and foot bones, and a kneecap. Many of them are covered in marks which suggest they were cut by stone tools. While such damage could be due to burial practices where modern people are concerned, such ritual behaviour seems unlikely for early humans, leading to the suggestion that these individuals were the victims of ancient cannibals, whether of their own or another group.

While some of the Gran Dolina teeth show ancestral features found in African fossils over 1.5 million years old, the frontal (forehead) bone suggests a rather large brain size for a *H. erectus* child, and the

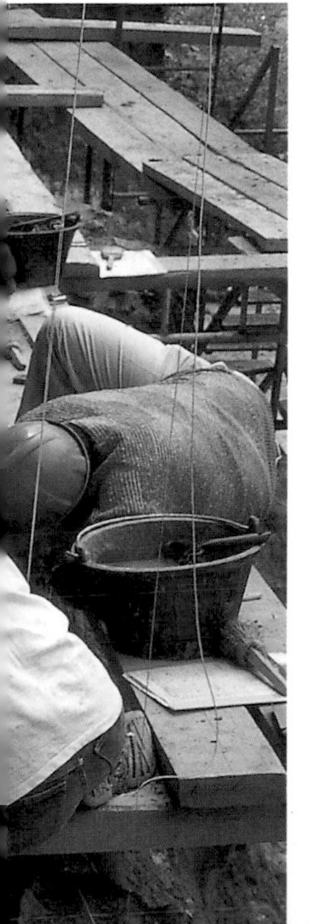

(Above and left) Excavations at Gran Dolina have revealed animal bones, stone tools and human fossils dating from about 800,000 years ago.

face looks remarkably modern in the shape of the nose and cheekbones. This has led the Spanish workers who have studied the bones to propose that they represent an entirely new species of human which they have called *H. antecessor* ('Pioneer man'). They believe that this species originated in Africa and then spread to Europe. In Africa it eventually gave rise to our species, *H. sapiens*, while in Europe it eventually gave rise to the Neanderthals. Thus they envisage antecessor as the last common ancestor of both our species and the Neanderthals, suggesting a very ancient split for the two lineages. As we shall see, there is other evidence that the split may not have been as ancient as this. Moreover, some researchers have noted that the teeth of *antecessor* closely resemble those of similarly dated jaw bones from the Algerian site of Tighenif (formerly Ternifine). As these have previously been referred to a species called *Homo mauritanicus*, both the Spanish and Algerian fossils could be called by that name if they do represent a distinct human species.

Homo heidelbergensis

(Right) Miner Tom Zwigelaar is pictured with the Broken Hill skull on the day of discovery in 1921. A human tibia and a fragment of thighbone were found on the same day in close proximity and may come from the same (male?) individual as the skull.

(Below) The Broken Hill Mining Company quarry in 1921.

In 1907, a strange-looking lower jaw was found by workmen in a sandpit at Mauer, near Heidelberg in Germany. The teeth were clearly human in number and shape, but the jaw was very thick, and it completely lacked a chin at the front. As was common in those days, it was assigned to a completely new species, *Homo heidelbergensis* ('Heidelberg Man'), but few scientists took this name seriously. As the century proceeded, it came to be regarded as probably representing a European form of *Homo erectus*, and from its associated fossil mammal remains, it was dated to about half a million years ago.

The Broken Hill fossils

Fourteen years later, in 1921, Africa produced its first significant human fossil, from the metal ore mine of Broken Hill (now Kabwe), in what was then the British colony of Northern Rhodesia, and is now

Zambia. The miners had been cutting through an extensive cave system, and had regularly thrown ore-coated fossil bones into the smelter. But on this occasion, the find was so remarkable that the workmen saved it, along with various other fossils. These included fragmentary human bones, some of which may belong with the skull, and others which do not. The Broken Hill skull was quickly brought to Britain, and was assigned to yet another new species, *Homo rhodesiensis*. Perhaps its most striking feature was the enormous brow ridges over the eye sockets, but although the braincase was long and low, the brain within must have been of modern human size. A shin bone found close to the skull was strongly built, but long and straight, suggesting a tall, but well-built, individual.

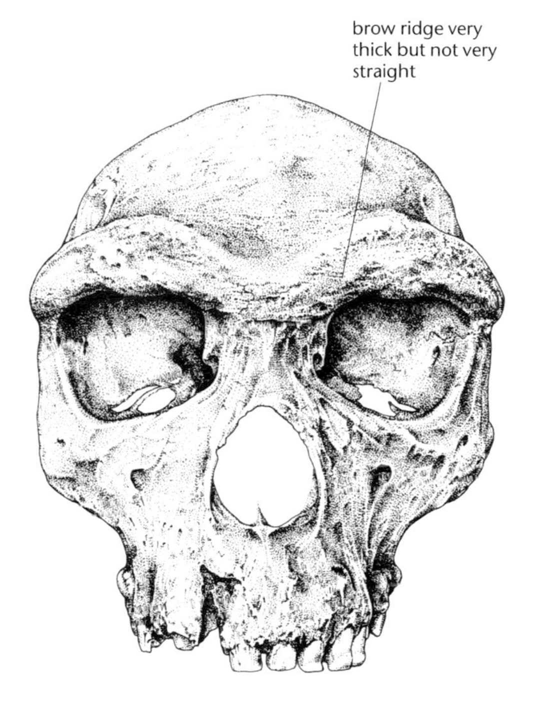

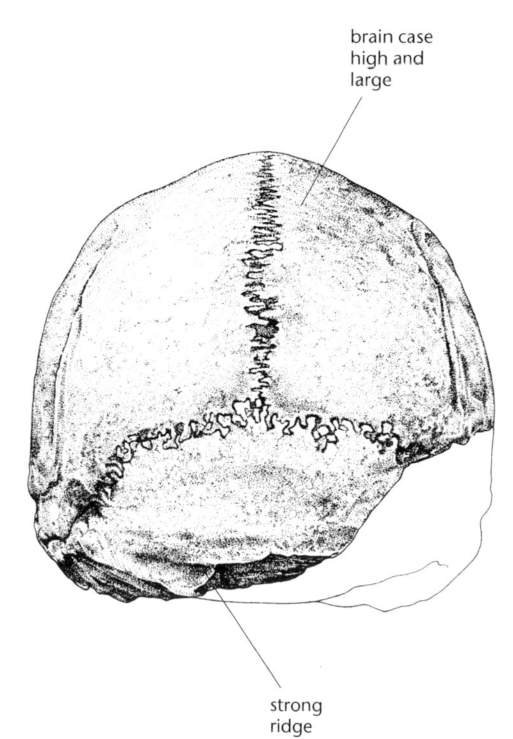

(Above and left) Broken Hill 1 is one of the best preserved of all ancient human fossils. It shares with H. erectus *fossils a massive bony browridge and ridge across the angled back of the skull, but it shares with later humans a relatively large brain size, reflected in its relative height and expansion of the parietal region. The face is relatively retracted and the base of the skull shows modern features. It also shows signs of disease in the temporal bone and the heavily decayed teeth.*

(Right) This comparison shows views of fossil human hipbones found at Broken Hill (left) and Olduvai Gorge (Olduvai Hominid 28). Both are large and strongly built and show a strong reinforcing ridge rising up above the hip socket. This feature seems to be present in H. erectus *and* H. heidelbergensis, *but is lost in Neanderthals and* H. sapiens.

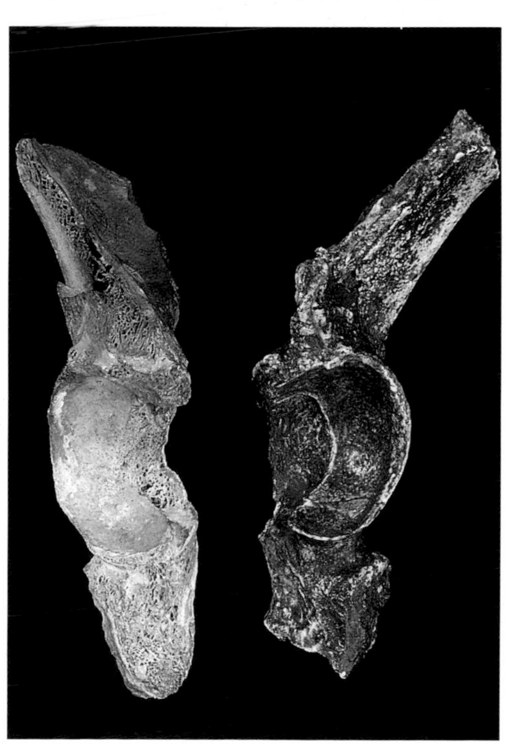

(Below) This reconstruction shows H. heidelbergensis *individuals driving hyaenas away from a rhinoceros carcass on the coastal plain at Boxgrove, southern England. Several rhinos were systematically butchered at Boxgrove but it is not known whether these potentially dangerous prey were hunted or whether they had died of other causes, and were then scavenged. However, the human inhabitants of Boxgrove certainly achieved primary access to the bodies for their meat.*

Other finds

Various other finds were made in Europe and Africa which were clearly more primitive than either the Neanderthals or modern humans. In Europe, these included specimens from Arago in France, Bilzingsleben in Germany, Vértesszöllös in Hungary, Boxgrove in England and Petralona in Greece. The Petralona fossil was particularly striking in its overall resemblance to the Broken Hill skull. In Africa, comparable finds included skulls from Bodo in Ethiopia, Ndutu in Tanzania and Elandsfontein in South Africa, and lower jaws from Baringo in Kenya. One view was that all these fossils represented late forms of *Homo erectus*, like the Mauer jaw, but gradually it was realized that there were enough distinctive features in these fossils to distinguish them from *erectus*. In particular, the braincase was higher and more filled out, especially at the sides, and this is a reflection of a larger average brain size, closer to that of living humans. In addition, the bony reinforcements of the skull, which are such a feature of *Homo erectus*, were reduced in this group of fossils. The face was also retracted under the braincase compared with the jutting face of *erectus*.

As a result of these comparisons, many workers have accepted the association of these African and European fossils with the Mauer find of 1907, and have, after all, recognized the validity of the species name *Homo heidelbergensis*. In this usage, the species shows derived features compared with *erectus*, but

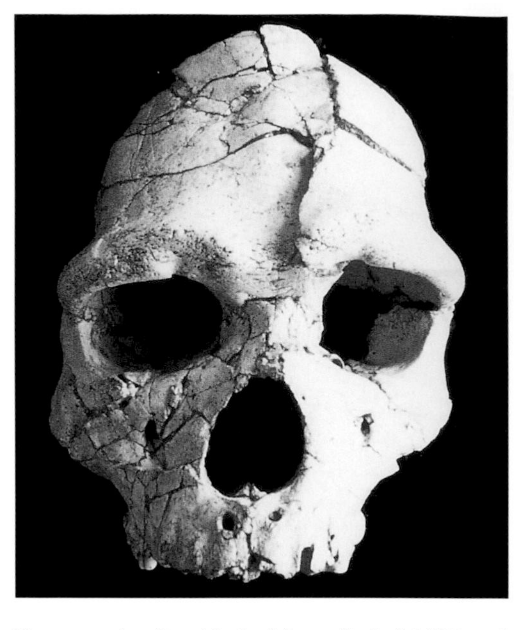

(Opposite) The Arago 21 fossil was found in 1971 in a cave near Tautavel, southern France. Represented by the somewhat distorted front and side of a skull, the specimen has sometimes been classified as H. erectus, *but more recently its similarities to specimens from sites such as Petralona and Atapuerca have led to its classification as* H. heidelbergensis *or even as an early form of* H. neanderthalensis.

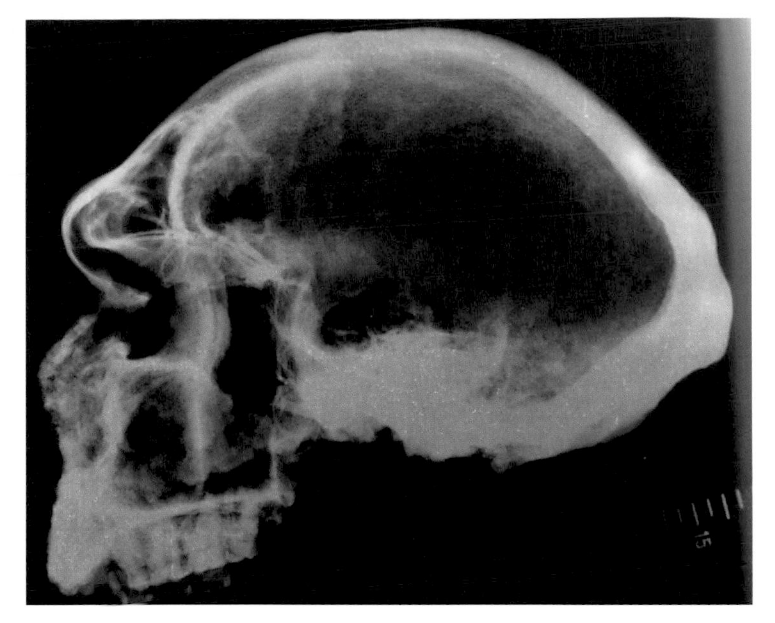

(Left) This radiograph (X-ray) of the fossil skull from Petralona, Greece, shows its interesting combination of features. While the long, low shape, angled and thick rear, and massive browridge are reminiscent of H. erectus, *the relatively high skull, retracted face and large air spaces in the brow and face are found in* H. heidelbergensis *and* H. neanderthalensis.

primitive features when compared with both Neanderthals and modern humans. It is regarded by its proponents as representing the last common ancestor of its two descendant species, Neanderthals and modern humans. Based on the evidence from Bodo, *heidelbergensis* must have originated somewhere in Africa, Europe or a region in between at least 600,000 years ago. The species spread across these regions, and then began to differentiate. Local 'racial' characteristics evolved and, accentuated by climate-induced geographic barriers, populations in Europe and Africa gradually split. In the north *heidelbergensis* eventually gave rise to the Neanderthals, and in Africa, to *Homo sapiens*. An alternative viewpoint recognizes *heidelbergensis* as only a European species, directly ancestral to the Neanderthals, in which case the contemporaneous African forms would be called *H. rhodesiensis*, after the 1921 find from Broken Hill. The relationship between the fossils assigned to *heidelbergensis*, and the earlier material called *H. antecessor* (Atapuerca Gran Dolina) and '*H. mauritanicus*' (Tighenif), is still unclear. Are the earlier fossils just more primitive versions of *heidelbergensis*, perhaps representing possible ancestors? This could certainly be the case judging by the shape of the more complete Ceprano skull from Italy, which is of similar age to the Gran Dolina and Tighenif fossils.

(Below) An improvised comparison of the original Bodo skull from Ethiopia (left) with a cast of Petralona (right). They are often both assigned to H. heidelbergensis *and share a broad and massive face and large nasal opening. However, it is possible that Petralona also lies close to the ancestry of the Neanderthals, as is suggested by the shape of its browridge and cheekbones.*

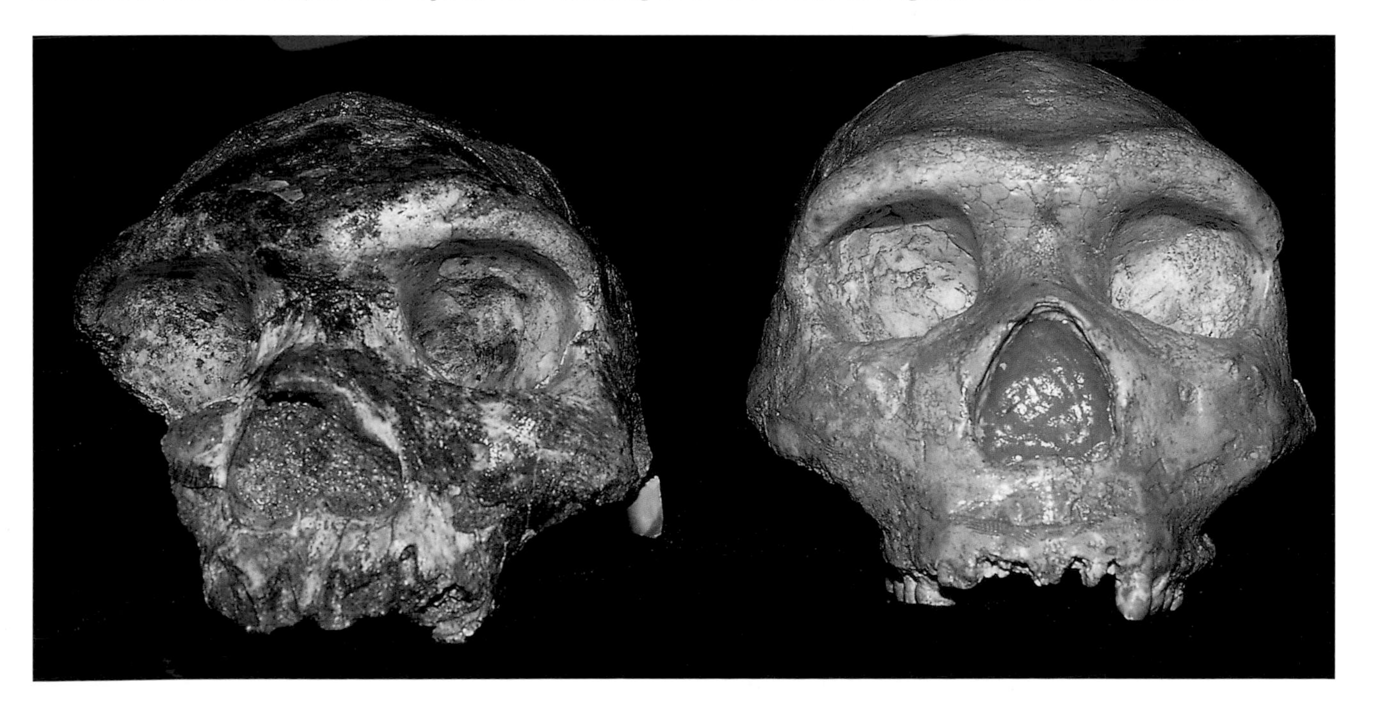

Atapuerca and the Origin of Neanderthals

The Sima de los Huesos (Pit of the Bones) at Atapuerca. This small chamber, deep within a cave, contains the richest concentration of ancient human fossils from any site in the world. It is still not clear how the human bones found their way into the chamber – were bodies intentionally thrown into the pit, or did the bones work their way down there through natural processes?

The Sierra de Atapuerca, near Burgos in northern Spain is hilly limestone country. As we saw on p. 145, it has produced a series of important finds concerning early human evolution in Europe. However, it has yielded another series of treasures from a later period which graphically illustrate the evolutionary transition from archaic humans of the species *Homo heidelbergensis* to the Neanderthals. Instead of an exposed series of sites which were once cave chambers or openings, such as Gran Dolina, the Atapuerca site concerned is deep within a cave system, at the bottom of a pit called, for good reason, the Sima de los Huesos – the Pit of the Bones. To get to this area now requires a long trek deep into the cave, at times crawling on the stomach, at other times using ropes. Eventually a 13-m (43-ft) vertical shaft is reached which is descended by a flimsy metal ladder. At the bottom of that is a small muddy and unprepossessing chamber – but that chamber contains the densest accumulation of fossil human bones ever found. Excavations are nowhere near finished, but already the Sima has yielded up over 2000 fossils of at least 32 individuals – men, women and children. What the bones are doing in this little chamber deep within the cave system is still an unsolved mystery, since it is evident from the lack of fire, food debris or stone tools that this inaccessible area was never a human occupation site. One beautiful handaxe made of pinkish rock was found amongst the bones, but it is unclear whether this had any special significance.

Discovering the Atapuerca cave system

The Sima must have been known to local villagers for centuries, because there is plenty of evidence of visits by young men out to impress their girlfriends by travelling deep into the cave to retrieve fossil cave bear teeth, which they gave as gifts on their safe return. In the search for these fossils, the sediments of the Sima were repeatedly and seriously

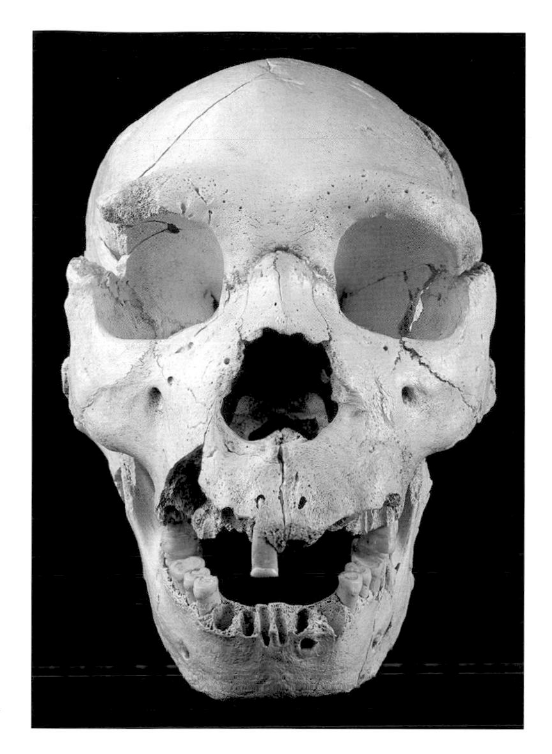

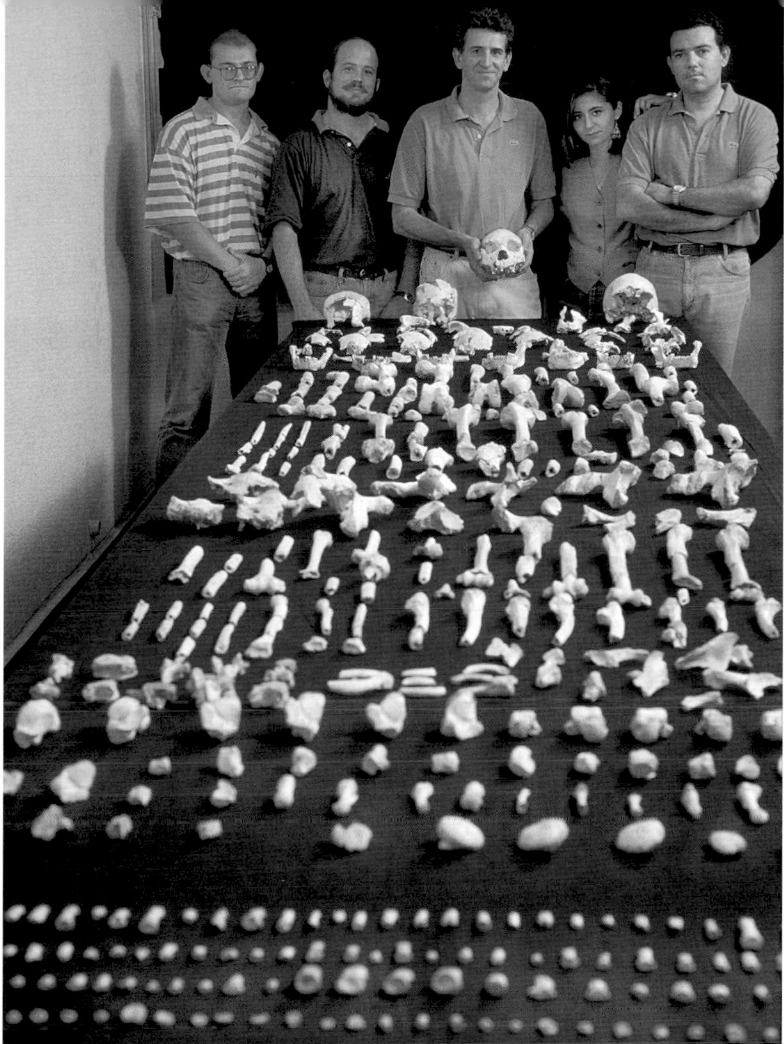

disturbed, and it was only in 1976 that a palaeontologist interested in cave bears finally visited the pit. It was then that a human jaw bone was recognized in the disturbed deposits, and proper scientific attention was paid to the site for the first time. It took many years of slow and laborious work to clear and sieve the disturbed sediments. There was only room for a few workers at a time in the cramped conditions, and oxygen was often in short supply. Moreover, every bag of excavated sediment had to be hauled up from the pit and carried through the cave to the surface. Numerous cave bear and human bones were recovered, often showing signs of recent breakage from the fossil hunters. Eventually, undisturbed deposits were ready for excavation, and in 1992, a rich collection of fossils started to be recovered, including nearly complete skulls of adults and children.

Studying the fossils

The fossil human material excavated from the Sima includes numerous examples of every part of the skeleton, down to the smallest hand and foot bones, but they are generally completely jumbled up in the sediment. The largest single element is the teeth, and these show that the majority of the individuals were teenagers or young adults. Some of the bones can be fitted together to show that they derive from the same person – for example where a lower jaw can be fitted to a skull – but in other cases, such as individual hand bones, it is much more difficult to associate them. Study of this enormous collection is still in progress, but already it is clear that the bones show a mixture of characteristics. For example, the most complete skull has a lower jaw associated with it which looks like a scaled-down version of the one from Mauer in Germany, the type specimen of *H. heidelbergensis*, while its face displays the projection around the nose which is typical of Neanderthals. The temporal bone (which includes the ear region) has a shape more like that of modern humans, while the back of the skull exhibits a small central depression which is found in all known Neanderthals.

Such combinations of ancestral and derived features occur throughout the whole collection in an almost random fashion, so that it is difficult to decide on how best to classify the sample as a whole. The Spanish team favour assigning the material to a late form of *heidelbergensis*, while we favour referring them to early Neanderthals – *H. neanderthalensis* – but the significance of these fossils is the same. They show that an evolutionary transition was taking place in European populations of about 400,000 years ago (the estimated age of the Sima material) which was leading on to the late Neanderthals, and that Europe records their evolution rather than that of *H. sapiens*.

(Above left) the skull of the most elderly member of the Sima assemblage, cranium 5 – nicknamed Miguelón, a rugged individual, but of uncertain sex. The teeth are very worn, and show signs of infection which had spread up the face and was perhaps the cause of death. The skull also shows a number of signs of injury.

(Above) The Sima has produced over 2,500 human fossils, representing at least 30 men, women and children who died over 300,000 years ago. The majority of the sample consists of adolescents and young adults, with relatively few very young or old people. Every part of the skeleton is preserved, providing the best data we have on variation and health in premodern humans.

The Neanderthals

The Neanderthals – *Homo neanderthalensis* ('Man from the Neander Valley') – are the best known ancient humans. There are two main reasons for this. Firstly, they lived in the region which has been explored more than any other for its prehistory – Europe. Secondly, many of them lived in caves, and they adopted the habit of burying their dead in the caves in which they lived. This has meant that Neanderthal bodies have had a greater chance of becoming fossilized, since they were protected by burial from destruction through erosion, trampling or scavenging. Moreover, caves have concentrated evidence of Neanderthal occupation much more than open sites, and so they have been excavated more intensively, leading to the recovery of numerous Neanderthal burials, as well as even greater numbers of occupation levels. Over 500 individual Neanderthals have been discovered, and although the vast majority of these are very fragmentary – often only a tooth, or a jaw fragment – about 20 of those men, women or children are represented by reasonably complete skeletons. Using these, we can build up quite a complete picture of a typical Neanderthal body. The burials also hint at complexity in Neanderthal minds and lives, since some appear to show particular care and treatment to the body. In Israel a man apparently had his skull removed after burial, while in Syria and France, stone slabs may have been placed in the graves of Neanderthal children.

(Above) Although some scientists dispute it, it is generally agreed that Neanderthals buried their dead, and some of the best examples have been found in Israel. This is the skeleton of a Neanderthal baby (Amud 7), apparently with the jaw of a deer laid next to it.

(Right) Amud means 'pillar' or 'column' in both Arabic and Hebrew, and this pillar of rock gave its name to the adjoining cave. Excavated by a Japanese-led team in the 1960s and an Israeli-led team in the 1990s, the site has produced rich evidence of animal bones, Middle Palaeolithic stone tools and several Neanderthal fossils. The most famous is an adult male burial (Amud 1).

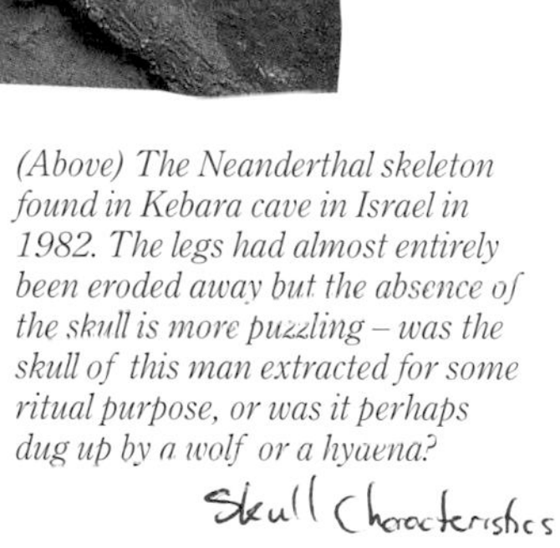

(Above) The Neanderthal skeleton found in Kebara cave in Israel in 1982. The legs had almost entirely been eroded away but the absence of the skull is more puzzling – was the skull of this man extracted for some ritual purpose, or was it perhaps dug up by a wolf or a hyaena?

(Above) This woman shows the typical short and stocky build of the Neanderthals, with wide shoulders and hips, a bulky trunk and relatively short forearms and shinbones. Differences in body size between male and female Neanderthals were somewhat greater than in people today, but Neanderthal women were certainly strong and muscular. It is not definitely known whether the Neanderthals wore clothing, but they certainly had the technology to process animal skins.

Neanderthal anatomy

These early humans had large brain sizes, housed in relatively long, broad and low braincases, with long faces, dominated by a voluminous nasal opening, and surmounted by a double-arched brow ridge. Judging from the inside of the braincase, the large Neanderthal brain was somewhat differently shaped from our own – somewhat smaller in the frontal region, and larger at the back (the occipital lobes) – but it is impossible to judge the quality of their brains from such limited data. The whole middle of the face was pulled forward, and the cheek bones were swept back at the sides. The front teeth were large, and often show severe wear, indicating that material such as skin, food or vegetable fibres were being regularly pulled across them. The lower jaw was long, with little sign of a chin at the front. Their skeletons suggest a relatively short, stocky physique, with powerful muscles, and the leg bones, in particular, show large joint surfaces

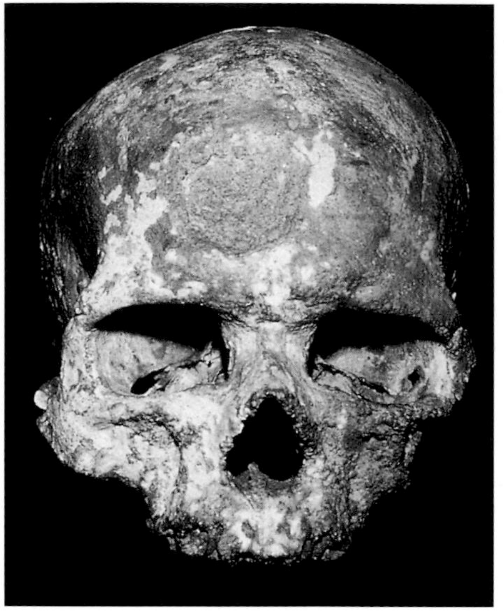

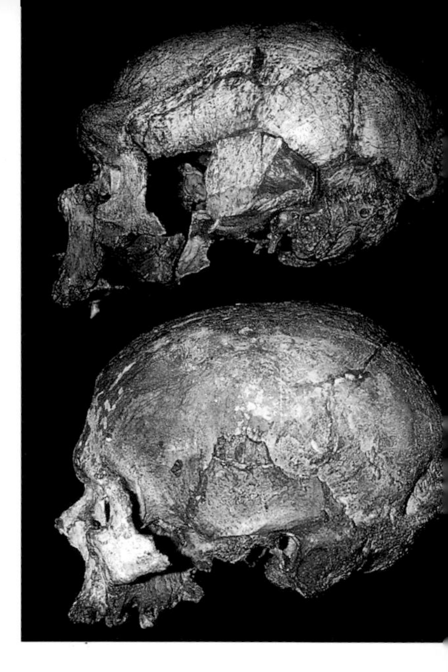

and thick walls. The best known or 'classic' Neanderthals lived from about 70,000 to 35,000 years ago, and were often associated with cold-adapted animals such as reindeer and mammoth.

A Eurasian species

These late Neanderthals were adapted to the climatic and physical rigours of life during the last Ice Age in Europe. Their bulky body shape would have conserved heat by minimizing the surface area of skin exposed to the cold, and the large internal volume of the prominent nose may have acted to warm and moisten cold and dry inhaled air. But Neanderthals also lived through warmer periods in Europe – for example, those from the Italian site of Saccopastore were associated with fossils of elephant and hippopotamus 125,000 years old. And although the best known Neanderthals are those from sites such as the Neander Valley (Germany), Spy (Belgium) and La Chapelle-aux-Saints and La Ferrassie (France), these people also ranged across western Asia into the present-day countries of Iraq, Syria, Israel, Georgia, Russia, the Ukraine and as far east as Uzbekistan. They characteristically made Middle Palaeolithic (Middle Old Stone Age, or Mousterian, from the site of Le Moustier in

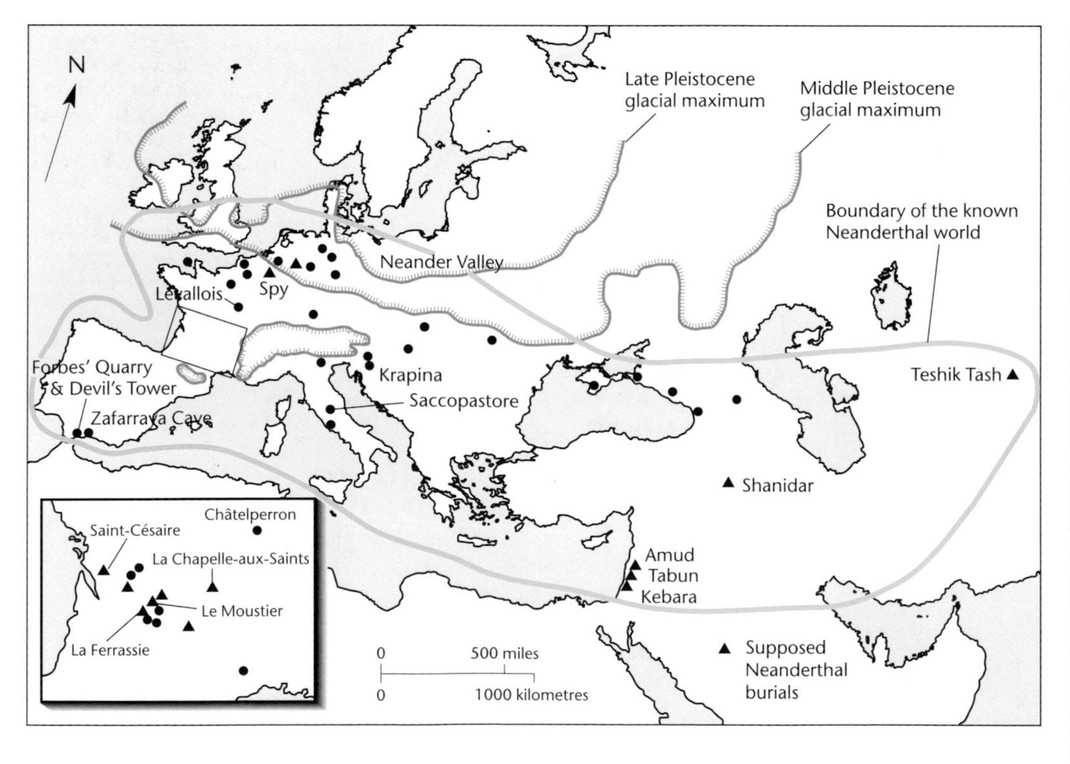

(Below) The (extrapolated) range of Neanderthal fossil remains. No Neanderthals are yet known from Africa or the Far East, but the wide geographical range of Middle Palaeolithic tools (possibly made by Neanderthals) mean that this map probably represents only their minimum extent in Eurasia.

(Opposite) *These comparisons show the skulls of the 'Old Men' of (far left and top right) La Chapelle-aux-Saints – a late Neanderthal – and of Cro-Magnon – an early modern human. They are perhaps separated by 20,000 years of time, but they show some of the most striking contrasts in the shape of the braincase and face found in fossil humans.*

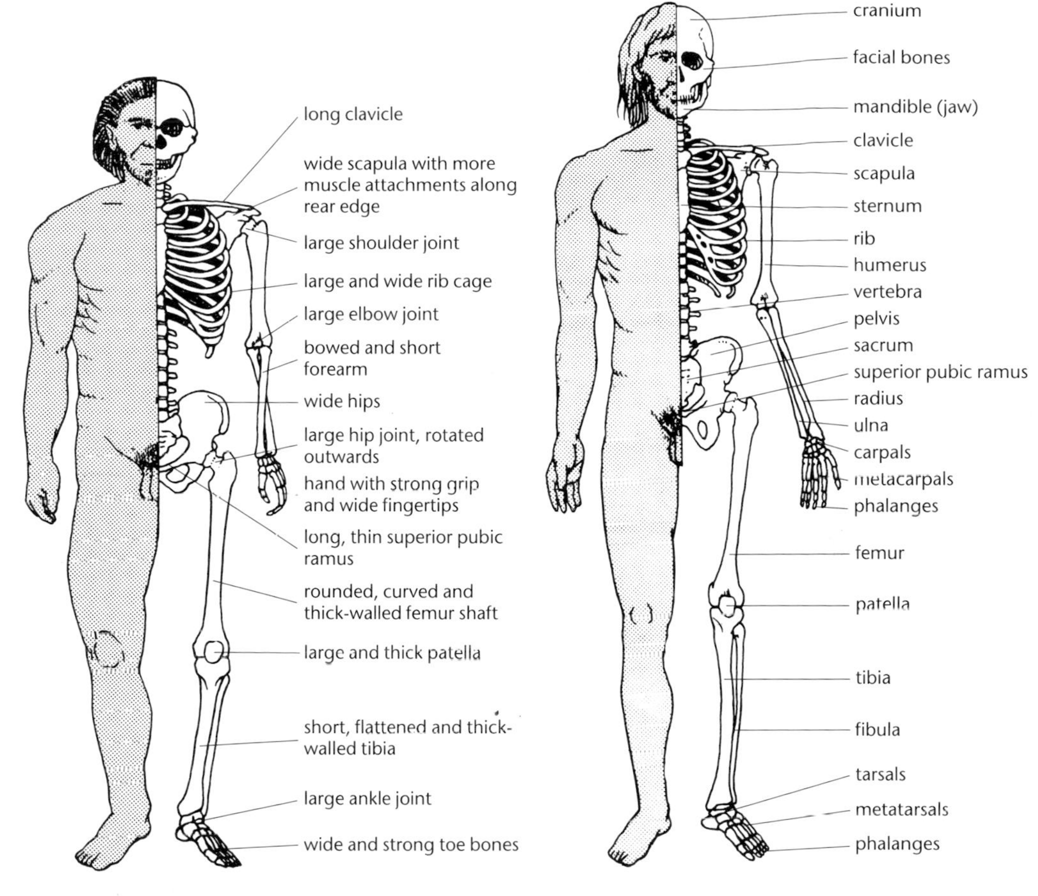

(Above) This comparison of the skeleton and reconstructed body shape of a European Neanderthal (left) and Cro-Magnon male shows their shared fundamental human form, but contrasts the short, wide body of the Neanderthal with the taller and more linear physique of the Cro-Magnons. However, both groups varied through time and space, with the Neanderthals of western Asia relatively taller and less robust, and the later Cro-Magnons shorter and stockier than their immediate predecessors.

France) stone tools, and these are found over even greater areas of Asia. However, there is no fossil evidence of Neanderthals from the Far East, nor from Africa – other peoples lived there, with their own, separate evolutionary histories.

Adapted for a hard life

Study of Neanderthal skeletons reveals the hard lives they led, and the way their bodies responded. Many show fractures and injuries, most of which had healed. These might have been caused by interpersonal conflicts, but it seems likely that some were from hunting accidents caused by close encounters with dangerous wild animals. Some injuries were serious enough to have severely disabled the individuals concerned – one man buried in a cave in Iraq may have been blind in his left eye, and was probably partly paralysed, with a withered handless arm on his right side. Nevertheless, he survived in this condition for at least a few months, leading to suggestions that he was being cared for by his fellow Neanderthals. In response to their demanding, and at times, dangerous lives, the Neanderthal skeleton was strongly reinforced with thick bone, particularly in the shape and strength of the leg bones. Their physique has been described as combining that of a powerful wrestler with the endurance of a marathon runner!

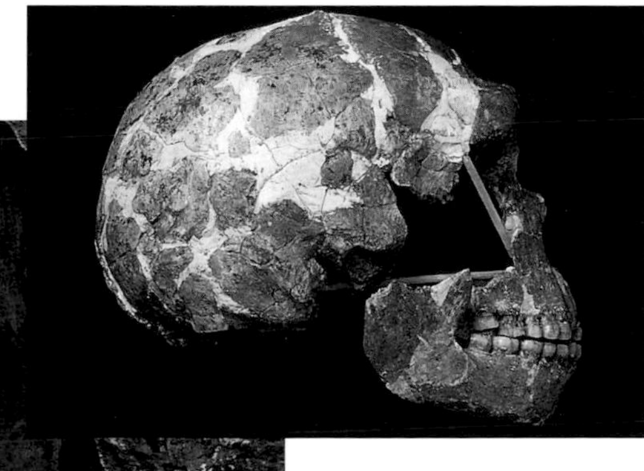

(Above) The oldest known burial of a Neanderthal (and perhaps the oldest known burial of any human) was excavated from the Tabun ('Oven') Cave (left) in what is now Israel, in 1932. It is from a small and rather lightly built woman, with small teeth but a strong browridge. Her pelvis was the first one in which the characteristic Neanderthal shape was identified. The skeleton has proved difficult to date, but direct ESR analysis of a tooth suggests a possible age of about 120,000 years.

Africa – Homeland of Homo sapiens?

The African continent has been rather slower to yield up evidence for the later stages of human evolution, compared with the relative profusion of fossils from the earlier stages. Moreover, Africa was for many years regarded as backward compared with the records of Europe and Asia. While archaeologists relied only on radiocarbon dating, it seemed that handaxe industries lingered on in sub-saharan Africa while the Middle Palaeolithic developed further north. Moreover, the subsequent Middle Stone Age of Africa was regarded as roughly contemporary with the Upper Palaeolithic of Europe, and it seemed technologically simpler, as well as lacking in features such as representational art and complex burials. The African fossil human record was sparse, but primitive looking specimens such as the skull from Broken Hill in Zambia, and the partial skull from Florisbad in South Africa were believed to be only about 50,000 years old, and thus contemporary with the Neanderthals.

Redating the African fossils

However, new discoveries and the application of new dating techniques that could reach back beyond the working range of radiocarbon dating have greatly changed our view of Africa's evolutionary sequence. We now believe that people similar to the European ancestors of the Neanderthals lived there about 400,000 years ago, as represented by fossils such as those from Broken Hill, Elandsfontein and Ndutu, and these are assigned to the species *Homo heidelbergensis* or *H. rhodesiensis* (see pp. 148–151). Thus the Broken Hill skull looked relatively primitive compared to the Neanderthals because it was actually much older than them. Between 400,000 and 130,000 years ago, an evolutionary transition to *Homo sapiens* ('Wise man') appears to have taken place in Africa,

(Left) This map shows the wide geographical range of African sites representing the later stages of human evolution. Nevertheless there is a dearth of material from the central and western zones of the continent.

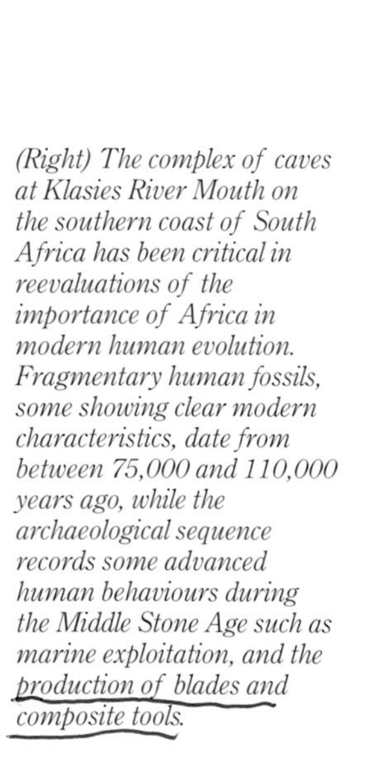

(Right) The complex of caves at Klasies River Mouth on the southern coast of South Africa has been critical in reevaluations of the importance of Africa in modern human evolution. Fragmentary human fossils, some showing clear modern characteristics, date from between 75,000 and 110,000 years ago, while the archaeological sequence records some advanced human behaviours during the Middle Stone Age such as marine exploitation, and the production of blades and composite tools.

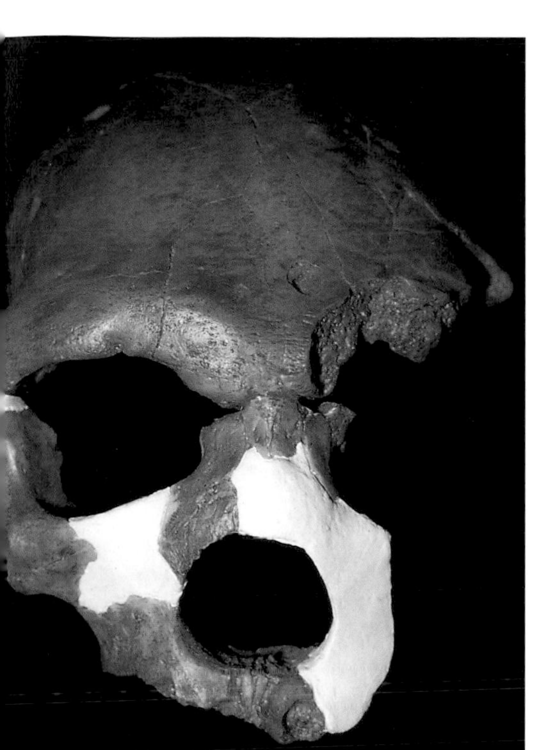

paralleling a similar transition in Europe to the Neanderthals.

New dating techniques suggest that fossils such as those from Florisbad, Ngaloba and Guomde, thought to be possible ancestors for modern humans, are much older than was formerly believed, and all could exceed 150,000 years in age. In fact, the Florisbad fossil was dated at about 260,000 years ago using electron spin resonance on the enamel of a molar tooth, and is associated with late handaxe industries, while the Ngaloba and Guomde specimens are somewhat younger, but were found in sites with Middle Stone Age tools. Thus the archaeological sequence of Africa can now also be redated, and no longer looks backward compared with records elsewhere. In fact, as we discuss on p. 211, some Middle Stone Age tools actually look more sophisticated than their Middle Palaeolithic counterparts in areas such as Europe.

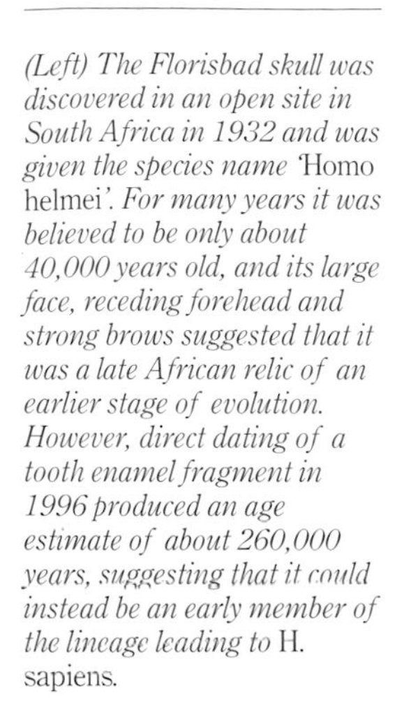

(Left) The Florisbad skull was discovered in an open site in South Africa in 1932 and was given the species name 'Homo helmei'. *For many years it was believed to be only about 40,000 years old, and its large face, receding forehead and strong brows suggested that it was a late African relic of an earlier stage of evolution. However, direct dating of a tooth enamel fragment in 1996 produced an age estimate of about 260,000 years, suggesting that it could instead be an early member of the lineage leading to* H. sapiens.

Early moderns span the continent

By 160,000 years ago, modern human specimens are known from opposite ends of the African continent – in caves in southern Africa, such as those at Klasies River Mouth, and Ethiopian sites such as Omo Kibish and Herto. The Klasies River Mouth

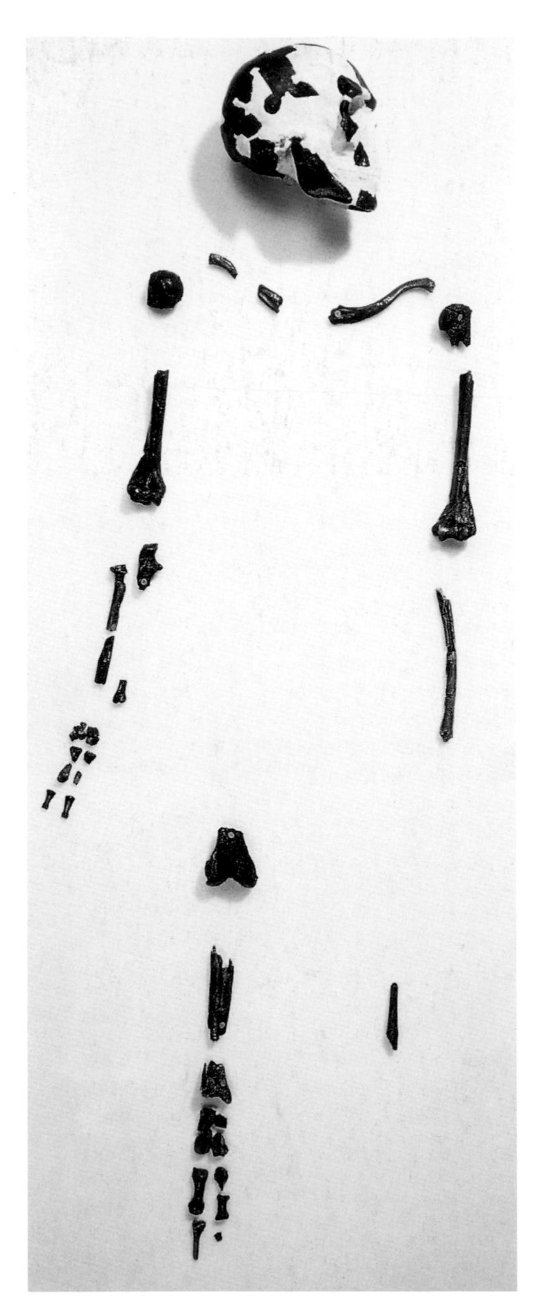

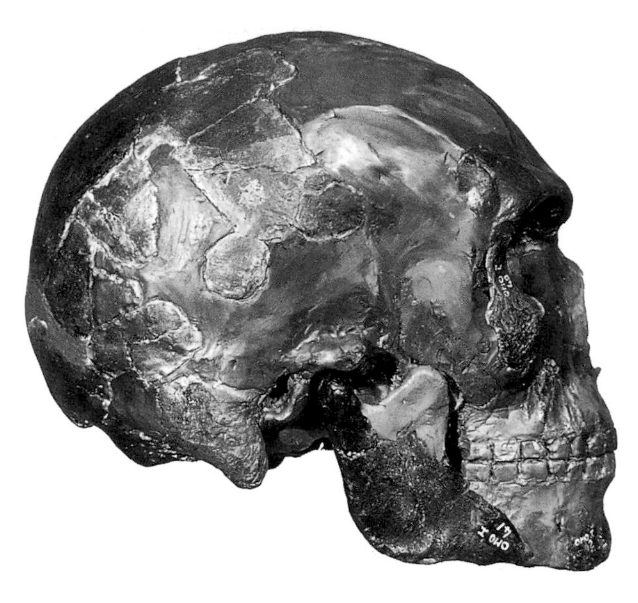

(Above) The skull bones of Omo 1 can be reconstructed to show a rather modern shape, with a clear chin on the lower jaw. The bones of the cranial vault are high and domed, and the cranial base is relatively narrow, not broad. The brow ridge is fragmentary but appears to be divided in a modern fashion.

(Above) In 1967 a team led by Richard Leakey recovered the partial remains of three fossil humans from the Kibish Formation in the Omo River region of southwestern Ethiopia. Omo 1 was a partial skeleton of a tall and well-built individual, apparently male, with clear modern characteristics in the preserved parts. The dating of this skeleton was controversially estimated at 130,000 years – and new research suggests this date is approximately correct.

caves are coastal and contain deep archaeological layers, including shellfish middens, hearths and fragmentary human fossils, some showing marks suggestive of cannibalism. Another famous South African site is Border Cave, which has rich Middle Stone Age levels, and several human fossils. One modern-looking mandible has been directly dated by ESR to about 75,000 years but there are doubts about the provenance of some of the bones, and it is possible that they were buried in the Middle Stone Age levels at a much later date. No such doubts surround the provenance of the fragmentary modern human skeleton from Omo, but a second isolated fossil braincase from the site looks more primitive, and probably derives from older levels. Recently, adult and child skulls showing modern features have been excavated from deposits at Herto in Ethiopia dated to about 160,000 years ago, and these may be the oldest definite examples of modern humans found so far. These, too, have cut marks, interpreted in this case as possible evidence of ritual behaviour. Further west, a peculiarly shaped fossil braincase was found at Singa, in the Sudan. Detailed study has shown that the

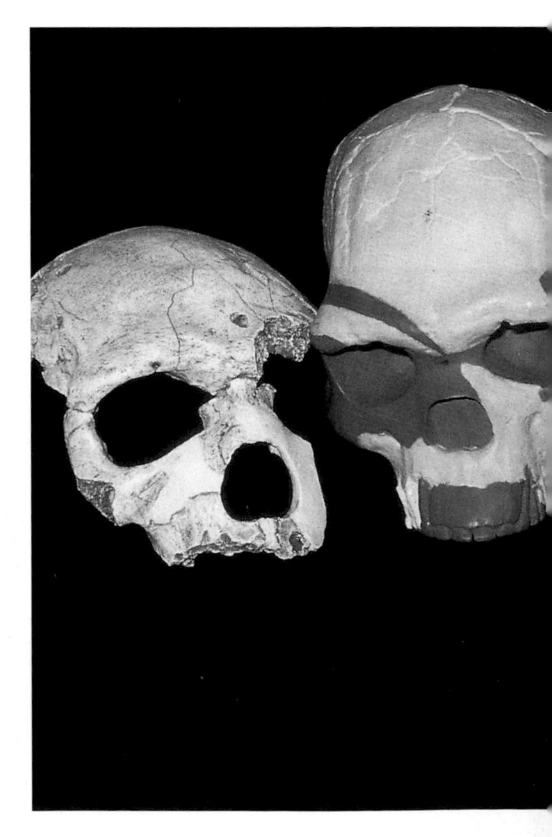

(Right) The endocranial cast from the inside of the Singa skull is modern in size, but its shape suggests that the Singa individual was unusual amongst fossil humans in probably being left handed. The strange shape of the skull and missing earbones on one side also suggest that the individual was suffering from a disease affecting bone growth.

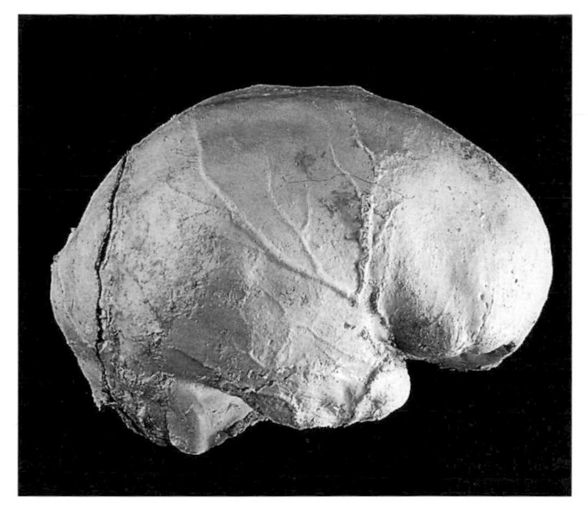

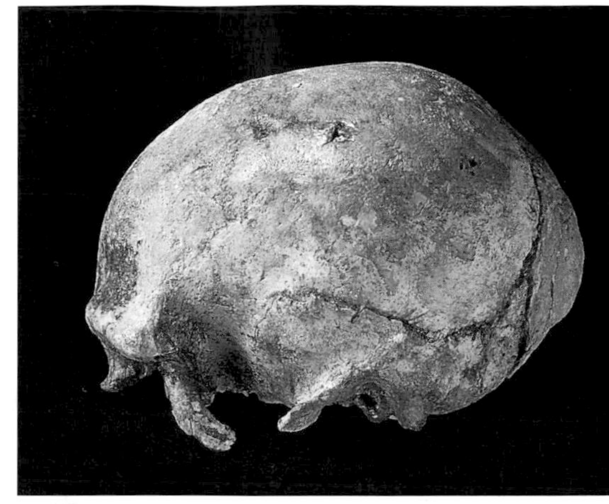

individual concerned probably suffered a severe infection which affected bone growth and caused deafness in at least one ear. New dating work shows that Singa has an antiquity of more than 130,000 years, and could therefore represent another proto-modern specimen.

North Africa has produced several primitive *Homo sapiens* fossils, including two adult skulls and the lower jaw of a child from Jebel Irhoud in Morocco, probably dating from about 150,000 years ago. More modern-looking specimens have been found at the Moroccan site of Dar es-Soltan, and at Taramsa Hill in Egypt. This recent discovery of the skeleton of a child, buried on the top of a hill above the River Nile about 70,000 years ago, is still under investigation.

Overall, the picture we have of Africa between 300,000 and 130,000 years ago is tantalizingly incomplete. This is when we believe our species originated, but many parts of the jigsaw are missing. Africa probably had the largest and most variable human population of this period, but our fossil samples are restricted to small portions of this vast continent. Handaxes and Middle Stone Age tools show that people were also living in central and west Africa, but of their physical appearance and evolutionary position we have absolutely no knowledge.

(Above) The Singa skull from Sudan was found in the riverbed of the Nile in 1931. Its strange mixture of primitive and modern features were noted in early studies, but until recently it was believed to be only about 20,000 years old. It is relatively low and very broad, with short and unusually thickened parietal bones. Yet the forehead is high and rounded and the browridge relatively small. In the last decade ESR and uranium series dating has shown that the skull is probably over 130,000 years old, and may be a very early relic of Homo sapiens.

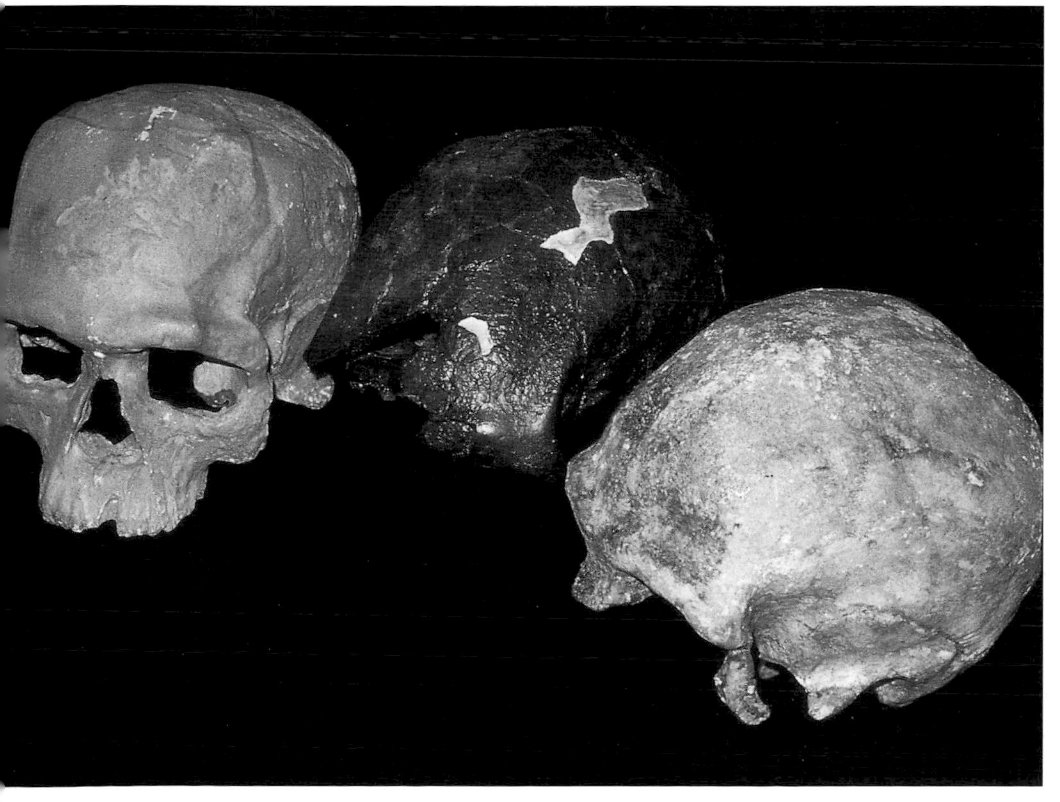

(Left) This array of fossil skulls from across Africa illustrates the range of possible modern human ancestors in time and space. The skulls and their estimated ages are (from the left) Florisbad (South Africa), 250,000 years old; Ngaloba – Laetoli Hominid 18 (Tanzania), 140,000 years old; Jebel Irhoud 1 (Morocco), 150,000 years old; Omo Kibish 2 (Ethiopia), 200,000 years old; and Singa (Sudan) over 130,000 years old.

Asia – Corridor or Cul-de-sac?

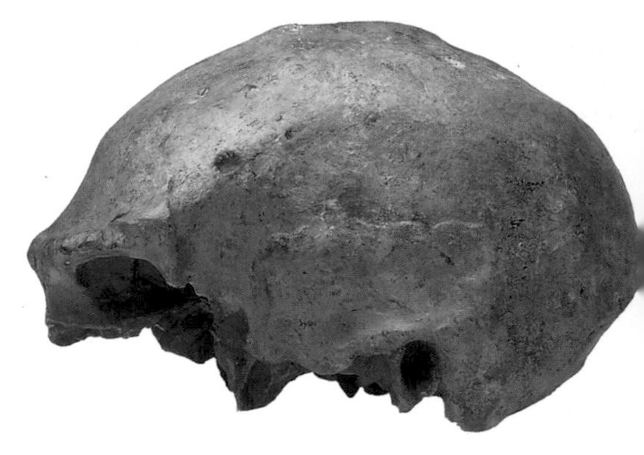

(Right) It is possible that the H. erectus *skulls from Ngandong, Java, are only about 50,000 years old. This skull is female.*

(Below) The Dali skull from China is perhaps 300,000 years old and appears to be distinct from Homo erectus. *Some believe that it forms a link between local* Homo erectus *populations and oriental peoples of the present day. However, it also resembles more ancient fossils from Africa and Europe, which has led some workers to classify it as* Homo heidelbergensis.

(Below right) Modern dating techniques have revolutionized our view of the sequence of Neanderthal and early modern fossils in the Middle East. From a situation which seemed to mirror that in Europe, where moderns succeeded Neanderthals, there is now a more complex scenario where these populations stretch back over 100,000 years and may have alternated or coexisted in their occupation of the region.

As we have seen, *Homo erectus* spread across the warmer regions of Asia over a million years ago, and is well known from fossils in regions such as China and Java. The species must have ranged even more widely to judge from artifact distributions, and new evidence suggests that *erectus* may have used rafts to reach the island of Flores, beyond Java, 800,000 years ago (see pp. 174–175). However, it does not look as though *erectus* ever reached Australia, despite occasional claims to the contrary, and both Australia and the Americas must have remained uncolonized until the arrival of modern humans. Scientists are divided over how isolated the east of Asia was, compared to the west. There is archaeological evidence for technological or behavioural differences between the peoples of the Far East and those of the western inhabited world, and these differences persisted for about a million years.

South and east Asia

It would be very helpful to have a good fossil record from the intervening regions, but there is little beyond a fossil braincase from Narmada in western India, which perhaps dates from about 300,000 years ago. This specimen may represent a more advanced species than *erectus*, and there are comparable fossils from Chinese sites such as Yunxian, Dali and Jinniushan. It is unclear whether these represent evolutionary developments from local *erectus* predecessors, or are evidence of a spread of a variant of *Homo heidelbergensis* into the region. But at the southern end of its range, in Java, it seems that *erectus*, as represented by fossils from Ngandong and Sambungmacan, persisted there with little change into the last 200,000 years. These fossils show somewhat larger braincases, but still display the characteristic robust *erectus* skull features of their more ancient predecessors.

Western Asia

In western Asia there is a very different story, for this region represented a corridor between the continents of Africa and Asia (and hence indirectly, Europe as well). There is evidence of human occupation in the Levant (the region bordering the eastern Mediterranean which includes Lebanon and Israel) about 1.5 million years ago at the Israeli site of Ubeidiya. A younger archaeological site at Gesher Benet Yacov in Israel contains a fragmentary human leg bone, many handaxes, evidence of elephant butchery, and even what look like worked slabs of wood, dated to about 800,000 years. There is also a possible *Homo erectus* skull fragment from Syria, and a possible *heidelbergensis* skull fragment

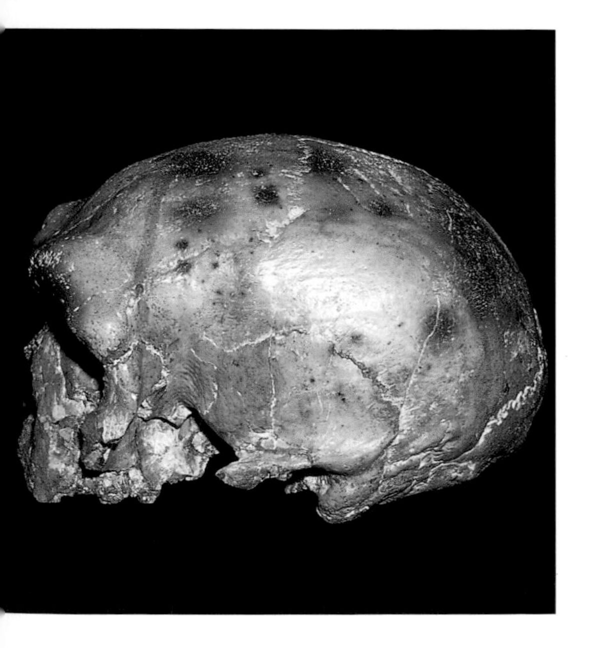

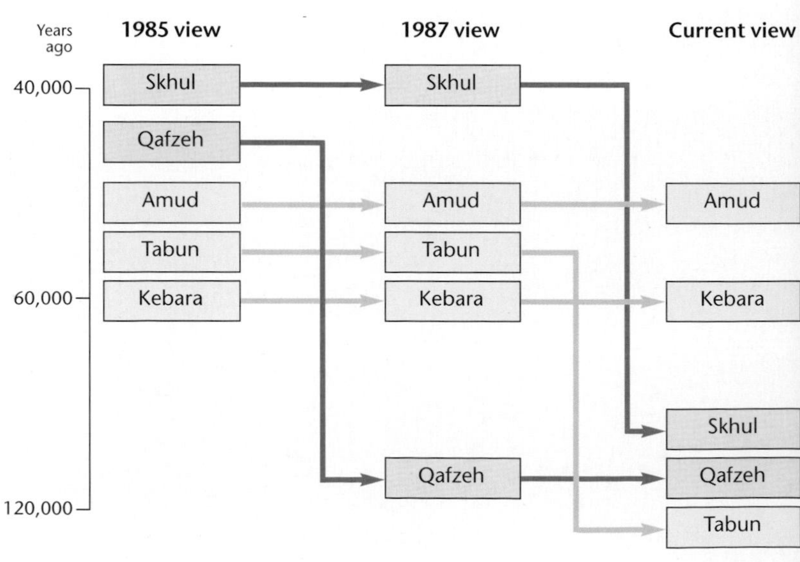

from Zuttiyeh in Israel. But the fossil record is sparse until the last 125,000 years, when more extensive evidence begins to appear in Israel.

Then, the region seems to have formed an overlap zone between the territories of the evolving Neanderthal lineage to the north, and the early *sapiens* lineage to the south. Recent research suggests that early modern humans were not only living in Israel long before the Cro-Magnons (modern humans) arrived in Europe, but that Neanderthal populations were there after them. Dating techniques applied to animal teeth and burnt flint tools found in the same levels as fossils of *Homo sapiens* individuals at the caves of Skhul and Qafzeh suggest that these individuals were intentionally buried between 120,000 and 80,000 years ago. The skeletons concerned show some primitive characteristics, but also many other modern features in skull shape, chin development and the slender build of the body, especially the legs. However, a Neanderthal burial from the Israeli site of Tabun has been dated to about the same period, while those from Kebara and Amud have been dated to only about 60,000–50,000 years, and it seems that the Neanderthals either overlapped with early *sapiens* populations, or that they alternated in occupation of the area, since modern people were present again from about 40,000 years ago.

One idea is that early modern people had emerged from Africa during the short interglacial period that began about 125,000 years ago. However, when the last Ice Age began, the moderns disappeared, or moved elsewhere, and the cold-adapted Neanderthals came into the region from further north, either because of environmental shifts which made the region more suitable for them, or because they were forced south by deteriorating conditions. Whether the Skhul-Qafzeh people represented just a brief and premature 'Out of Africa' excursion which never went any further, or whether populations like them actually managed to move deeper into Asia and Arabia to give rise to later moderns such as the ones who reached Australia, remains unclear.

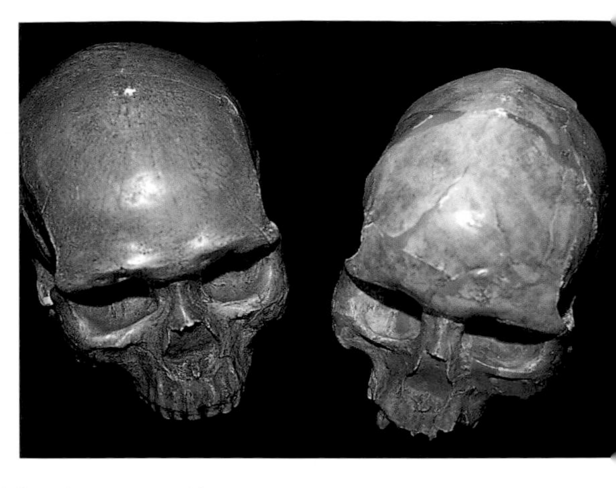

(Right) These early modern skulls are from the Upper Cave at Zhoukoudian, China, and may be about 30,000 years old. Probably those of a man (left) and woman (right), they show significant variation, which led some researchers to argue that they were related to different present-day 'racial' populations. However, they are more likely to represent an ancient human population of East Asia which has either died out subsequently or had not yet developed the regional features found in populations today.

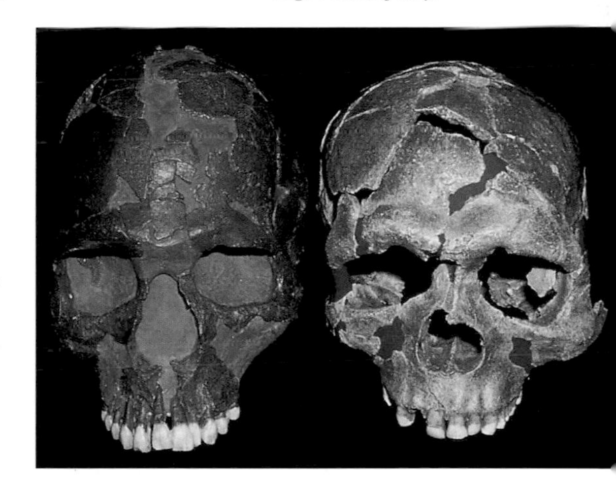

(Below) These skulls from Qafzeh show typical early modern features such as a broad flat face, rather low and square eye sockets, and a very large brain capacity.

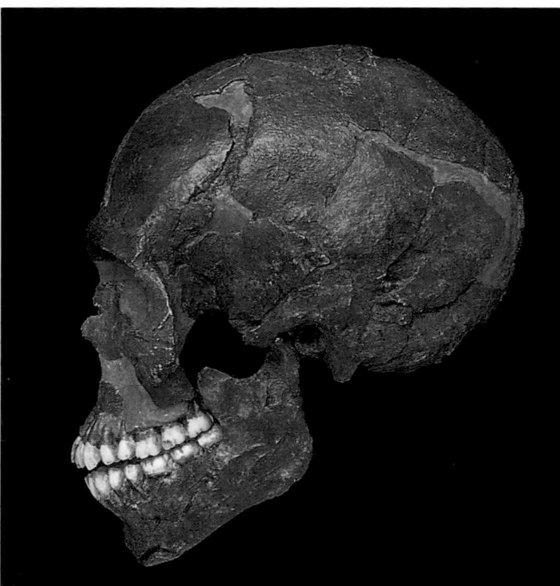

(Below left) The cave site of Qafzeh, near Nazareth, Israel. The Qafzeh 9 individual (below right), of which this is the skull, had a tall and rather lightly built skeleton, which was excavated with the skeleton of a child (Qafzeh 10) at its feet.

What Happened to the Neanderthals?

(Below left and right) Zafarraya Cave, in a mountainous region inland from Malaga in Spain, may have been one of the last refuges of the Neanderthals. The evidence suggests that modern humans may have migrated through central Europe from the east while Neanderthals were able to maintain their numbers longer in peripheral areas such as Britain, southern Iberia and southern Italy longer before they went extinct, or were absorbed into the populations of the newcomers. However, the radiocarbon and uranium series dates of about 27,000 years were not made directly on the human bones, and it is possible that they are, in fact, older than this date.

We have very little data on the reasons why early human species became extinct. But in the case of the Neanderthals, the fossil and archaeological record is good enough for us to make some informed guesses. As we have seen, their lineage can be traced through increasing specializations in the braincase, face and even ear bones over about 300,000 years in Europe. But they disappeared over a period of less than 20,000 years. They lived in western Asia until at least 50,000 years ago, and in western Europe they may have survived in marginal regions until less than 30,000 years ago. Between 40,000 and 30,000 years ago, it appears that the last Neanderthals and the first modern people in Europe (the Cro-Magnons – see pp. 166–169) actually coexisted, but it is unclear whether they regularly encountered each other, were in conflict, or at the other extreme, perhaps even interbred. Our dating methods are still too imprecise to map whether these peoples coexisted in separate regions, the same regions but in different locations, or perhaps in the same places at the same time. The fossil evidence does not show an evolutionary transition occurring between the last Neanderthals and the first moderns, but it is disputed whether there is evidence of hybridization.

Late-surviving Neanderthals

Excavations at a Spanish site not far northeast of Gibraltar, called Zafarraya Cave, recently revealed new fossil evidence of the Neanderthals, their stone tools and food debris, and indicated that they had lingered on in this region much later than had previously been suspected. Evidence from France suggested that the Neanderthals had disappeared from there by about 32,000 years ago, dated by the radiocarbon method. But the same method dates the last Neanderthal occupation of Zafarraya Cave to only about 27,000 years ago. If this disputed date is accurate, it means that the Neanderthals seemingly survived in southern Spain long after they had disappeared further north, and even longer after the appearance of the first modern people in Europe, some 40,000 years ago. Further dates from other sites in Gibraltar, southern Spain and Portugal appear to support this scenario of late Neanderthal survival in pockets, and in this context, areas like mountain ranges and even Britain (not then an island) could have provided refuges. There are comparably late dates from regions such as Croatia, the Crimea and the Caucasus.

While the Neanderthals of southern Iberia continued to make typical Middle Palaeolithic artifacts (p. 210), other Neanderthal groups did start producing more advanced stone tools, bone tools, and apparently even necklaces and pendants during their period of coexistence with the Cro-Magnons – for example, the makers of the Châtelperronian in France and northern Spain, and of the Uluzzian in Italy. Some experts believe that this is a reflection of contact or trading between the two populations,

(Right) Comparison of a late Neanderthal skull from Saint-Césaire in France, about 37,000 years old, and an early Cro-Magnon skull and mandible (not from the same individual) of about the same age from the Czech Republic. It is unclear to what extent late Neanderthal and early modern populations overlapped and interacted. Some archaeologists believe that late Neanderthals show evidence of behavioural change, indicating contact with contemporaneous modern humans.

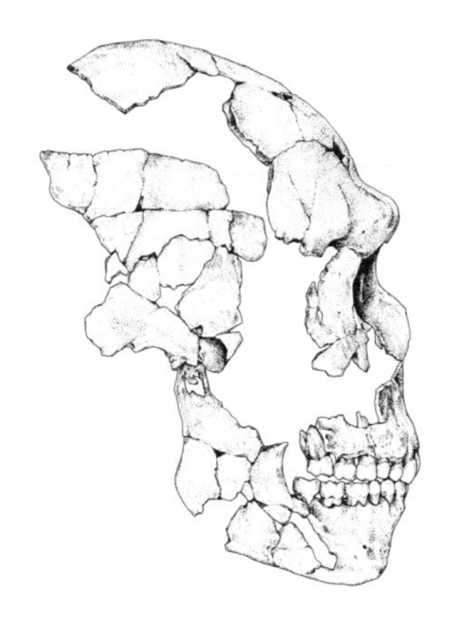

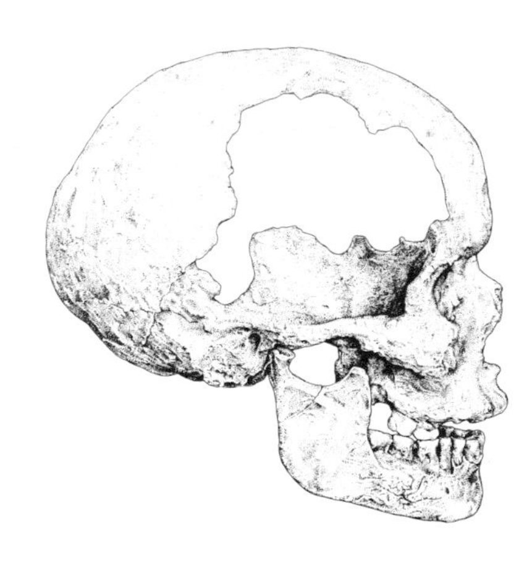

while others believe it indicates technological competition between them for the available resources, competition which drove the Neanderthals to innovation and social change. Yet others argue that the Neanderthals were developing these innovations independently, even before the Cro-Magnons arrived.

Neanderthal – Cro-Magnon relations

Within an area as large as Europe, and over such a long time span, many different kinds of interactions would have been possible, ranging from warfare, through avoidance, to peaceful coexistence, trade and even interbreeding. Given what we know of human behaviour today, any of these could have occurred, and indeed perhaps they all did at certain times and in certain places. Whether the two populations could have interbred successfully, we cannot say. Even if they were separate species, their genetic differences would not have been large, and they were probably as closely related as distinct mammal species today which can interbreed. So the main factors controlling possible hybridization were probably behavioural, cultural and social. Perhaps offspring had reduced fertility, or were shunned as prospective mates by members of their parent populations. The evidence of DNA studies, both on recent Europeans and on a Neanderthal fossil (see pp. 180–181), speak against Neanderthal genes surviving in Europe today, but whether their replacement was absolute remains to be seen.

The disappearance of the Neanderthals may in the end have been the result of a combination of factors: the rapidity of climatic oscillations at this time and the resultant constantly fluctuating environments, coupled with the presence of newcomers who were perhaps more flexible and innovative in adapting to these rapid changes. While the Neanderthals had survived such environmental stresses many times before by retreating to more sheltered refuges and then recovering when things improved, this time there may not have been enough space for both them and the Cro-Magnons to survive in the long term. By 25,000 years ago, this well established and successful human lineage had gone forever, leaving *Homo sapiens* as the sole surviving human species on Earth.

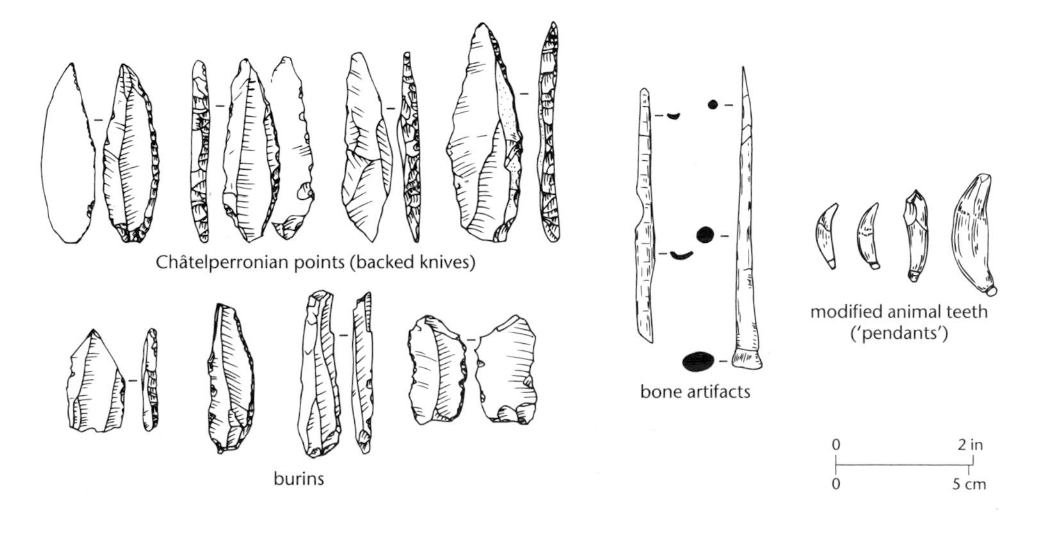

(Left) The Reindeer Cave (Grotte du Renne) at Arcy-sur-Cure in France is one of the most interesting and controversial in Neanderthal studies. Artifacts of the Châtelperronian industry are associated there with Neanderthal fossils, as at Saint-Césaire, and apparently also with stone hut structures, bone tools and the remains of necklaces made of animal teeth, dating from about 35,000 years ago. Some archaeologists have argued that the Châtelperronian is a late Middle Palaeolithic industry, others that it is genuinely Upper Palaeolithic, created with or without Cro-Magnon influence on its Neanderthal manufacturers, while others question its integrity as a real assemblage.

The Cro-Magnons

In 1868, three human skulls and parts of skeletons were found in the rock shelter called 'Cro-Magnon' ('Big hole' in old French). They were found near stone tools which we now assign to the Aurignacian and Gravettian industries of the Upper Palaeolithic (Late Stone Age), about 30,000 years old, and shells which had been pierced to make a necklace. In time, these fossils gave their name to the whole human population of the European Upper Palaeolithic. Earlier Cro-Magnons are known from sites in countries like Romania, Germany and the Czech Republic, dating from about 35,000 years. Because of their proximity in time to the last Neanderthals, some workers have seen Neanderthal traces in their anatomy, such as larger brow ridges, bigger faces and teeth, and bulging backs to their skulls. However, most evidence suggests that the first Cro-Magnons were quite distinct from the Neanderthals. Their faces, for example, were relatively short, flat and broad, and their noses, although wider than those of modern Europeans, were much smaller than those of Neanderthals. Moreover, there is both archaeological and physical evidence that they were intrusive in Europe. The Aurignacian industry seems quite distinct from its local Middle Palaeolithic predecessors, and many archaeologists believe it came into Europe from Asia.

(Below) These two skulls are of French Cro-Magnons, probably both female. That on the left is from Abri Pataud, perhaps 22,000 years old, associated with a 'Protosolutrean' industry, and that on the right is from Cro-Magnon itself, perhaps 30,000 years old, associated either with Aurignacian or Gravettian tools. Both skulls appear fully modern in the shape of the face and vault, but the Pataud skull has robust jaws with large and unusually shaped teeth.

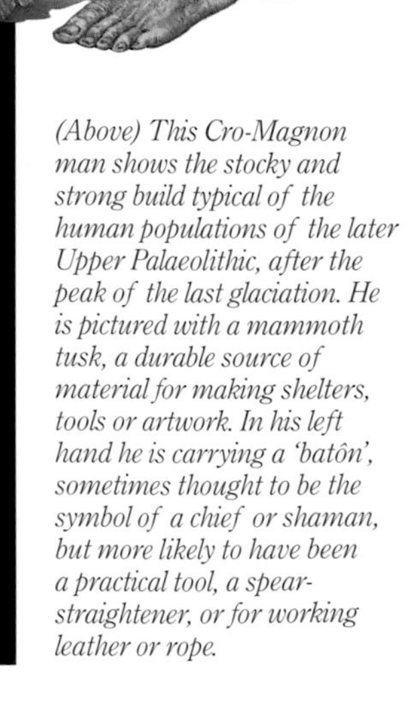

(Above) This Cro-Magnon man shows the stocky and strong build typical of the human populations of the later Upper Palaeolithic, after the peak of the last glaciation. He is pictured with a mammoth tusk, a durable source of material for making shelters, tools or artwork. In his left hand he is carrying a 'batôn', sometimes thought to be the symbol of a chief or shaman, but more likely to have been a practical tool, a spear-straightener, or for working leather or rope.

A warm-adapted species?

Early Cro-Magnon skeletons show that they had linear body shapes like those of modern people who come from hot climates, in contrast with those of the Neanderthals, who were shorter and wider, more like modern cold-adapted peoples such as the Inuit. It seems that in both their body shape and face shape the Cro-Magnons started off contrasting strongly with both the Neanderthals and recent Europeans, and they only began to change away from that distinctive non-local pattern after the climax of the last Ice Age, about 20,000 years ago. When the Cro-Magnons entered Europe, over 40,000 years ago, the climate was relatively mild, but several short but severe cold periods soon followed. The fact that these people apparently came in with, and maintained, warm-adapted body shapes indicates that they must have had excellent protection from the cold, in the form of clothing and insulated shelters. In fact, there is evidence that Cro-Magnons in northern Europe even burnt mammoth bones and dug out coal to use as fuel.

Cro-Magnon life and art

As we will discuss later (p. 212), the Cro Magnons displayed the complexity of life which we find in modern hunter-gatherers. They hunted, fished, traded, produced art and apparently even recorded time. We think that they marked bones to indicate

(Above) A spearthrower works by adding extra length to the throwing arm. In skilled hands, a spear can be propelled up to 4 times the distance of an unaided throw.

(Left) The Maz D'Azil spearthrower. Made of reindeer antler between 13,000 and 17,000 years ago, it represents a young ibex with two birds perched on its rear. The hook on which the butt of the spear was placed is cleverly formed by one of the birds. Its delicacy suggests it may not have been intended for actual use.

(Right) The bone plaque from Les Eyzies, France. It is between 25,000 and 30,000 years old, and is marked with rows of holes and notches – it has been argued that such plaques are ancient calendars, recording days, lunar months, and seasons.

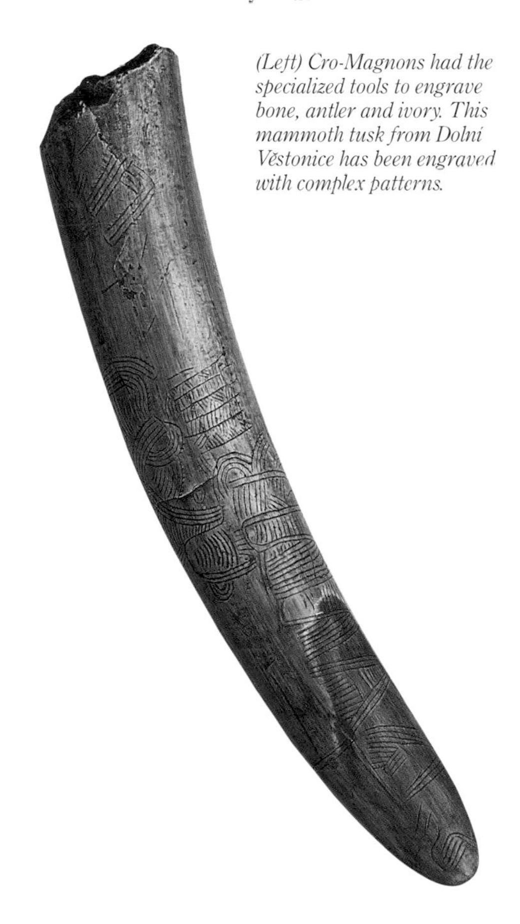

(Left) Cro-Magnons had the specialized tools to engrave bone, antler and ivory. This mammoth tusk from Dolní Věstonice has been engraved with complex patterns.

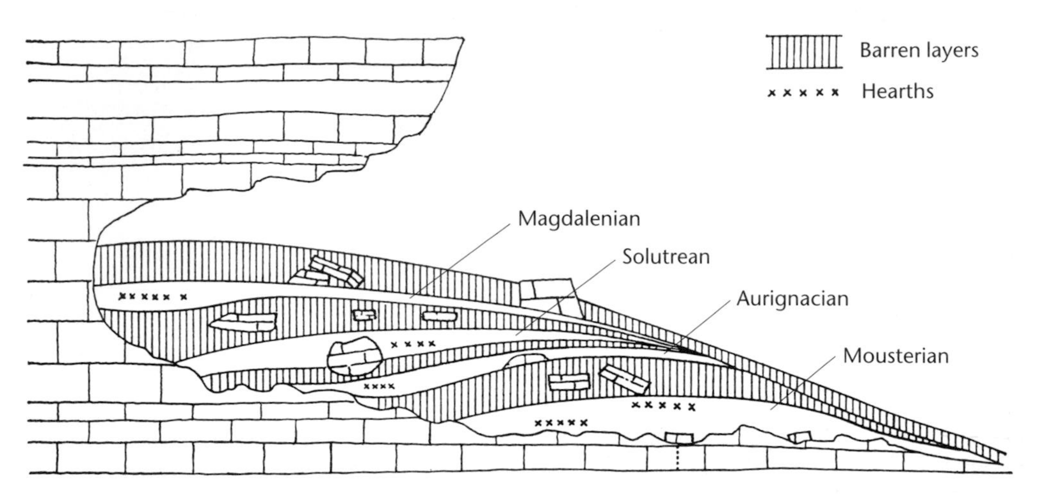

(Left) This schematic section of a French abri (rock shelter) shows an idealized archaeological sequence of Middle and Upper Palaeolithic industries spanning some 40,000 years. Generally, intermittent occupation or erosion prevents the accumulation of such a complete sequence.

days, lunar months and the changing seasons, and this no doubt helped them to plan for the future, whether for hunting, fishing (for example when salmon would be passing through the rivers), gathering plant resources, storing food, moving camp or carrying out ceremonies. These resourceful people survived in Europe for about 25,000 years, and it is commonly assumed that they represent the ancestors of modern Europeans, although as we have seen above, they did not look exactly like them. While some genetic data calibrates the age of local European populations to between 40,000 and 30,000 years ago, their origin does not have to lie with the known Cro-Magnon fossils. We have no direct evidence of the physical appearance of these people beyond what we can reconstruct from the fossils

(Right and opposite) One of the most famous features of the Cro-Magnons was their art, sometimes in the form of engravings or paintings on the walls of caves, at other times in the form of statues of animals or the human form. Three famous Upper Palaeolithic statuettes are shown here. On the left, the Venus of Galgenberg from Austria, carved from green slate and about 30,000 years old. It apparently depicts a dancing woman. Centre is the Lespugue Venus from France, carved from ivory but unfortunately damaged during excavation, and perhaps 25,000 years old. On the right, the Venus of Willendorf, found in 1908 near the town of Willendorf in Austria. It was carved from an exotic limestone and is about 25,000 years old.

(Left) Facial representations are rare in Palaeolithic art. This tracing of an engraved plaquette from La Marche shows the face of a bearded man which has been extracted from a mass of other markings.

(Below) This nearly complete skeleton from Předmostí in the Czech Republic (Předmostí 3) was discovered in 1894 in a large multiple burial site of the Gravettian period, dated to about 27,000 years ago. This individual shows the tall and linear body shape of many of the early Cro-Magnons. The burials were associated with numerous decorated objects and tools, many of them made of mammoth ivory, and the whole collection was perhaps the richest such record ever found. Tragically, it was all lost in World War II.

and from their art. Women were often depicted in the so-called Venus figurines as plump, with large breasts, stomachs and buttocks, but facial features are usually poorly marked. There are a few realistic engravings and paintings of faces from the later Upper Palaeolithic, and some seem to show men with dark beards, long hair and prominent noses.

People like the Cro-Magnons also lived in North Africa and the Levant, and probably further afield too. Fossil material found with Upper Palaeolithic-style artifacts has been found in both Sri Lanka (Batadomba Lena) and China (Upper Cave, Zhoukoudian), both sites dating to about 30,000 years ago. In the latter case, bone needles, shell necklaces and red ochre were also found, items strongly reminiscent of the European Upper Palaeolithic. But Java and Australia were probably colonized by distinct peoples who arrived there before the Cro-Magnons, and the Upper Palaeolithic, appeared in Europe. Southern Africa, too, had different populations at this time, some of whom seem to represent forebears of the present-day Khoisan peoples.

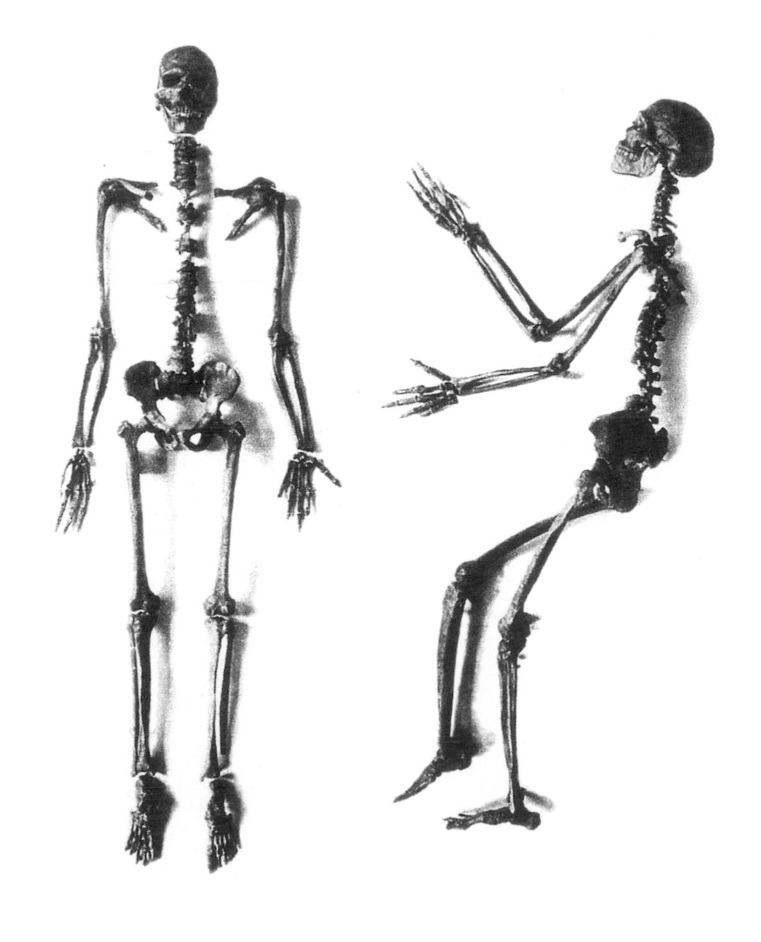

The First Australians

(Below) *Despite much lower sea levels in the past, there was never a landbridge between Java and Borneo, and Greater Australia. The first Australians must have used boats to make repeated sea voyages between islands to reach the new continent.*

Even at the times of lowest sea level during recent ice ages, there has never been a land bridge connecting Australia with the islands of Indonesia, including Java, although New Guinea and Tasmania were part of an enlarged Australian continent at such times. As we have already seen, humans had colonized the islands of Indonesia by the early Pleistocene (1.5 million years ago), but seem not to have reached Australia until the late Pleistocene, less than 70,000 years ago. There is good evidence that *Homo erectus* persisted through the lower and middle Pleistocene in Java without very significant evolutionary change. Even more remarkably, recent dating work suggests that *Homo erectus* might still have been living in Java as recently as 50,000 years ago from fossil evidence at the sites of Ngandong and Sambungmacan, on the banks of the River Solo. If these dates on associated animal teeth are accurate, then this ancient species survived in the southeast of the inhabited world as long as Neanderthals did in the northwest.

Arrival by water craft

People could only have arrived in Australia on water craft, and this would have entailed repeated

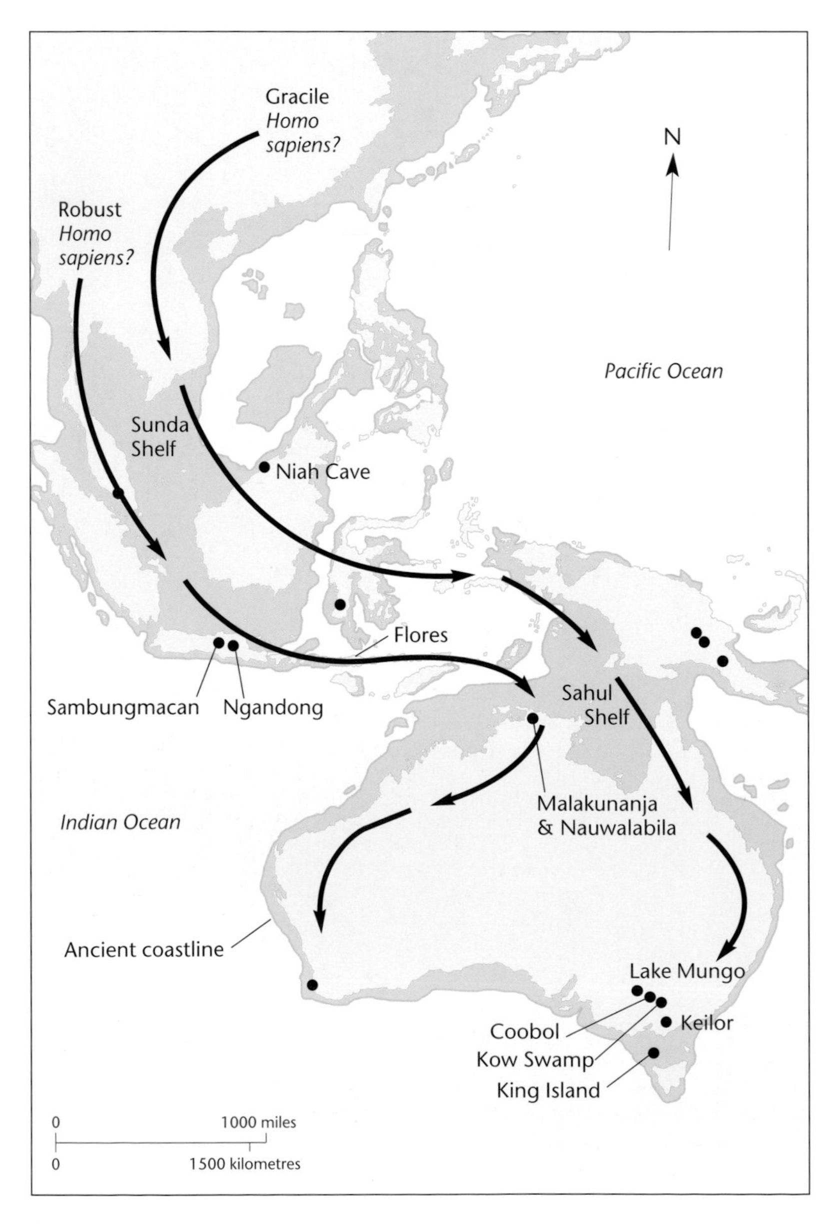

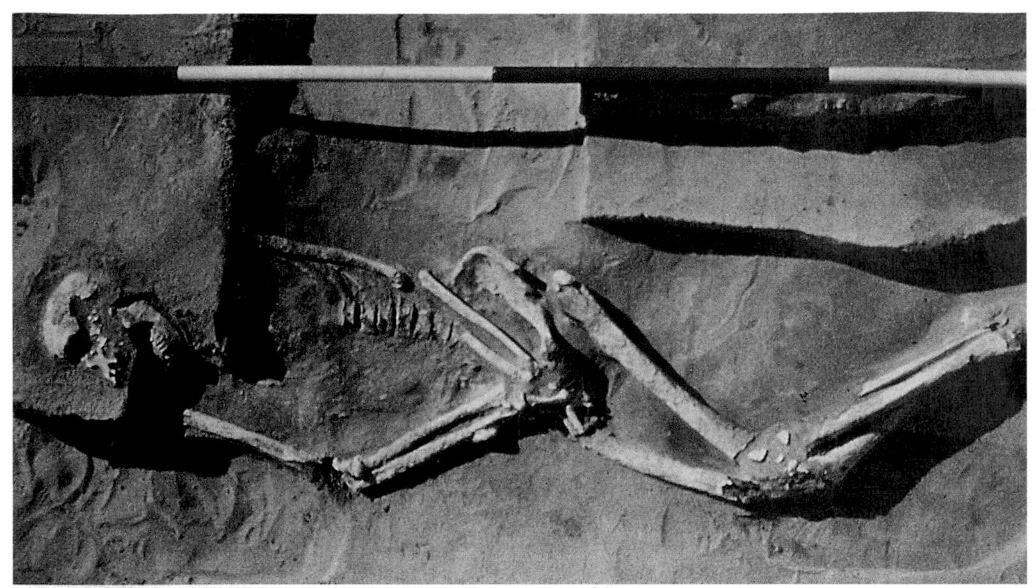

(Left) The Mungo 3 burial in the Willandra Lakes region of southeastern Australia is at least 40,000 years old. Its gender is uncertain, but red ochre powder was found on the upper part of the skeleton, perhaps originally painted on the body, perhaps added afterwards. Reconstructed below, this is the oldest known ceremonial burial using red ochre. Now an arid desert region, this was a fertile series of lakes 40,000 years ago, around which some of the first Australians lived, hunting kangaroo, and collecting fish and shellfish from the lakes.

A Worora youth paddles a mangrove log raft on George Water, Glenelg River district, western Australia. Larger rafts made of bamboo may have been used in the first voyages to New Guinea and Australia.

island-hopping journeys of at least 50 km (30 miles) over open seas. The first Australians were probably unwilling colonists, carried by unkind winds or seas off course from an island they wished to visit to one they had never seen before. The exact route taken by these first colonists is unknown. There were at least two feasible routes – an eastern one via Timor to New Guinea, or a western one via Java to northwest Australia itself. Archaeological sites in New Guinea have been dated to about 30,000 years ago, but more recently, human occupation of inland rock shelters in northern Australia, Malakunanja II and Nauwalabila, has been dated to at least 50,000 years ago. In southeast Australia, the Willandra Lakes region also has human occupation stretching back to about 50,000 years.

Theories of colonization

We have no evidence yet of the physical appearance of the first colonizers of Australia, since the earliest known fossils come from sites in the southeast of the continent, in the Willandra Lakes region, and date from a slightly later stage of colonization. The large sample from there contains a cremated individual and a partial skeleton, both lightly built and possibly female, dated to at least 40,000 years ago. However, there are also much larger and more robust individuals, and this has led to different ideas to account for this marked physical variation. The simplest idea is that only one founder population reached Australia; it then dispersed across a continent empty of people, and in doing so began to develop the extensive physical variation which is found later on.

A more complex scenario linked with the now largely abandoned model of multiregional evolution suggested that there were two distinct founding populations. The more robust one colonized from western Indonesia, derived from late *Homo erectus* people, such as those known from Ngandong. The other more gracile arrivals came on the eastern route via New Guinea, and ultimately derived from Chinese *Homo erectus* ancestors. It would be helpful in testing these ideas to know the age of all the important fossils, but many are currently undated, and it is therefore impossible to know the degree and direction of changes in robusticity. However, what evidence there is, suggests that gracile populations preceded robust ones. Moreover, there are problems with the interpretation of the fossil material itself because it appears that some of the most robust specimens have been altered by intended or unintended artificial deformation of the head during life. This has exaggerated their flat frontal bones, increasing resemblances to their supposed ancestors, the late *erectus* people of Java. More difficult still, many of the most critical specimens in the debate about human origins in Australia have now been returned to aboriginal custodians for reburial, thus making them unavailable for further study.

Some fossil and recent skulls from Australia do show features that resemble those found in more ancient peoples, such as a relatively broad, flat face, with low orbits and nose and a relatively long skull. However, these features are also found in the earliest moderns of Africa and Israel, and it is possible that this reflects the early arrival of modern people

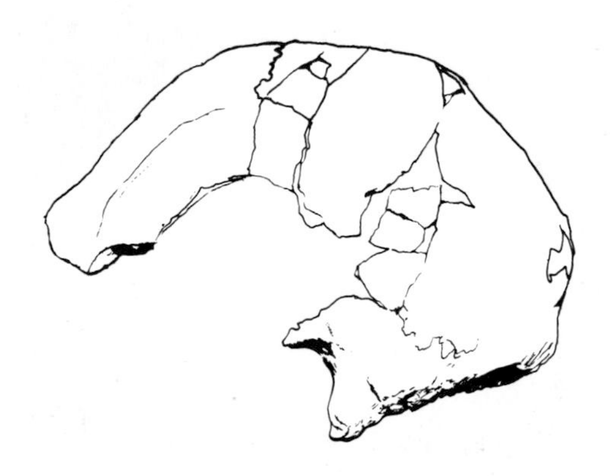

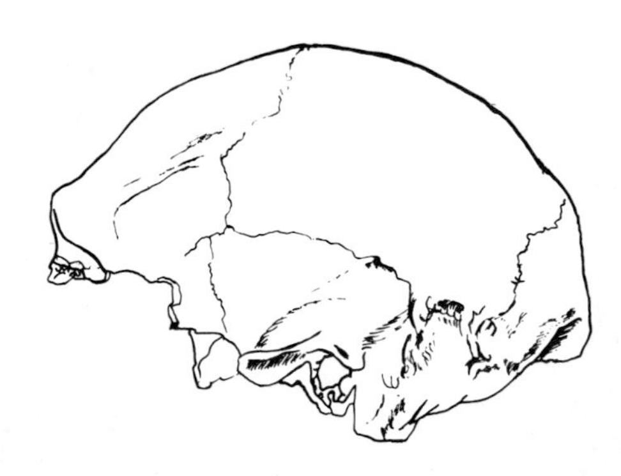

This comparison is of the Willandra Lakes 50 skull (left) from Australia and Ngandong (Solo) skull XI from Java. Some workers see evolutionary connections between these skulls, suggesting a Javanese origin for some early Australians. However, comparative analyses suggest that the Australian skull is a large and robust modern human, while the Ngandong specimen is fundamentally similar to Homo erectus.

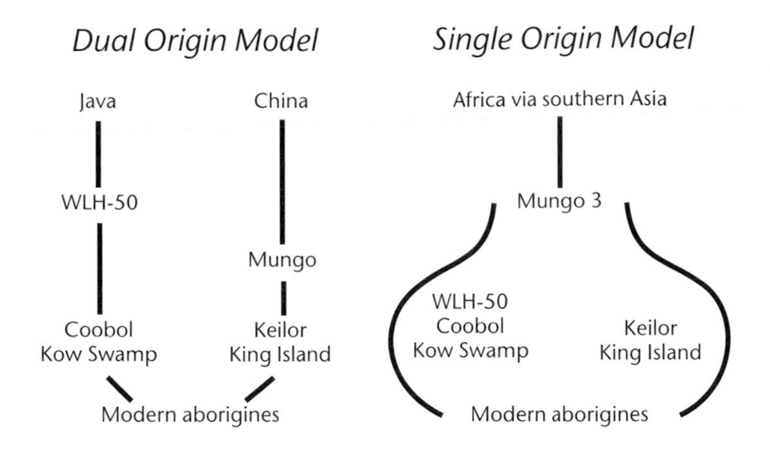

(Right) One model for the origin of the Australians (left) suggests that they had a dual origin – a 'robust' lineage derived from Javanese ancestors, while a 'gracile' lineage descended from Chinese antecedents. These then blended to produce recent aboriginal populations. However, dating of some of the early Australian fossils shows that the robust forms are relatively late in time, not early, and this favours a single origin model, where diversification developed within Australia (right).

in Australia, followed by their relative isolation. While the arrival of modern people in Europe apparently coincided with the first appearance of Upper Palaeolithic technology there, the archaeological record of Australia appears not to show comparable developments in stone tools until the last 10,000 years. Yet, other signs of 'modern' behaviour such as the use of boats or rafts, cremation of the dead, the production of art and body decoration, and the working of bone, were certainly present early on in Australian prehistory, and demonstrate the behavioural modernity that must have arrived with the first Australians at least 50,000 years ago.

(Below) 'The Walls of China' are crescent-shaped dunes which stand along the eastern shore of dried-up Lake Mungo in Australia.

Homo floresiensis

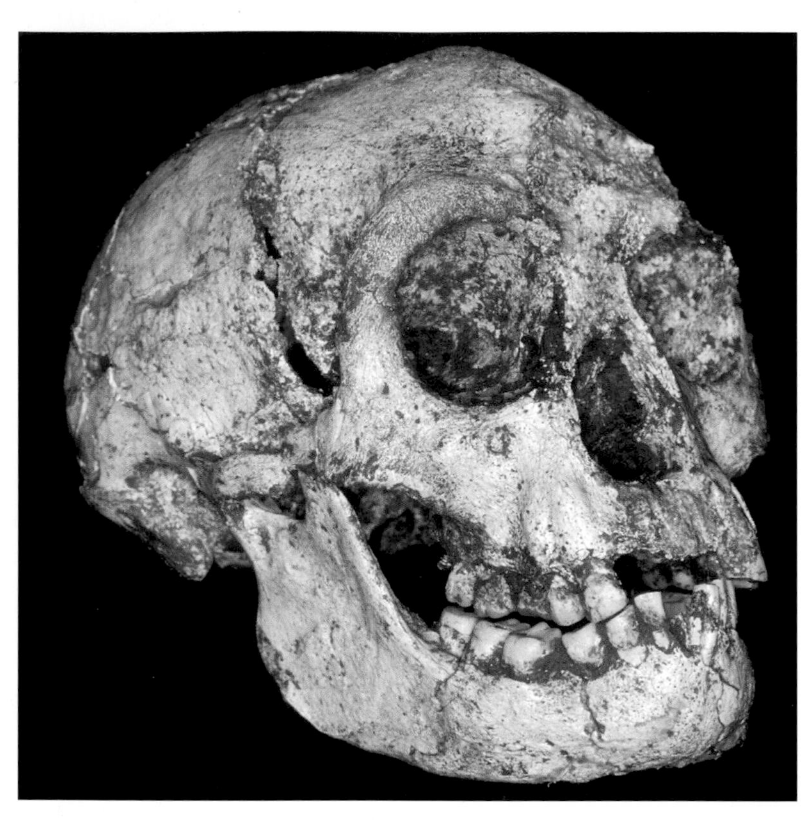

The skull of Homo floresiensis*, believed to be a female specimen.*

It is usually assumed that only one species of early human – *Homo erectus* – lived in Southeast Asia before modern people arrived there. Furthermore, up to now, *Homo erectus* fossils have only been identified in the region from the island of Java, in Indonesia. East of Java, towards New Guinea and Australia, it was thought that deep water had kept people from venturing any further until the ancestors of Australian Aborigines used boats to hop across the intervening chain of islands, some 60,000 years ago. This simple picture was challenged when it was reported a few years ago that 800,000 year-old stone tools had been found on the island of Flores, some 300 miles east of Java, but most experts wanted more evidence to back up the claim that ancient humans had migrated that far.

That evidence has now turned up in an extraordinary discovery from Flores. The skeleton (including a well-preserved skull) of a metre tall 'human' with a brain size of about 380 ml (about the same as that of a chimpanzee) has been excavated from the Liang Bua cave, together with stone tools, and remains of a pygmy form of an extinct elephant called *Stegodon*. There are also bones of smaller animals, some of which have been burnt. Remarkably, the level in which the skeleton was found has been dated to only about 18,000 years ago, so modern people must have actually encountered this strange creature. What was it, what was it doing on Flores, and what happened to it?

Naming the new species

The Flores find is so unexpected that deciding what kind of creature it represents is not easy. The possibility that it was a single abnormal individual can be discounted because other similar remains have already been found in the cave. Although the legs and hipbone suggest that it walked upright in a fundamentally human manner, in details of shape as well as size, the hipbone resembles those of the prehuman australopithecines, who lived in Africa over 2 million years ago. Together with the very small brain size, this might suggest that this is actually some kind of australopithecine that migrated out of Africa long before the spread of *Homo erectus*. Yet details of the skull, the shape of the face, the small teeth, the evidence of tool making and, perhaps, hunting all suggest that the creature was fundamentally human. Thus the describers of the skeleton have named a new human species *Homo floresiensis* ('Man from Flores'), after its island home. They suggest it might be a descendant of *Homo erectus* that arrived early on Flores, perhaps using boats, and under completely isolated conditions evolved a very small size – a phenomenon known from other mammals, called island dwarfing. Alternatively, the dwarfing process could have occurred along the route to Flores, on one of the islands nearer to Java, such as Lombok or Sumbawa. As the skeletal remains are hardly fossilized, there is the possibility that DNA can be extracted from them, which could provide valuable insights into our relationship with both *Homo floresiensis* and its presumed ancestor *Homo erectus*.

It is also possible that DNA might be recovered from the cave sediments, preserved droppings and, if it survives, ancient hair.

Questions for future research

This remarkable discovery raises many questions for future research. One of these is how *Homo floresiensis* got to Flores. Could its ancestors really have made watercraft (perhaps of bamboo) to reach the island? This would certainly be surprising, because such behaviour is thought to be exclusive to *Homo sapiens*. But the alternatives – a short-lived land bridge that allowed very few species to cross it, or accidental transport on natural rafts of vegetation – seem even more unlikely.

A second question concerns the behavioural evidence from the Liang Bua cave. Some of the excavated stone tools are small and sophisticated, and there is evidence of the use of fire, and possible predation on young *Stegodon*. Was *Homo floresiensis*, with its ape-sized brain, really capable of such behaviours? The answer to that question may only come from further excavations to exclude the possibility that early modern humans were also using caves on Flores, at least after 60,000 years ago, and could be responsible for some of the archaeological evidence left behind.

A third and especially intriguing question is what happened to *Homo floresiensis*. Climatic changes at the end of the Pleistocene may have affected its habitat, or modern humans could have killed it off directly, or by consuming the resources on which it lived. There is also evidence of a massive volcanic eruption which devastated Flores about 12,000 years ago. However, there is the fascinating possibility that it (or species like it) lingered on, and form a source of the widespread legends of 'wild-men' living in the jungles of Southeast Asia. Whatever the truth, its very existence shows how little we still really know about human evolution in Asia.

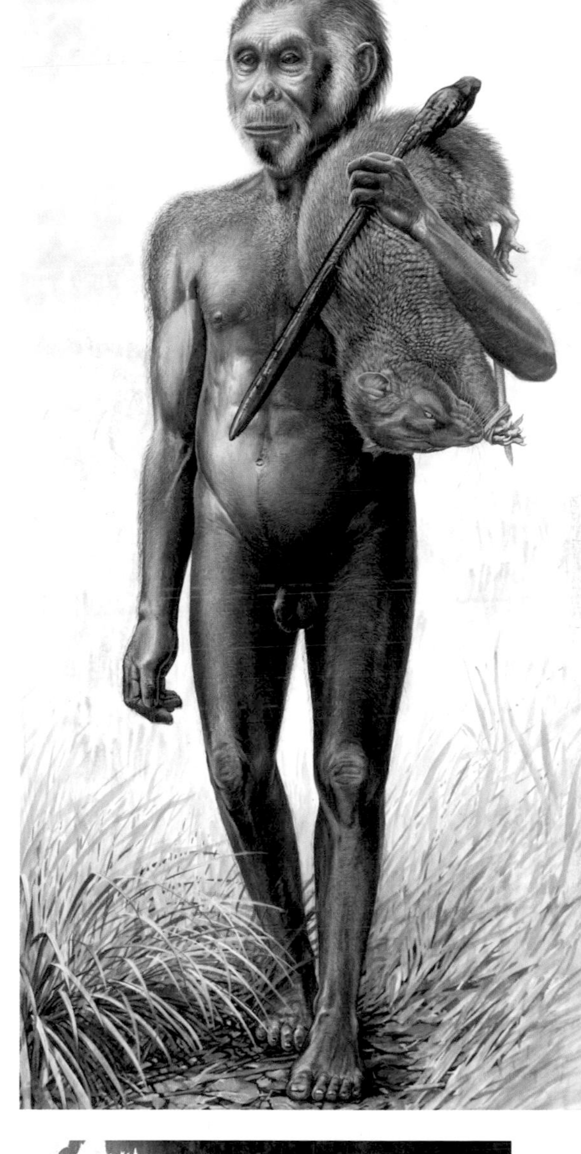

(Right) Homo floresiensis *returning from the hunt. The animal draped over his shoulder is the Flores giant rat* (Papagomys armandvillei), *which still lives on the island, but, as it is also hunted by modern* Homo sapiens, *it is vulnerable to extinction.*

(Below) Work in progress in the Liang Bua cave. The little skeleton of Homo floresiensis *was found in the deep excavation area on the far right against the cave wall. Excavations here reached a depth of 11 m (36 ft), which meant that safety shoring, multiple platforms at different depths and hard-hats had to be used.*

(Opposite) The co-directors of the archaeological excavations at Liang Bua, R. P. Soejono (left) and Mike Morwood (middle) discuss survey results with technical officer Sri Wasisto.

Genetic Data on Human Evolution

A great deal of inferential information about human evolution can be reconstructed from the DNA (deoxyribonucleic acid of living humans. Because DNA is repeatedly copied, especially when it is passed on from parents to their children, copying mistakes are made, and if the changes are not lethal, these mutations are then also copied on. They thus accumulate through time and allow us to follow particular lines of genetic evolution, and also to estimate the time involved in their accumulation. For our purposes, there are three kinds of DNA which can be studied.

(Below) Each of us has an evolutionary history locked up in our DNA, which has a characteristic double-helix structure.

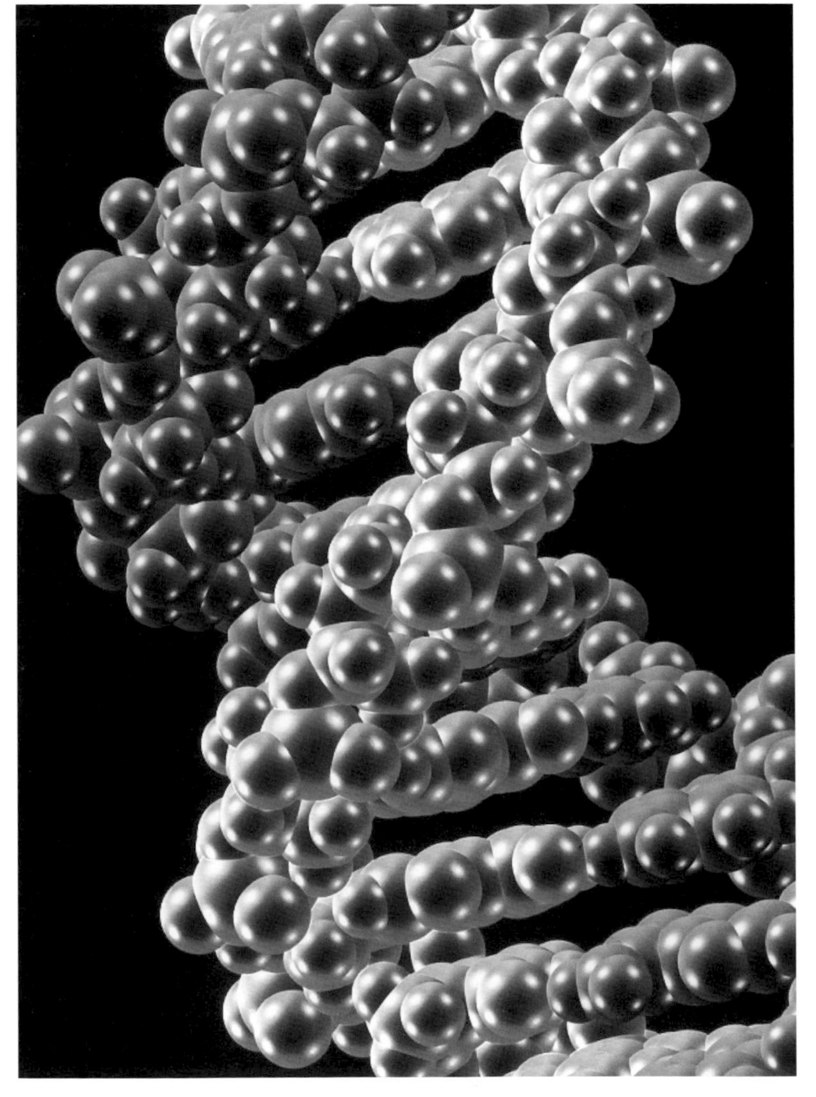

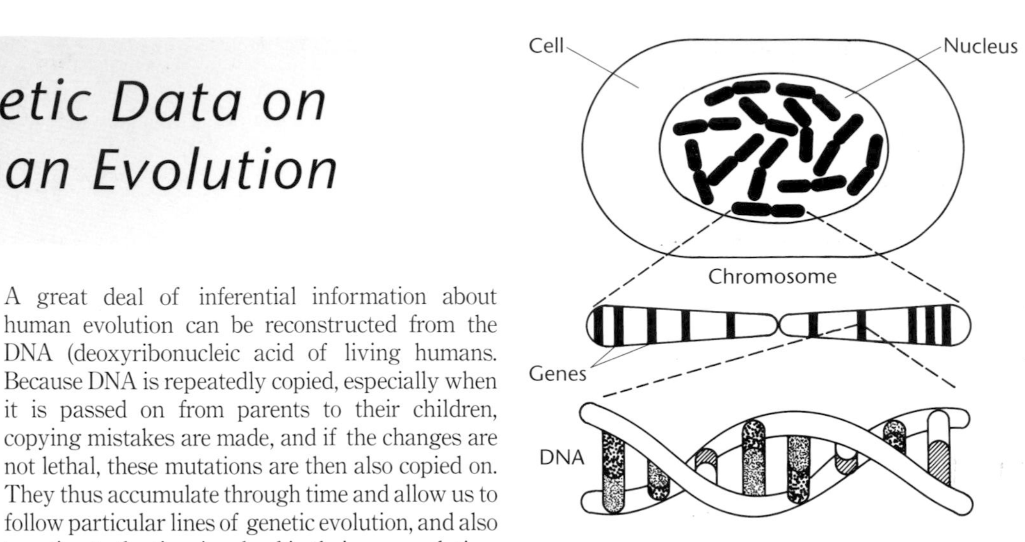

(Above) These diagrams show the relationship between cells and their DNA. Nuclear DNA is contained within the chromosomes, and active (coding) regions of DNA on the chromosomes are called genes.

Three types of DNA

The first type is the DNA which makes up the typical chromosomes contained within the nucleus of our body cells – called nuclear DNA. This DNA contains the blueprints for most of our body structure, and we inherit a combination of it from both of our parents. Nuclear DNA also contains many segments of so-called 'junk DNA', which do not code for features such as eye colour or blood group type. They nevertheless get copied along with the coding DNA and mutate through time, and can therefore also give us information on evolutionary relationships. The second type is Y-chromosome DNA, which lies on the chromosome which determines the male sex in humans. The DNA on this chromosome can be used to study evolutionary lines in males only, without the complication of inheritance from two parents which comes with the study of normal nuclear DNA. The third type is mitochondrial DNA (mtDNA), which is found outside the nucleus of cells and which is inherited through females only (p. 178). Although it is the last type of

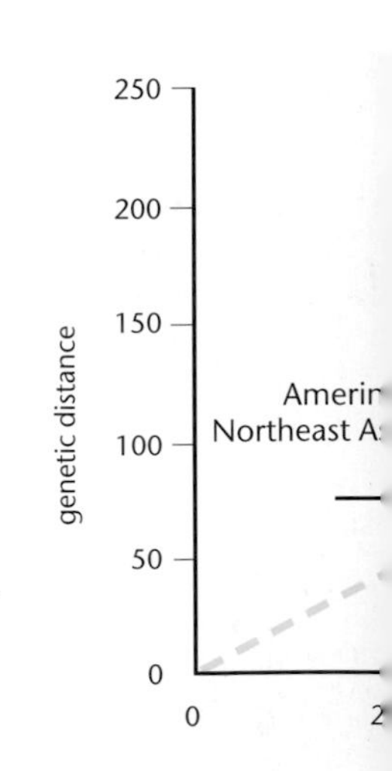

(Right) Differences between the genes of human populations may reflect their histories. Measures of 'genetic distance' roughly match the separation times of populations according to anthropological and archaeological data.

DNA that has attracted the greatest recent attention, the analysis of nuclear DNA and its products (most of our bodies' chemicals) has a much longer history in evolutionary studies.

Nuclear DNA studies

For example, it was a study of ape and human blood proteins which led, over 30 years ago, to the first suggestion of a late divergence between humans and African apes. Now, studies are able to use combinations of data from many different gene systems, or they can look at variation in a particular segment of DNA in great detail. For example, analysis of global variations in a nuclear DNA strand called the CD4 locus on chromosome 12 shows that African populations display many different patterns of variation, while those from the rest of the world have basically only one pattern. The results suggest that non-African populations are descended from ancestors who emerged from North or East Africa about 90,000 years ago. Many other nuclear DNA studies have given similar support to an 'Out of Africa' model, but there are exceptions, such as some recent research on a blood chemical called beta-globin, the DNA coding for which is found on chromosome 11. In this case, there are beta-globin types found in modern Asians, but not Africans, which appear to have been evolving their distinctive patterns for at least 200,000 years. This would suggest local continuity in Asia which goes back well beyond the time of the supposed 'Out of Africa' dispersal.

Y-chromosome DNA studies

In the case of Y-chromosome DNA, it has taken longer to establish evolutionary patterns, but very detailed results have now been obtained. Recent research suggests that variation in Y-chromosome DNA is relatively low, and a hypothetical 'Adam', forefather for present-day males, may have lived at an even younger date than the hypothetical mitochondrial 'Eve'. However, although it seems established that the 'Adam' and 'Eve' in question both lived in Africa, the precise region(s) involved are not yet clear.

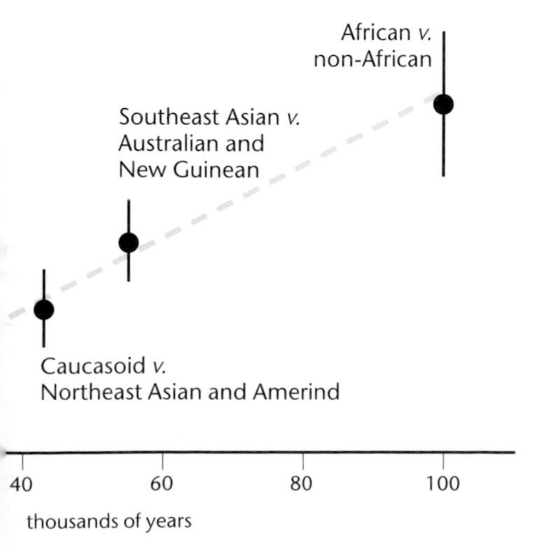

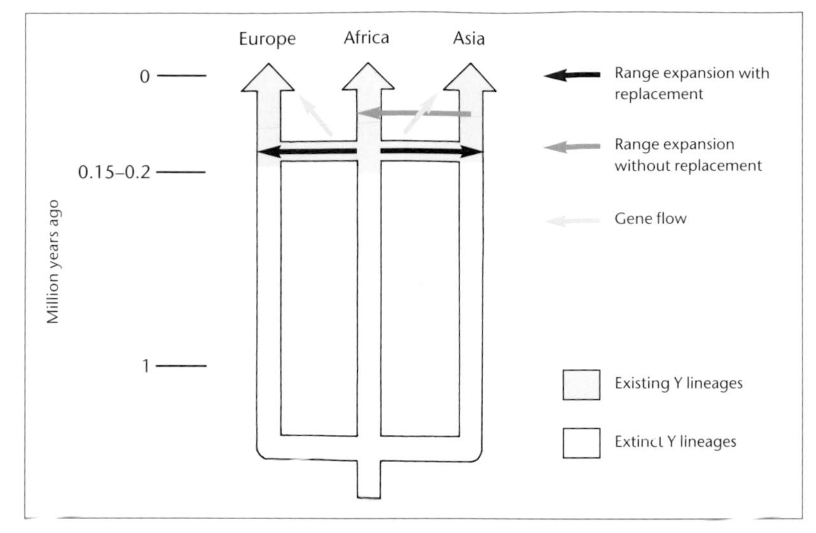

A diagrammatic representation of the evolution of Y-chromosome diversity in recent men. Within the last 150,000 years, the Y-chromosomes of African populations spread from their place of origin to replace older lineages inside and outside of Africa. However, there were later back migrations of Y-chromosome DNAs from Asia into Africa.

Population histories

The degree of DNA variation within modern populations can also be used to reconstruct something of ancient population histories. For example, a Japanese scientist called Naoyuki Takahata, using estimates for the age of common ancestors for particular DNA types, argued from our relatively high nuclear DNA variation that the breeding size of the human population through most of its evolution was about 100,000 individuals. This may not seem a large number compared with the enormous world population now, but this was a significant size for a large-bodied mammalian species. However, from the much lower variation inherent in human mtDNA, Takahata has argued that this number must have dropped to about 10,000 individuals in our recent evolutionary past, producing a 'bottleneck' which filtered out some of our previous genetic variability.

This contraction in numbers may reflect divisions in the formerly large and widespread common ancestral population of Neanderthals and modern humans. Vast distances, repeated climatic extremes, and the intrusion of geographical barriers such as ice-caps and deserts during the last 400,000 years, may have progressively isolated human populations from each other, leading to their increasing differentiation, and eventual separation as species. Perhaps it was only one of these isolated groups, about 10,000 in number and restricted to Africa, which gave rise to all living *Homo sapiens*. It seems likely that the most important physical barrier was the Sahara Desert, increasing in its extent and impact through each cold period over the last 500,000 years, and that our origins were influenced by its growth, in regions to the south.

Mitochondrial DNA

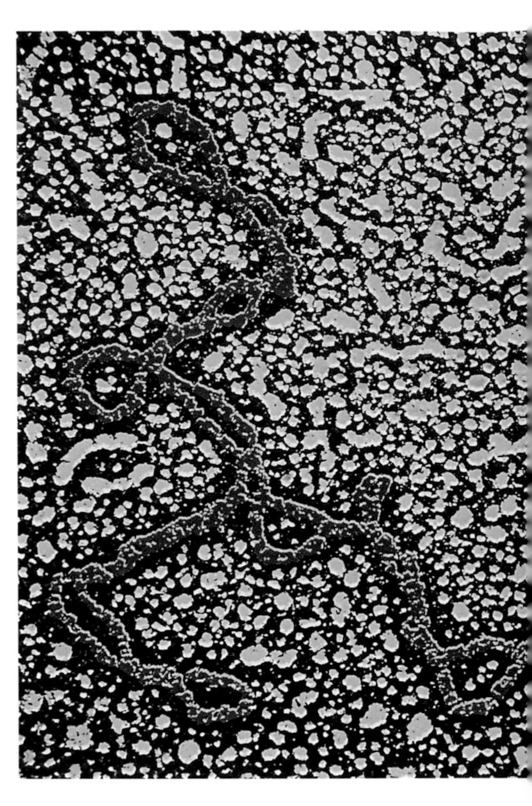

(Right) False-colour transmission electron micrograph (TEM) of mitochondrial DNA.

As its name suggests, mitochondrial DNA (mtDNA), which is found outside the nucleus of cells, is found in the mitochondria. These are little bodies which provide the energy for each cell. Their DNA is passed on in the egg of the mother when it becomes the first cell of her child, and little or no DNA from the father's sperm seems to be incorporated at fertilization. This means that mtDNA essentially tracks evolution through females only (mothers to daughters) since a son's mtDNA will not be passed on to his children. The molecule of mtDNA is shaped in a loop, and consists of about 16,000 base pairs. Only some of these are functional – that is, contain genetic code to produce specific proteins such as cytochrome – and the rest of the DNA is therefore much more prone to mutation. MtDNA seems to mutate at a much faster rate than nuclear DNA, allowing the study of short-term evolution.

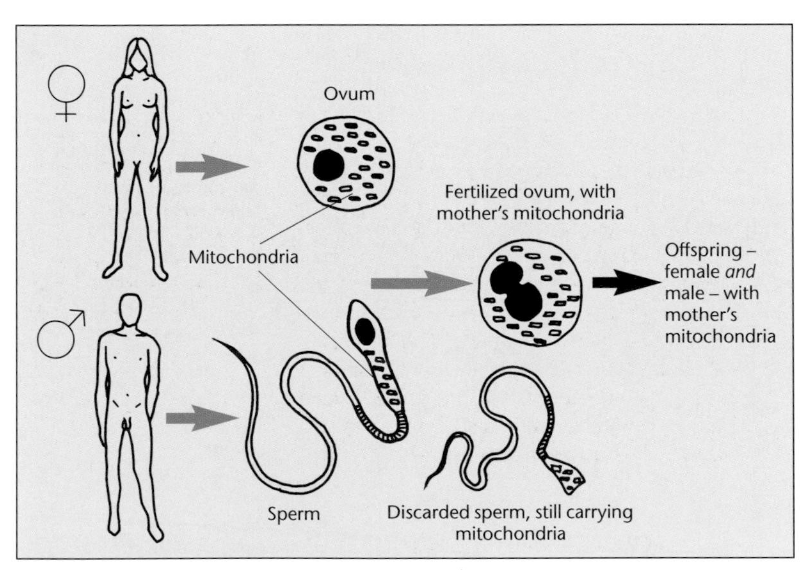

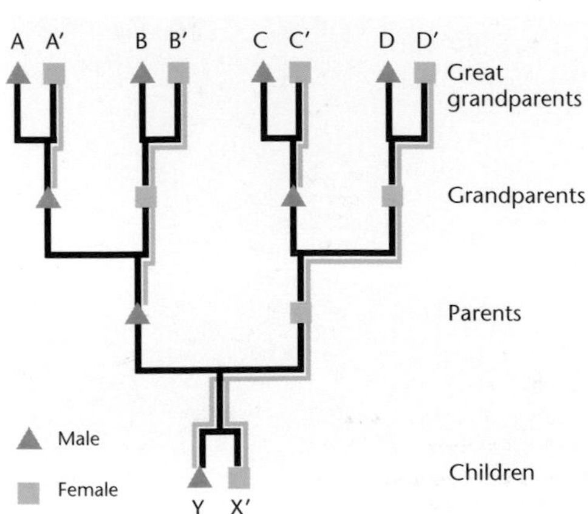

(Above and left) Mitochondria are unusual in that they have their own distinct DNA which is passed on separately, through the mother. Thus children will inherit nuclear DNA approximately equally from both parents, but mitochondrial DNA only from their mother. This means that tracing mitochondrial lineages and their evolution through unique mutations is readily possible through maternal ancestors.

The birth of 'mitochondrial Eve'

Prior to the recovery of Neanderthal DNA (p. 180), the biggest single impact of genetic data on research on modern human evolution came in 1987, with the publication of a study of mtDNA variation in modern humans. About 150 types of mtDNA from around the world were investigated, and their variation determined. Then a computer program was used to connect all the present-day types in an evolutionary tree, reconstructing hypothetical ancestors. In turn, the program connected these ancestors to each other, until a single hypothetical ancestor for all the modern types was created.

The distribution of the ancestors suggested that the single common ancestor must have lived in Africa, and the number of mutations which had accumulated from the time of the common ancestor suggested that this evolutionary process had taken about 200,000 years. This, then, was the birth of the famous 'mitochondrial Eve' or lucky mother, since the common mitochondrial ancestor must have been a female. These results seemed to provide strong support for the 'Out of Africa' model of modern human origins, because the research suggested that a relatively recent expansion from Africa had occurred, replacing any ancient populations living elsewhere, and their mtDNA lineages.

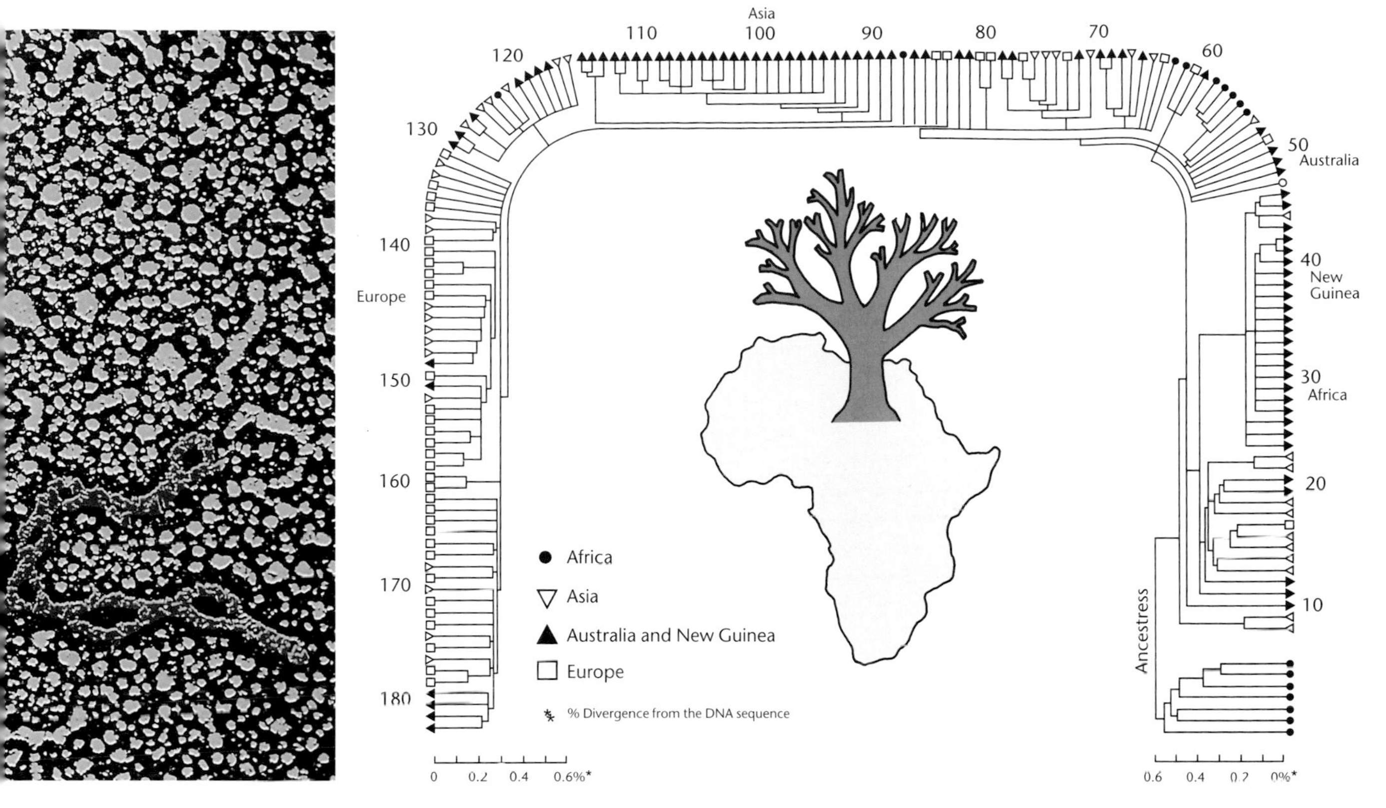

(Above) Evolutionary tree based on maternally inherited mitochondrial DNA. The tree structure suggests the ultimate ancestor – necessarily a woman – lived in Africa. Calculations of mutation rates suggest that this 'African Eve' lived about 200,000 years ago.

Eve under attack

However, the work was soon heavily criticized. It was shown that the kind of computer program used could actually produce many thousands of trees which were all more or less as plausible as the published one, and not all of these alternative trees were rooted in Africa. Moreover, other workers criticized the calibration of the time when Eve lived, while yet others questioned the constitution of the modern samples analysed (for example, many of the African samples were actually from African-Americans). The team involved in the original work admitted that there were deficiencies in their analyses, but they and many other workers have continued to use mtDNA to reconstruct recent human evolution.

The more detailed results obtained since, suggest that even if the 1987 conclusions were premature, they were essentially correct, and a recent African origin for our mtDNA variation is established – indeed, some calculations place the last common ancestor as closer to 150,000 years ago. Moreover, the mtDNA of humans varies across the world far less than is the case in the species of our closest relatives, the great apes, leading to the idea that a recent bottleneck – a drastic drop in population – pruned the variation previously found in our species.

Other mitochondrial DNA studies

The mtDNA of living people is now being used to study a host of unanswered questions in recent human evolution. These include the timing of the first colonization of the Americas, the origin and dispersal of the peoples of Polynesia, and the factors behind the present genetic patterns of the peoples of Europe. In this latter case, the mtDNA results have challenged the prevailing theory that present European patterns mainly reflect the spread of farming populations, and perhaps the associated Indo-European languages, over the last 10,000 years. Instead, the new research suggests that European variation is mainly a result of population growth during the Upper Palaeolithic, at least 20,000 years ago. This is being further tested by attempts at retrieving DNA from Cro-Magnon fossils.

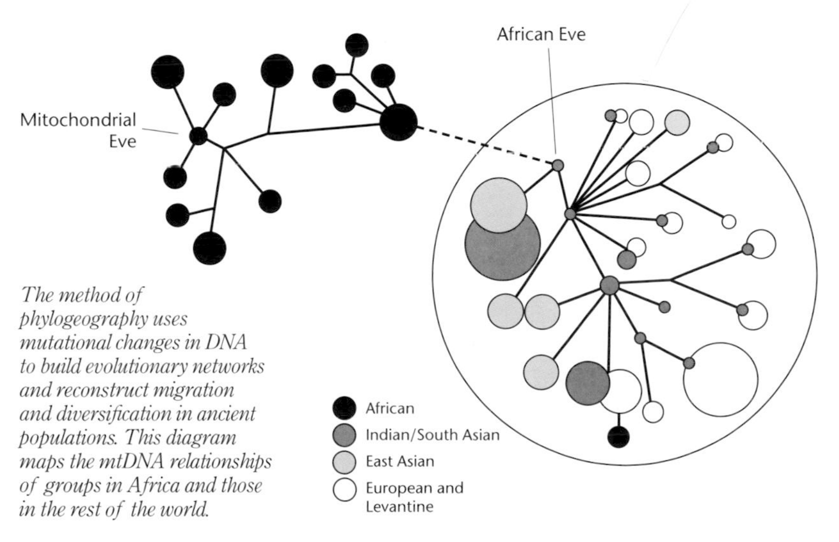

The method of phylogeography uses mutational changes in DNA to build evolutionary networks and reconstruct migration and diversification in ancient populations. This diagram maps the mtDNA relationships of groups in Africa and those in the rest of the world.

Neanderthal DNA

In 1983 a complete human skeleton was found in a cave at Altamura, near Bari in southern Italy. It was covered in stalagmite and apparently represents the remains of an individual who died in the cave during the Pleistocene. Such a complete and associated skeleton of an ancient human is of great importance for reconstruction of the whole body and its functional anatomy. However, sampling of the skeleton for dating and possible ancient DNA has not yet been permitted.

In 1997, teams of scientists working in laboratories in Munich and Pennsylvania contributed to the solution of the Neanderthal problem – the problem of the place of Neanderthals in human evolution – without studying the anatomy of a single fossil. They recovered DNA from the arm bone of the original Neanderthal skeleton found in 1856, and the results of their analyses support the idea that the Neanderthals were a separate human lineage or even species, that died out some 30,000 years ago. The DNA in question was mtDNA, which has proved a powerful tool in reconstructing recent human evolution.

Running the DNA tests

The DNA extraction was painstaking work, carried out in the face of several previous disappointments, because so much previous work on ancient DNA, whether from dinosaurs, fossil leaves or insects preserved in amber (the supposed source of dinosaur DNA in the film *Jurassic Park*) had proved to be questionable for one reason or another. The possibility of fragile DNA chains surviving over many thousands or millions of years had been challenged by some scientists, while others had highlighted the considerable difficulties of distinguishing possibly genuine but minute fragments of ancient DNA from large amounts of more recent contamination, which can even come from the atmosphere in the laboratory, or the skin flakes of scientists handling the fossils or carrying out the DNA extraction. However, the teams of researchers concerned had excellent track records in both successful DNA extraction from extinct species, such as the mammoth and giant ground sloth, and in the careful re-examination and refutation of previous ancient DNA claims which did not stand the test of time. The apparent Neanderthal DNA was replicated independently in both laboratories concerned, and the team conducted every conceivable test to exclude the possibility of recent contamination, particularly from recent human DNA.

Results

The team of scientists first managed to piece together about a fortieth of the whole mtDNA sequence, and have since recovered even more from the fossil. They compared its pattern of genetic coding with those found in about 1,000 people from around the world, and as a more distant comparison, with those found in chimpanzees, our nearest living relatives. The Neanderthal DNA lay closer to the human sequences, but still clearly distinct from them. The Neanderthal pattern was equally distinct from those of each modern population, from whichever continent. So the Neanderthal was no closer to a living European than to an African, Asian or Australian.

This result certainly did not support the idea that Neanderthals were specially linked with Europeans, as partial or complete ancestors. But the researchers were also able to use the differences

(Right) Ancient DNA has been recovered from skin, hair, bones, teeth and sediments. Sterile laboratory conditions must be used to minimize the possibility of contamination by recent DNA, as this illustration of a fossil bone being sampled shows.

between the living human, Neanderthal and chimpanzee DNA sequences to estimate the likely time depth of the Neanderthal line of evolution. Although the Neanderthal fossil has been dated to about 40,000 years old, its separation time from the modern human line is estimated at about 500,000 years. Genes begin to diverge before populations and species do, but this date is long before the estimated start of the divergence of modern human mtDNA types, 200,000–150,000 years ago, and certainly indicates that this Neanderthal could not have been one of our ancestors, given its late date and its genetic and physical distinctiveness. However, on its own, it does not necessarily prove that the Neander Valley individual was from a different species, since the variation it shows from living people in mtDNA is at a level which can be found both within and between living primate species.

This was only one sequence from one Neanderthal fossil. Did it really settle the fate of the Neanderthals? The authors cautiously said that the Neanderthal mtDNA sequence supported a scenario in which modern humans arose recently in Africa as a distinct species and replaced Neanderthals with little or no interbreeding. But they pointed out that other genes might tell somewhat different stories. This is certainly possible because mtDNA is only inherited through females. So any genetic heritage passed on from Neanderthal males to present-day populations, for example, would not be recorded in that particular DNA. Nevertheless, when put together with a growing body of fossil research showing the uniqueness of the Neanderthals, it seems that they could have had only a minimal genetic input, at best, on the modern populations that succeeded them.

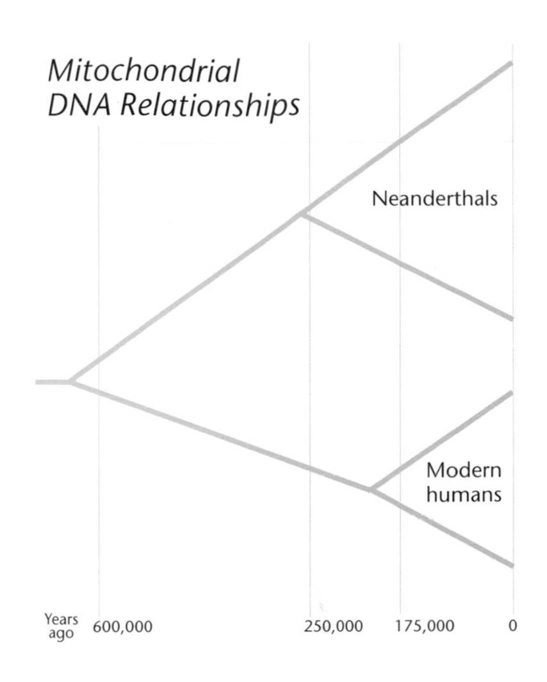

(Right) This schematic diagram of Neanderthal and modern human mtDNA evolution shows that the projected mtDNA diversity of Neanderthals was comparable to that of humans today, but that the populations represent distinct lineages that may have begun to separate in the Middle Pleistocene, perhaps 600,000 years ago. However, genetic separation would have begun within a common ancestral population, so diversification and possible speciation would have occurred after this date.

Further Neanderthal DNA studies

This achievement was a great step forward in human evolutionary studies, and has led on to yet other successes. At least four more Neanderthals have now yielded mtDNA, demonstrating that they had their own population variation, comparable with, but quite distinct from, that of recent humans. However, work on identifying Cro-Magnon DNA has been more difficult, as the molecules involved appear similar to, or identical with, that of recent Europeans, giving real problems of authentication. Similar problems of authentication surround claims for the recovery of mtDNA from early Australian fossils.

(Below) A histogram of the extent of DNA mutational differences between modern humans, between modern humans and Neanderthals, and between modern humans and chimpanzees. In their mtDNA Neanderthals are certainly more closely related to modern humans than are chimpanzees, but variation within chimpanzee populations exceeds that so far recorded between modern humans and Neanderthals.

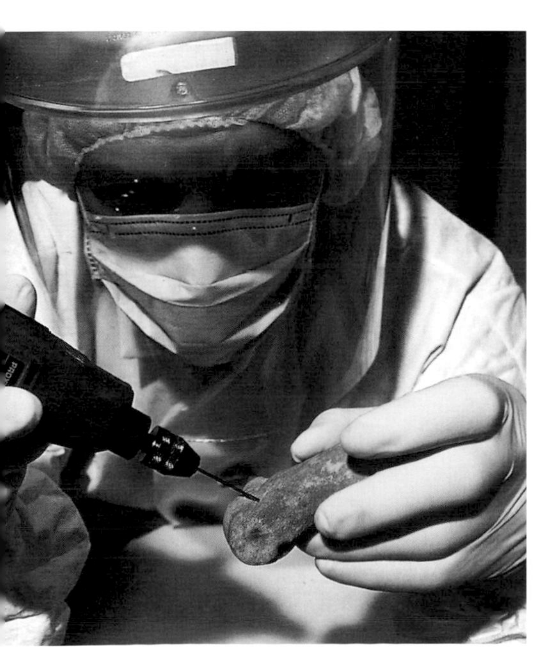

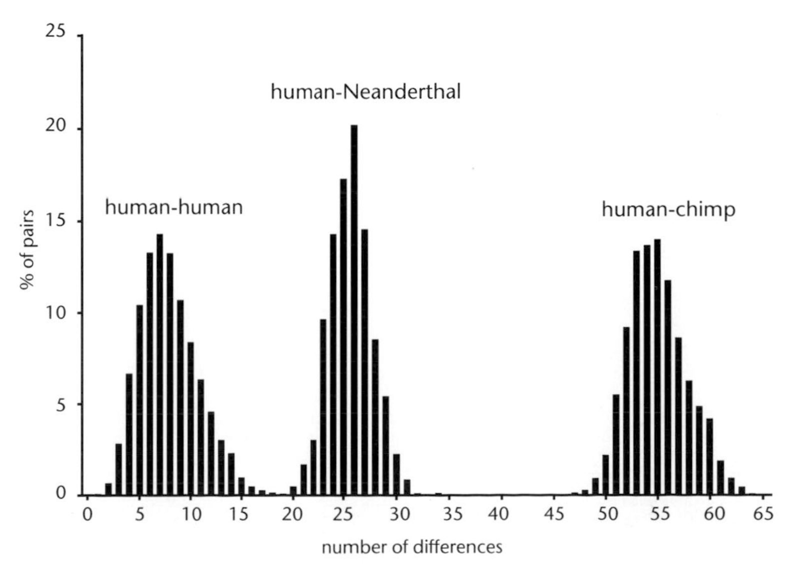

So far in this book we have described the fossil evidence for our ape and human ancestors, and the environmental context in which they have been found. Now we are going to try and put these together in order to interpret all lines of evidence for human evolution. First we will consider the evolution of locomotion, both because this is a key adaptation and because it is uniquely important in human evolution. Hominoids (apes and humans) generally have extreme locomotor adaptations and none more so than humans, for bipedalism or two-legged walking is extremely rare in mammals. We will go on to consider diet, which is also important, for we must eat to live. In this case some of our hominin relatives also show evidence of extreme specialization, with enlarged thick-enamelled teeth, suggesting very abrasive foods, although in this case the feature is shared with some extinct apes. These adaptations evolved in ancestral forms that were considerably smaller than modern great apes and humans, living in tropical woodland/forest environments.

We saw above that the earliest evidence for both ape and human evolution has been found in tropical Africa. In the case of the apes, they must have quickly diversified into a variety of environments ranging from wet lowland tropical forest to woodlands, and eventually to subtropical forests and more open seasonal woodlands. Possible hominin ancestors have been found associated with woodland environments that are as yet poorly defined, although the one feature that seems to be present in all is that the environments were variable, diverse as opposed to uniform, encompassing many different habitats over short distances. Such was the situation in the Afar Depression, for instance, in Ethiopia, and at Laetoli in Tanzania. When the locomotor and dietary adaptations of fossil apes and hominins are interpreted in terms of these environmental constraints, a distinctive pattern for human evolution emerges.

Our knowledge of the evolution of humans is both constrained and illuminated by the development of culture. We can reconstruct something of the probable mental and physical capabilities of early hominins by studying our closest living relatives, the Great Apes. However, no living apes show the ability to make complex tools and to teach those skills to other apes, so the development of sophisticated tools and language must have been a powerful catalyst in later human evolution. By using cultural adaptations, humans have been able to transcend their physical limitations and have eventually colonized every continent and every terrestrial environment on Earth.

The Panel of the Horses in Chauvet Cave, France, which dates from around 30,000 years ago. The horses are given movement and expression with the careful use of shading. Some areas are also emphasized with engraving, and in others it is possible to see where the black pigment has been carefully spread out using a finger.

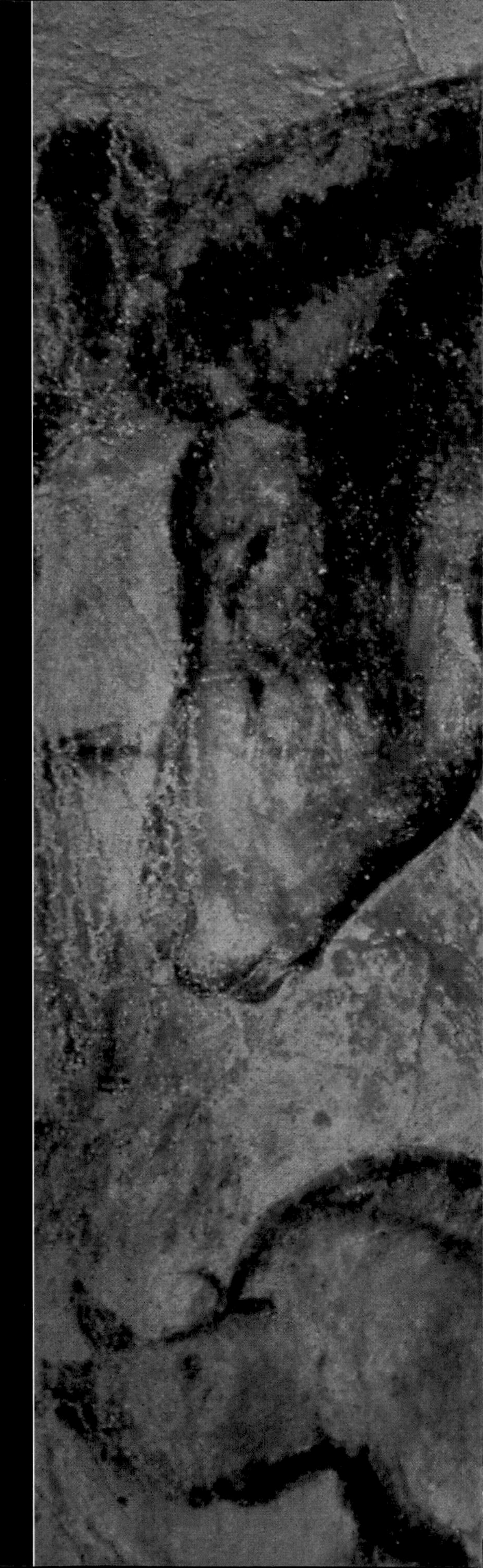

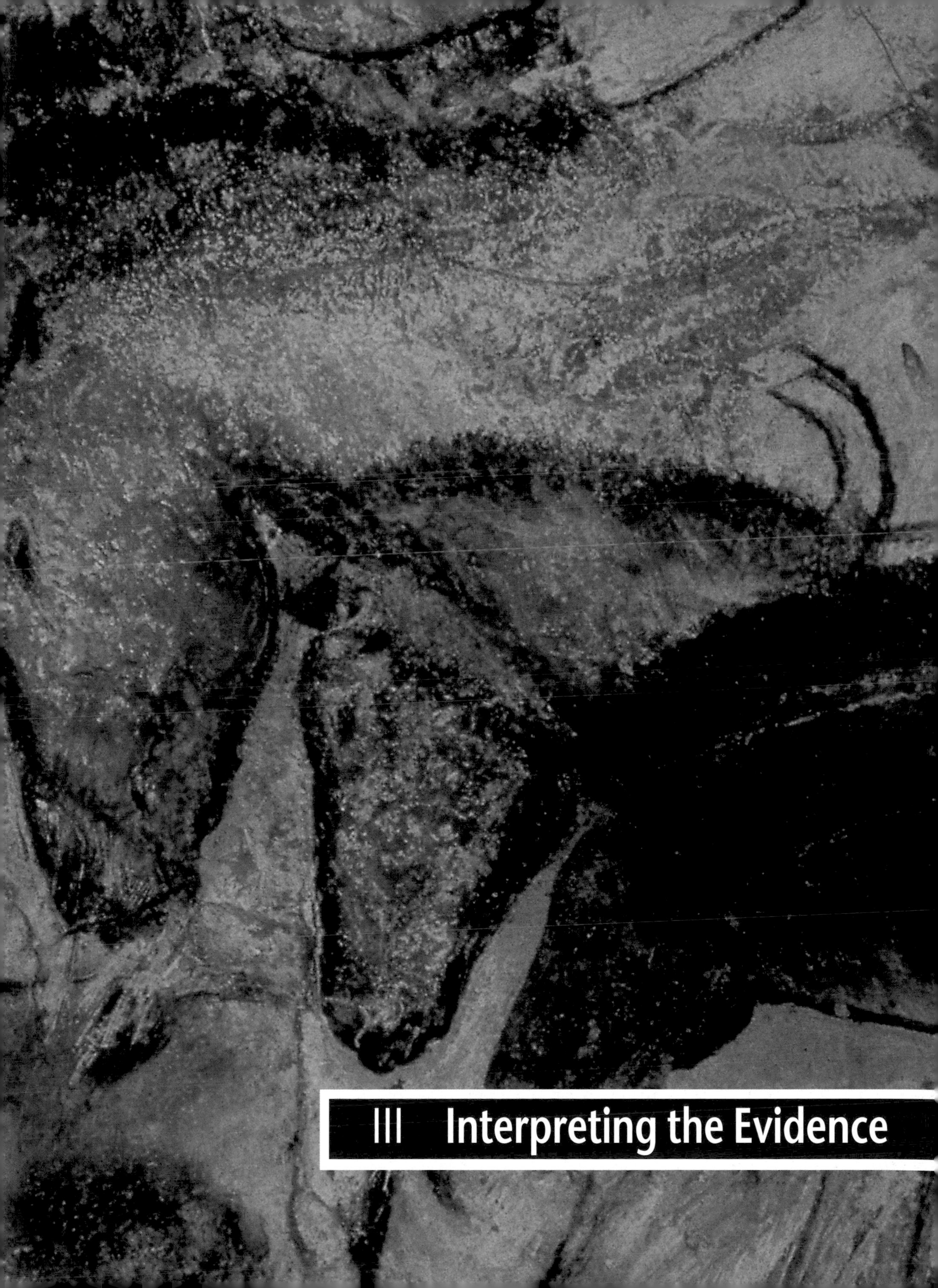

III Interpreting the Evidence

Evolution of Locomotion in Apes and Humans

Apes and humans display an extraordinary variety of forms of locomotion, and all are different from other primates and most other animals. Gibbons have long muscular arms by which they hang beneath branches of trees and move through their forest homes by swinging from one branch to another, sometimes leaping long distances in the process. This form of locomotion is called brachiation and is unique to gibbons. Other apes and a few monkeys, such as the spider monkeys of South America, may hang by their arms, but none uses brachiation as its normal form of locomotion.

Whereas gibbons move extremely rapidly through the forest canopy, orangs move slowly. Orang utans also have long arms, but partly this is due to their large size. As apes and monkeys get larger, their arms get longer relative to their legs. They also have very mobile and agile legs, and this enables them to move around in trees as if they had four arms, and on occasion they can even hang upside down from branches holding on with their legs. Normally, however, they move cautiously from one branch to another holding on with at least three of their 'arms'.

Knuckle-walking in chimps and gorillas

Chimpanzees and gorillas share another unique form of locomotion called knuckle-walking. Their arms are also long because of their large size, and they extend the length of their arms by supporting their weight on the knuckles of their hands, which are also elongated. This raises the front parts of their bodies, so that when you see them walking on the ground they can be in a semi-upright posture with their shoulders higher than their hips. They also spend much of the daylight hours on the ground, with the mountain gorilla being almost entirely confined to the ground.

The knuckle-walking posture of chimpanzees (and gorillas) has the effect of lengthening the arms, already longer than the legs, so that they walk in a semi-upright position.

Different postures seen in upright bipedalism, knuckle-walking and quadrupedal locomotion. Knuckle-walkers are semi-upright, a trend that is seen in all hominoid primates, which are intermediate in this respect between quadrupeds and bipeds.

Human bipedalism

All these forms of locomotion are unusual in the animal kingdom, and humans are equally unusual, walking upright on just two legs. This is called bipedalism, and while a few other mammals have also preferred two legs, they use their legs for hopping, like the kangaroo, not walking. Bipedalism is unusual because it slows us down and at first sight makes us vulnerable to faster-moving predators. Anyone who has taken their dog for a run on the beach or park and tried to have a race with it will have found that even the smallest of dogs can run faster than we can. So in adopting such a restricted form of locomotion humans must have had other means of defence against attack, and yet it seems that bipedalism was one of the first, if not the first, distinctively human attribute to evolve. The adaptations for bipedalism are many and extend to all parts of the body. The head has to be balanced on top of the backbone instead of being slung in front; the backbone has developed curves to withstand stresses and to function as a spring; the hip has broadened and wrapped around the sides of the body to give better leverage to the muscles that maintain us in an upright position; the legs become longer and angled inwards to keep the centre of gravity along the midline of the body; and the feet have developed arches and the big toe has rotated in line with the rest of the foot to provide additional leverage.

One thing is held in common by all the living apes, with some left over effects in humans. This is a series of adaptations of the arm and shoulder for suspending the weight of the body beneath branches of trees rather than walking on the tops of branches. This type of suspension necessitates an upright posture that is quite different from the prone position adopted by most other primates and other mammals. It appears late in hominoid evolution, however, and most fossil apes lack any of the specializations that distinguish brachiation, knuckle-walking or bipedalism. *Proconsul*, for example, was a generalized four-footed ape that lived in trees in the tropical forests of East Africa, and the adaptations present in *Proconsul* were retained with only minor changes in many of the later apes.

Adaptations for ground-living among fossil apes

One modification that was important for later fossil apes was for ground-living. *Kenyapithecus* in East Africa and the closely related *Griphopithecus* in Turkey, both living 14 to 15 million years ago, showed changes in their limbs that showed that they had some adaptations for life on the ground while still retaining much of their tree-living *Proconsul*-like ancestral characters. This change was almost certainly because the habitats they were living in were more seasonal and therefore more variable, and the forests or woodlands they occupied were more open and less rich in food and structure. In such environments it is difficult for animals larger than 3–5 kg to move through the tree canopy and they have to come down to the ground. Similar patterns were maintained in some later fossil apes as well, particularly in the probable orang utan ancestor, *Sivapithecus*, from India and Pakistan. *Sivapithecus* is thought to be ancestral to the orang utan because of certain skull and facial features, but it did not have any of the adaptations

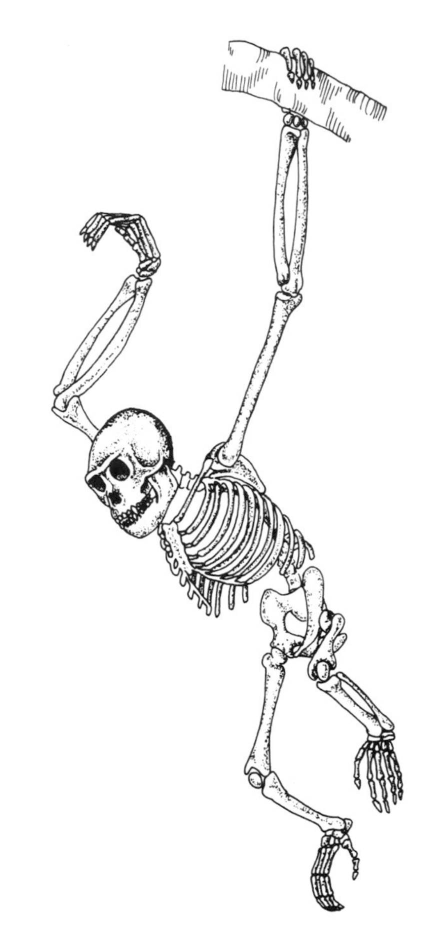

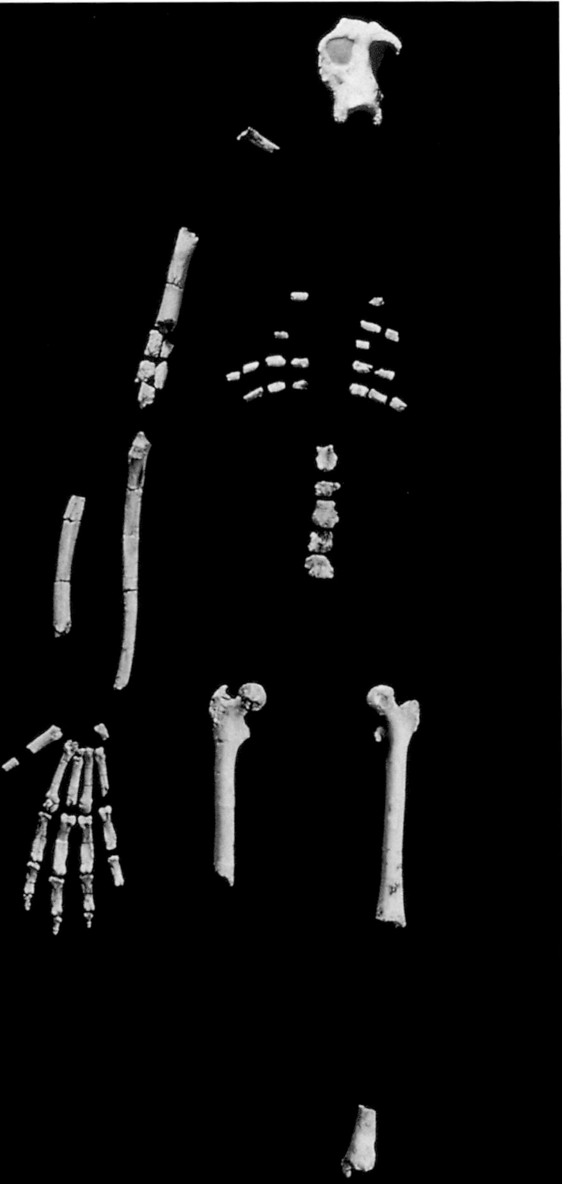

(Above) The skeleton of Sivapithecus *with known parts shown in black. Its quadrupedal posture and many aspects of its anatomy are little changed from that of* Proconsul *living some 10 million years earlier, but both differ from the suspensory adaptations of* Dryopithecus *and* Oreopithecus.

(Right) The limb proportions of Oreopithecus *and its short trunk show it to have had a degree of suspensory behaviour, moving through its forest habitat by suspending itself below branches with its long and powerful arms.*

(Left) The skeleton of Dryopithecus *from Can Llobateres in Spain is one of the best known. Like* Oreopithecus, *it had long and powerful arms, and the shape of the hip joint indicates that it had high mobility of the legs. In both respects it is similar to the living orang utan.*

of the limb bones that today distinguish the orang utan. This is puzzling, for *Sivapithecus* lived from 12 to 7 million years ago, long after the time when the orang utan is thought to have diverged from the other apes, and this could indicate that in fact *Sivapithecus* was not an orang ancestor.

As well as these ground-living fossil apes, there existed two fossil apes, probably related to each other, that had developed some degree of suspensory behaviour. They were *Dryopithecus* and *Oreopithecus*, and both had adaptations of the upper arm and shoulder which are very similar to those of the orang utan and which first appear in the fossil record between 10 and 9 million years ago. These characteristics are close to the condition from which all the living apes and humans must have evolved, but neither they nor any other fossil ape have the particular sets of characters that would indicate an early stage of evolution of brachiation (in gibbons) or knuckle-walking (in chimpanzees and gorillas). It has been suggested that *Oreopithecus* may have been partly bipedal, although this evidence is still controversial and is not directly relevant to the emergence of human bipedalism. For this there is evidence of a most striking kind, for not only are there fossil foot bones that are similar to our own, but there are also fossilized footprints that show beyond any doubt that early humans were walking upright on two feet.

(Right) The fossil hominin footprints at Laetoli. The only hominin known from this locality is A. afarensis, *which on skeletal evidence walked bipedally, and it is thus the most likely culprit. The weight transfer of* A. afarensis *when walking is shown below by contouring the depth of the impressions made by the feet. Like human walking, the main stress is on the ball of the foot and the heel, with secondary stress on the outside of the foot in line with the little toe.*

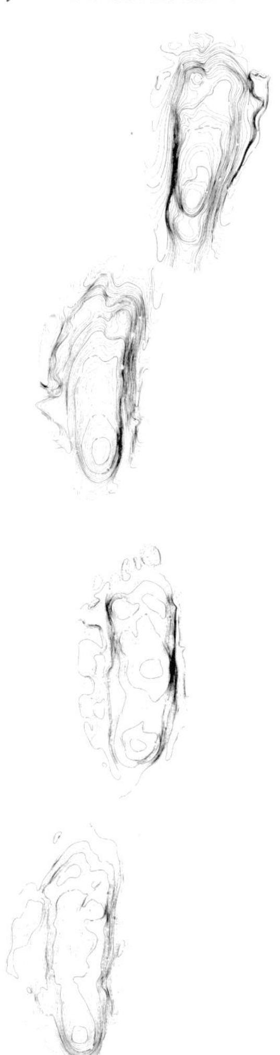

The Laetoli footprints

The evidence from the footprints comes from Laetoli, a fossil site in Tanzania nearly 4 million years old. They show the tracks of three of our human ancestors (identified as *Australopithecus afarensis*) walking across an area of open ground at a time when a nearby volcano was emitting clouds of ash, which settled on the ground in successive fine layers. Almost as soon as the footprints were made in the ash they were covered over and filled in by more ash slightly different from the earlier ash, and the whole lot became hardened by a series of light rainfalls coinciding with the ashfalls. This difference in texture of the ash has enabled archaeologists today to excavate out these footprints to show the trackway illustrated here, along with scores of other footprints of animals living at the same time to produce a unique record of life in Africa for a few days nearly 4 million years ago. The ash even records the rainfall at the same time.

The Laetoli footprints show that *A. afarensis* undoubtedly walked upright, but unfortunately there are no actual fossil bones of the feet of these fossil humans to show what they looked like. There are pieces of jaw and teeth, but they do not help much when it is their form of locomotion we are interested in. The lack of fossils and disagreements over the interpretation of the footprints have led some anthropologists to conclude that the form of bipedalism practised at Laetoli was different from that of living humans, while other anthropologists conclude the opposite, that it was exactly the same. At an earlier site in northern Kenya, Kanapoi, there are some limb bones that appear more modern than those of *A. afarensis*, and while undoubtedly bipedal, the exact form this took is not known at present.

Bipedalism in Lucy

The earliest fossil remains complete enough to investigate bipedalism comes from the partial skeleton of Lucy from Hadar in Ethiopia. This is also attributed to *A. afarensis*, and it shows that Lucy had different body proportions from modern humans, with short legs proportioned more like apes. On the other hand, her hip bone was broad and flared, more like humans than apes. The hip is important in bipedalism, because as every child knows, one of the main difficulties in learning to walk is keeping your balance. This is achieved both by developing the muscles connecting the hip bones to both the legs and the backbone and by improving their leverage by increasing the distance of their position of origin from both legs and backbone. The first is done by buttressing the hip with crests and pillars, and the second by broadening the upper part of the hip, the iliac blade. Both adaptations are present in australopithecines to some extent, and although both the buttressing and the broadening are less than in modern humans, there is some evidence of increase in later species.

The knee joints in even the earliest australopithecines are very similar to modern humans, showing that they were definitely bipedal. Most australopithecines, however, have femora (upper leg bones) with small heads and long necks in marked contrast to both humans and apes, and this seems to

(Below right) The angle of the upper leg is similar in Lucy (A. afarensis) and modern humans, and both differ from chimpanzees. This angle brings the centre of gravity of the upper body down the mid-axis of the hips instead of being offset as in chimpanzees.

(Below) Lucy is very small when compared with a modern human (right). Both are fully upright but differ in the shape of the pelvis and the length of the legs, which suggest that Lucy could not stride out like modern humans but walked with more of a shuffle-like gait.

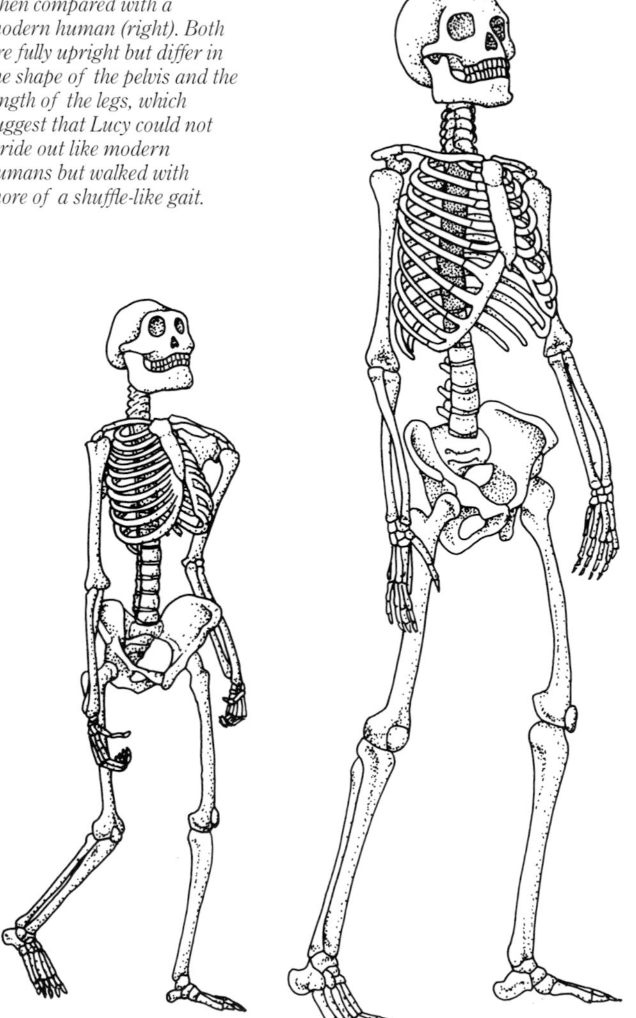

Lower Limb Geometry

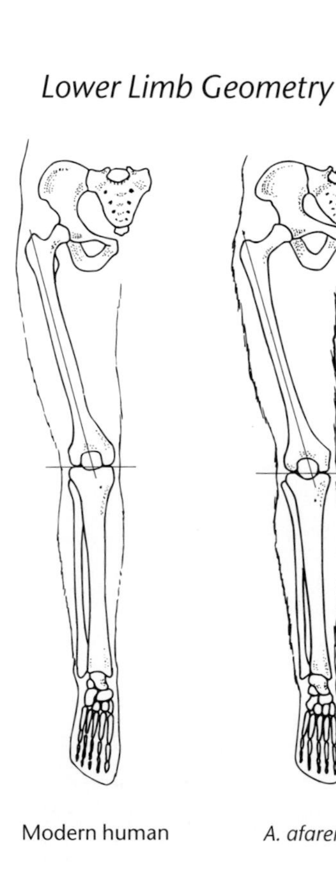

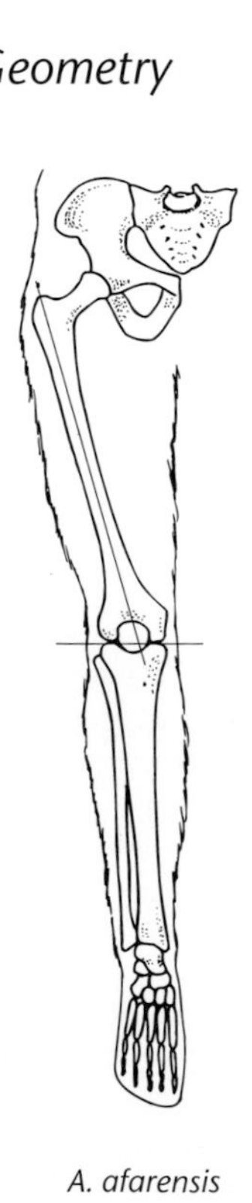

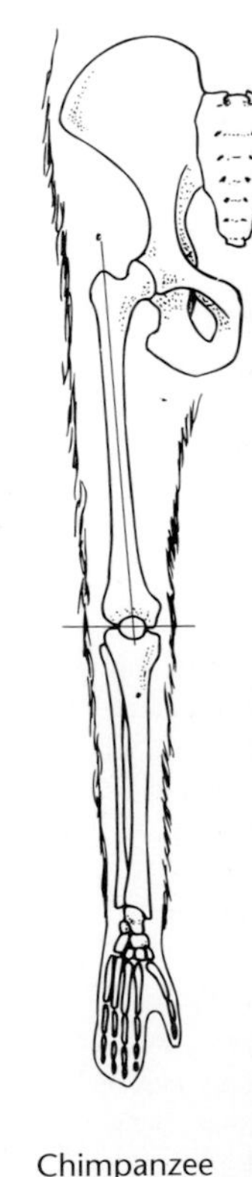

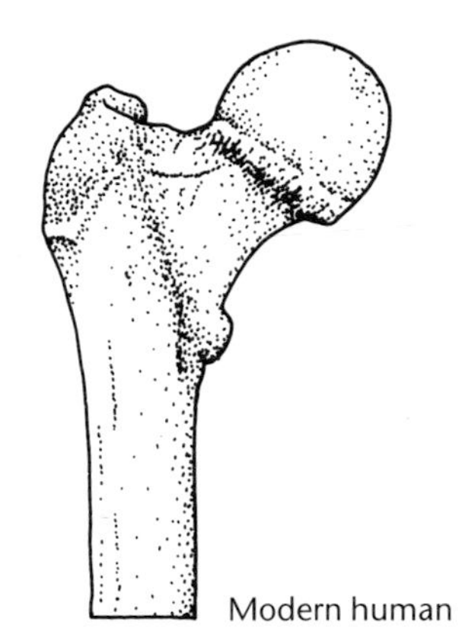

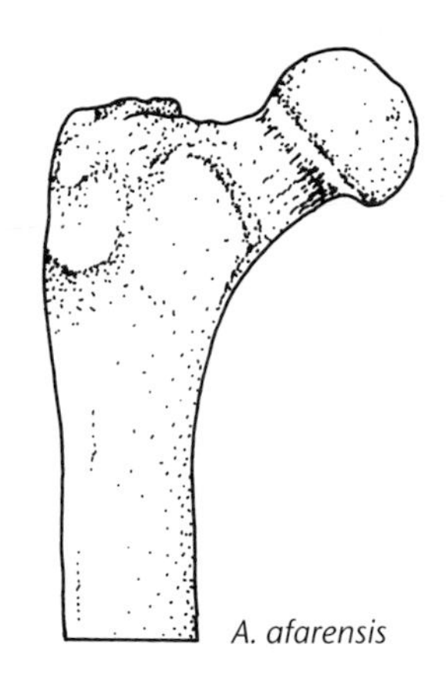

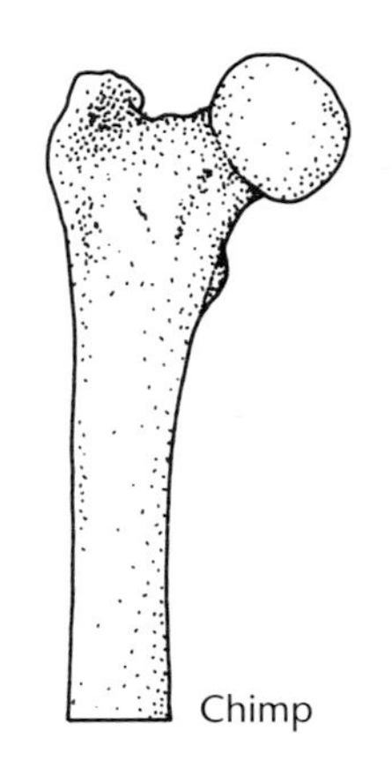

Proximal femora of human, A.afarensis *and chimpanzee. The neck of the femur is more elongated in* A.afarensis, *unlike both chimpanzees and humans, but it retains other characteristics similar to chimpanzees (and some Miocene apes).*

indicate a different form of weight support carried by the legs. It is interesting that *A. afarensis* lacks the extreme development of these later australopithecine characters, and it could be that it is intermediate in this respect between them and an earlier chimp-like ancestor. This ancestor may well have been found in the shape of *Orrorin*, 2–3 million years earlier than *A. afarensis*, for this fossil had a femur with a large head and short neck, just like later Miocene apes and chimpanzees. Finally, there are features of the hand, such as long curved finger bones producing a grasping hand, and the shoulder, the joint of which is directed upwards for suspension from branches as in apes, that suggest that *A. afarensis* retained a significant adaptation to tree climbing inherited from its ape ancestors.

Australopithecus afarensis: ground-living and tree-climbing abilities

What appears to have happened at this early stage of human evolution was the development of adaptations for bipedalism in the legs and the retention of ape-like characters of the arm for suspension in trees. In other words, *A. afarensis* had the best of both worlds: a new form of locomotion on the ground combined with retained abilities for climbing trees. In evolutionary terms, it is the new adaptation that is important, for that shows relationship with later humans, but in functional terms it is significant that tree-climbing was clearly still important to early humans, for that formed their best line of defence against predators. Two aspects of these adaptations will be explored later in this book: retention of tree-climbing ability only makes sense when there are trees to be climbed, and significantly it will be seen that early humans are found associated with woodland or even forest environments, not open plains; and the development of bipedalism was probably associated with behavioural and social changes by which the disadvantages of bipedalism as a form of locomotion were more than countered by advantages resulting from freeing of the hands for tool use and toolmaking.

Locomotion in early *Homo*

The early members of the genus *Homo* showed some advances in the way they moved compared with the australopithecines. Limb proportions were more similar to modern humans in most species, although not apparently in *Homo habilis* if the identification of OH62 from Olduvai is correct (see p. 135). The thumb joint was typically human, although the hand was still capable of powerful grasping. The legs had lengthened, the lower leg and foot were very like the modern human condition, the broadening and flare of the hip joint was also more like modern humans, but the neck of the femur (thigh bone) was proportionally as long as in australopithecines. It may only have been in the last 1 million years that essentially modern locomotor adaptations evolved in the human line.

(Below) Two A. afarensis *individuals make footprints in the recently fallen ash at Laetoli nearly 4 million years ago. The previously vegetated region has been decimated by the volcano, and the two australopithecines would have been very exposed to predators whilst in the open like this.*

The Evolution of Feeding

We have just seen the remarkable variety of methods of locomotion practised by apes and humans (pp. 184–189). In contrast to this, the feeding behaviour of these animals share many similarities. All apes and humans are primarily fruit-eaters, what scientifically are called frugivores, with just a few species eating leaves as well. Chimpanzees and humans also eat animal flesh, but neither has any biological adaptations for catching and eating other animals.

(Below) Chimpanzees use their large front teeth in preparing food for ingestion. Large incisors are characteristic of fruit eating primates since there is greater need for initial preparation of fruits than there is for vegetation or insects.

(Below right) Microwear on the surface of an orang utan molar. There are many large pits resulting from the hard fruit diet of this ape, similar to the pattern seen in some fossil apes such as Griphopithecus alpani *(see p. 102).*

Gibbons

Among the living apes, gibbons are frugivorous except for the largest species, the siamang which is partly folivorous, which means leaf-eating. Its teeth have more projecting cusps and ridges which are adaptations for cutting up the tough leaves, whereas other gibbons have low-crowned teeth with thin enamel covering for crushing soft fruits.

Orang utans, chimpanzees and gorillas

Orang utans are also almost entirely frugivorous. Their teeth are also low-crowned, but they are curiously ridged with thicker enamel than is present on gibbon teeth. The function of the ridging on orang teeth has never been satisfactorily explained, but it does not seem to be related to its diet. Chimpanzees again are frugivorous with diets quite similar to the orang utan, but their teeth are much more simple and lack both the ridging and thickness of the enamel. Both chimpanzees and orang utans eat some leaves, flowers and buds when fruit is not available, but gorillas are different in that they have a much greater leaf component in their diet throughout the year, with mountain gorillas eating almost wholly leaves. As would be expected with such a big difference, gorilla teeth are adapted for cutting up the tough leaves by having more strongly developed ridges, as in the siamang.

Fossil apes

We know quite a lot about the diets of fossil apes because teeth are what we use for chewing food, and teeth are the best-preserved body parts found as fossils. The types of teeth found in living apes have been described above, but there is other evidence as to the nature of diet as well: when we eat, food particles leave microscopic scratches on the surface of our teeth. By examining similar scratches on the teeth of fossil apes with a scanning electron microscope, it is often possible to identify what they had been eating. Analysis of carbon isotope differences in tooth enamel also provides information about diet.

As for the living apes, so for the fossil apes: they mainly ate fruit. This was the case for *Proconsul*, for example, although another ape living at the same time and place was at least as folivorous as gorillas. This was *Rangwapithecus* which in other respects was similar to *Proconsul*. It differed in that its teeth had more developed ridges and analysis of tooth wear indicates that it ate leaves rather than fruit. Many later fossil apes were similar to *Proconsul* in their diets, such as *Dryopithecus*, but later in the Miocene another form of fossil ape appeared. The first examples of this new form, *Afropithecus* and *Kenyapithecus*, appeared in Africa and their relatives

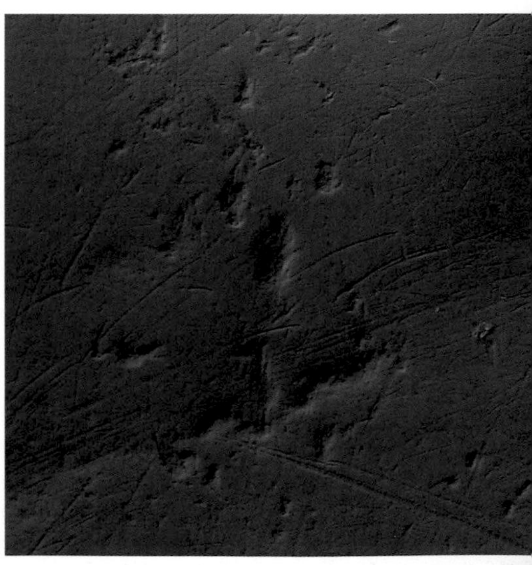

were the ones that first left Africa and moved into Europe and Asia. What these fossil apes have in common is enlarged teeth with low crowns and thick enamel, and in this respect their teeth appear human-like. In fact several of these thick-enamelled apes have been claimed as human ancestors in the past. The wear on their teeth indicates that they had a diet of hard fruits, nuts and seeds.

The fossil apes with large teeth and thick enamel may have given rise to humans, but one branch more certainly was ancestral to the orang utan. *Sivapithecus* is generally thought to be related to the orang utan, and it had teeth and diet very similar to those of *Kenyapithecus* and *Griphopithecus*. All these fossil apes had feeding habits that were probably very similar to the orang utan – in this sense the orang utan is very conservative and has changed little from its ancestral diet.

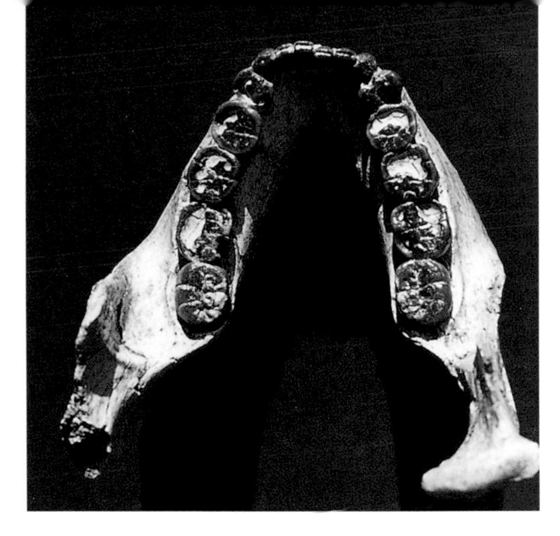

The mandible of Paranthropus boisei, *with huge molars and premolars for grinding up food, and greatly reduced front teeth. This is a pattern seen in primates that eat small hard objects like seeds or smaller fruits that do not need any initial preparation.*

The australopithecines

Some of the apparent human ancestors like *Australopithecus afarensis* and *Orrorin tugenensis* had thin enamel on their teeth, whereas others such as *Australopithecus anamensis* had teeth that were just like those of the thick-enamelled apes: they were large, with flattened crowns and thick enamel on their surfaces. The wear patterns on their teeth resembled those of the thick-enamelled apes. The later australopithecines took this adaptation further by exaggerating several of these features – for example, by increasing enamel thickness still further – and they reduced the size of their front teeth and expanded their back teeth. They even increased the sizes of their milk teeth, so that an ability to eat the coarse, hard food on which they lived must have been of enormous importance to them.

The diet of *Homo*

If the trend in earlier stages of human evolution was towards larger teeth and thicker enamel, the later trend was just the reverse. *Homo habilis* had teeth that were not markedly different from those of the australopithecines, but they did not exhibit the heavy wear from hard, coarse foods that was present on the australopithecine teeth. They were evidently eating a higher-quality food, like fruit and probably meat, which they may have scavenged from predator kills. *Homo erectus* had smaller teeth still, and it had probably become an accomplished hunter so that meat was an increasing part of its diet. This trend continued to the time of the agricultural revolution around 10,000 years ago, with further tooth reduction spurred on by man's development of cooking, which renders tough or hard food soft and more easily chewed. For a while after agriculture was developed, tooth wear increased, reversing the trend, because even when cooked, cereal grains have abrasive particles in them, but now we eat processed food our teeth are once again reducing.

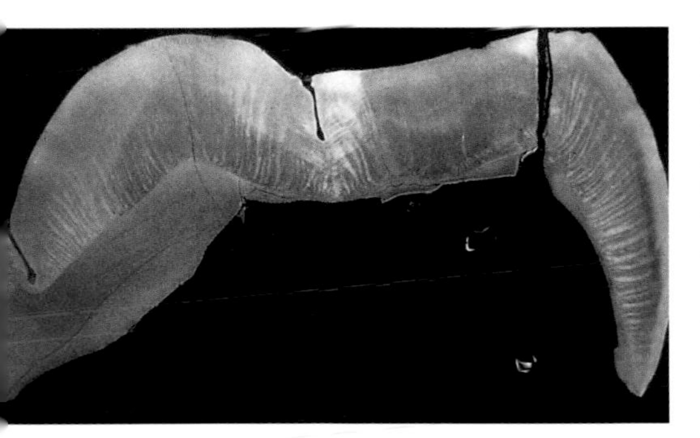

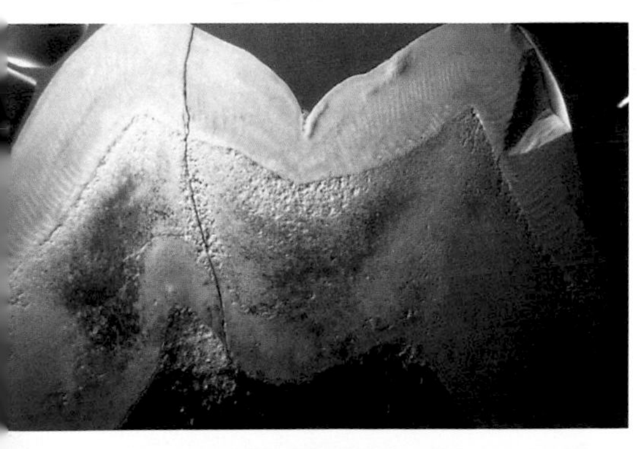

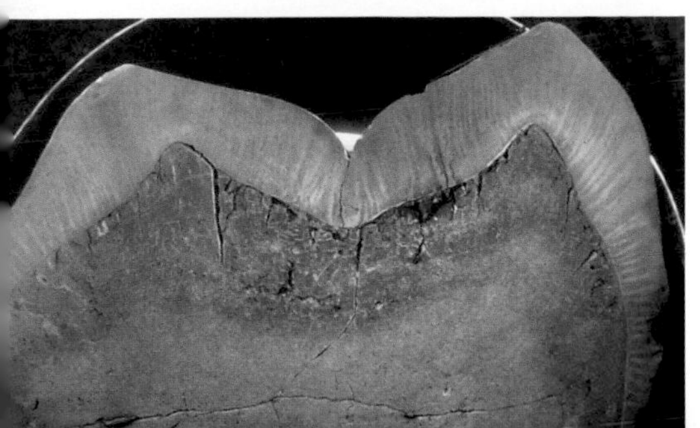

(Left) Sections through teeth of Graecopithecus, Australopithecus *and* Sivapithecus. *All have thick enamel and similar cusp morphology of the molars, and it is likely that they all had similar diets.*

(Below) Two different diets, one based on fruit (frugivorous) and one on leaves (folivorous) can produce major differences between species: for example, fruit has more concentrated food value, especially carbohydrates, resulting in shorter guts and different tooth types, but the source of the fruit is also usually more scattered, so that fruit eaters have to remember the locations of fruiting trees and when they are likely to be in season.

Fruit eaters v. leaf eaters

	frugivores	*folivores*
protein	low	high
fat	low (seeds high)	low
carbohydrates	high	low
gut dimensions	small	large
food properties	deformable hard or soft	tough fibrous
tooth structure	flat tooth surfaces thick enamel	teeth with ridges thin enamel

The Geographical Spread of Apes and Humans

Since the second half of the last century, it has been recognized that humans probably first evolved in Africa. There was no fossil evidence for this, but it was recognized by Darwin that chimpanzees and gorillas are the hominoid primates most closely related to humans, and since these apes live in Africa, this was likely to be the place where humans evolved. This view was championed and elaborated by 'Darwin's Bulldog', the zoologist and anatomist T.H. Huxley, who first made explicit both the relationship of humans and apes and their origin in Africa.

Curiously enough, there actually was one fossil ape known at the time Darwin published his *On the Origin of Species* in 1859. This was *Dryopithecus*, discovered in the south of France in 1856, but it was not until the end of the century and the early years of this century that additional fossil apes were found, and then they were discovered in India. These fossil apes are all quite late in time, in deposits 12 to 8 million years ago. It has been suggested that the anthropoid primates, the group including all monkeys and apes, may have originated in Asia, for early Eocene fossils with apparent anthropoid affinities have been found in China in recent years. No fossil anthropoids are known from Africa during this time, but they occur in North Africa during the late Eocene and Oligocene nearly 30 million years ago, from Oman and northern Egypt (see pp. 84–87). The absence of deposits in Africa makes it difficult to determine if the earlier absence of anthropoids is real or an artifact of collection.

Apes evolve in Africa

The first apes from Africa were found in western Kenya, and when the specimen was shown to A. Tyndall Hopwood, then a palaeontologist from the Natural History Museum in London, he visited Kenya in 1931 where he collected several additional specimens from a site called Koru. This was the first indication that there were fossil apes in Africa, but later work by Louis Leakey and his assistant Donald MacInnes was to show that there was a great proliferation of fossil apes in Africa long before any were present in Europe and Asia. In other words, Leakey confirmed that apes evolved in Africa and were common there for at least the first 10 million years of their evolutionary history, from 26 to 14–16 million years ago. Many species of *Kamoyapithecus, Proconsul, Rangwapithecus, Nyanzapithecus, Kenyapithecus, Equatorius, Morotopithecus* and *Afropithecus* are now known from this time in East Africa, and they were probably as abundant and widespread as monkeys are today in Africa.

Fossil ape migration from Africa

When fossil apes did come to leave Africa, they probably did so at least three times. The first emigration was for the line leading to the gibbons, and nothing is known of this except for the present-day existence of gibbons in eastern Asia. No fossils are known until the Pleistocene, and molecular evidence suggests that the divergence of living species took place in the late Miocene. The second emigration was by the thick-enamelled apes, first represented outside Africa by the kenyapithecine *Griphopithecus* from middle Miocene deposits in Germany, the Czech Republic and Turkey. The date for these fossils was close to 15 million years ago, similar in age to African representatives of the Kenyapithecinae. Later Miocene genera like *Sivapithecus, Gigantopithecus, Ankarapithecus* and *Graecopithecus* spread across eastern Europe, the Middle East and Indo-Pakistan, and it is likely that part of this lineage led to the orang utan.

The third emigration, which seems to have been quite separate from the first two, was the one leading to *Dryopithecus* and *Oreopithecus*. Like the first, this was a dispersal of tree-dwelling apes, whereas the thick-enamelled apes probably had some degree of ground-living capability as well. There may have been other emigrations from Africa, but in the meantime it is assumed that evolution was continuing in Africa with the line leading to the African apes and humans. Unfortunately there is a long gap in the fossil record in Africa from 14 to 6 million years ago, which is just the time

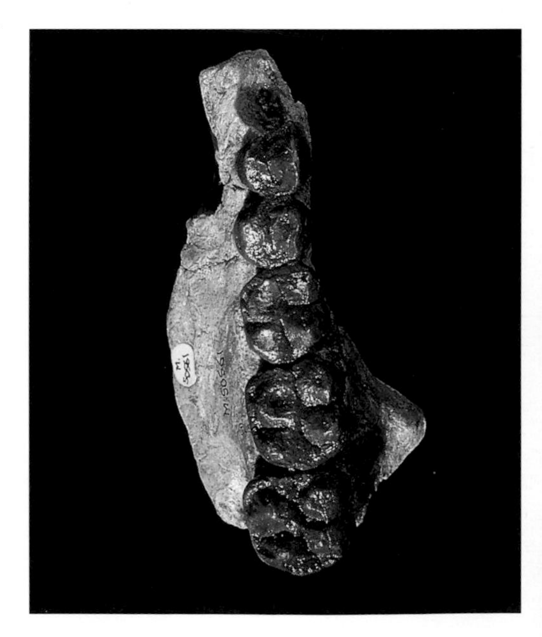

Maxilla of Samburupithecus kiptalami *from the Samburu Hills in Kenya. The teeth of this fossil ape are very similar to those of gorillas and differ from those of chimpanzees and humans. Unfortunately this is the only specimen known at present, and the evidence is too weak to establish whether this similarity has evolutionary significance.*

There is evidence of at least two emigrations of fossil ape species out of Africa during the Miocene, neither of which account satisfactorily for the gibbons and orang utan in Asia. The earliest emigration for which there is evidence is that of the thick-enamelled apes entering Europe and later extending their range eastwards around 15 million years ago, and the other is the migration of suspensory hominids into Europe about 3 million years later.

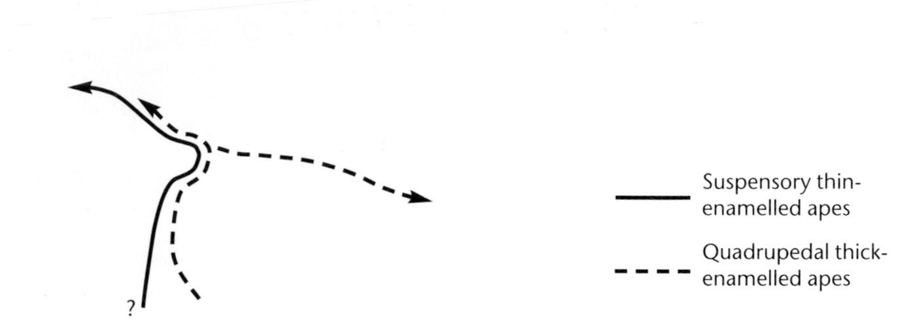

when humans and apes would have been diverging, and so we have little evidence of what was happening. There is one fossil ape called *Samburupithecus* (ape from Samburu) from 6-million-year-old deposits that is very gorilla-like, which indicates that apes were present in Africa, although it is too incomplete to identify its relationships. There is also the similar aged fossil hominid from the Kenya rift valley mentioned above, *Orrorin tugenensis*, which is generally ape-like although there is some evidence of hominin affinities. The same can be said of the recently discovered skull from Chad, *Sahelanthropus*, definitely a hominid but its hominin affinities still uncertain. It is clear that there was some diversity of fossil hominids 7–6 million years ago in Africa – three fossils representing three genera – but how they are related to each other or to living hominids is far from clear.

Human ancestors evolve in Africa

Be that as it may, we do know that the earliest close relatives of humans came from Africa, ranging in age from 5 to 2 million years ago. They occupied large parts of Africa, and they are known from East Africa, South Africa, Malawi and Chad during this time period. They are best known from eastern Africa, particularly Ethiopia, Kenya and Tanzania, with a fragment of mandible from Lothagam at about 5 million years ago and the material from Aramis slightly later at 4.4 million years ago. By 4 million years ago the first australopithecines are known from Kanapoi, Kenya, and between 4 and 3 million years ago they are known from Chad in the north, Ethiopia, Kenya and Tanzania. By 3 million years ago the australopithecines are present in South Africa at several sites. The major factor that was probably limiting their distribution in the Pliocene was the environment, for at this stage of their evolution the early hominins were still adapted for woodland conditions. Hominins were probably in other parts of Africa as well, but no fossils have yet been found.

'Out of Africa'

After 4 million years ago, the australopithecines evolved into at least two distinct groups, the gracile and robust forms, as well as variants such as *Kenyanthropus* and *A. garhi*. By about 2.5 million years ago, the first putative species of the genus *Homo* had appeared, although it is still unclear what they evolved from. Both *Homo* and the australopithecine evolutionary lines were still restricted to their original homeland, Africa. Soon after that date, however, people made the first dispersal from Africa, an event sometimes known as 'Out of Africa I', although in fact this was at least the fourth emigration of hominoids from Africa. These humans may have belonged to the species *Homo erectus* (for some scientists the more primitive ancestral form *H. ergaster*) or perhaps even species related to *H. habilis* and *H. rudolfensis* (see pp. 132–135). Initially, they must have remained in the tropical or subtropical environments to which they were adapted. Their initial move out of Africa must not be seen as a purposeful pioneering colonization of new territories, but rather as a gradual extension of their foraging ranges, tracking plant or animal resources into previously uninhabited territories. Later, as human capabilities grew, their descendants adapted to new environments, such as temperate grasslands and woodlands. Later still, after about 500,000 years, populations managed to

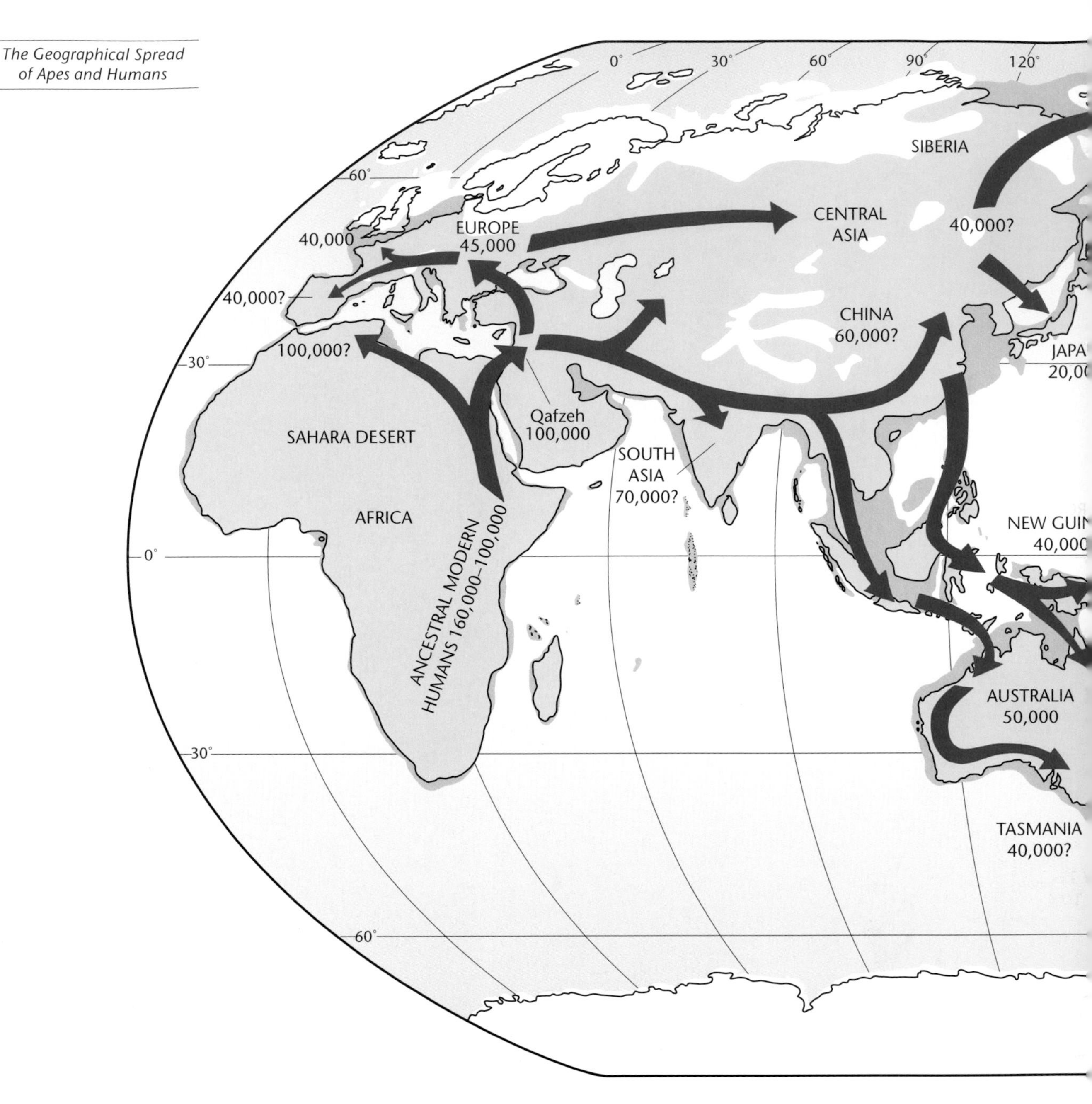

First colonization of the world by modern humans, with approximate dates (years ago) and ice sheets/low sea levels of c. 18,000 years ago. Modern humans first emerged from Africa at least 100,000 years ago, when their fossil remains appear in sites in Israel such as Qafzeh. By 12,000 years ago Homo sapiens *was established in regions as far apart as Tasmania and Chile.*

persist in northern Europe and Asia during cooler climatic stages, gradually adapting to steppe conditions. However, it seems that even the Cro-Magnons were not able to sustain their occupation of northern Europe at the peak of the last Ice Age, about 20,000 years ago, and regions of northern Europe and Asia must have been completely depopulated at such times.

During periodic warm stages, when the climate was like our present interglacial (the Holocene), human numbers probably grew into the hundreds of thousands, and regions such as central Africa, southern Europe, India, and Southeast Asia, were probably relatively densely populated. But at that time, the Earth's ice caps were small, and global sea levels were high. Thus regions like the British Isles, Sicily and Japan were periodically cut off as islands from neighbouring larger land masses, to be rejoined with them again as ice age conditions returned. These corridors for human movement were periodically created and destroyed. Similarly, during glacial stages the sea level fell up to 100 m (330 ft), producing new coastlines and land bridges, but this increase in land area was balanced by the spread of ice caps, snowfields and deserts. Some regions have never been interconnected during human history, however low the sea level fell. Thus Madagascar has remained separate from Africa,

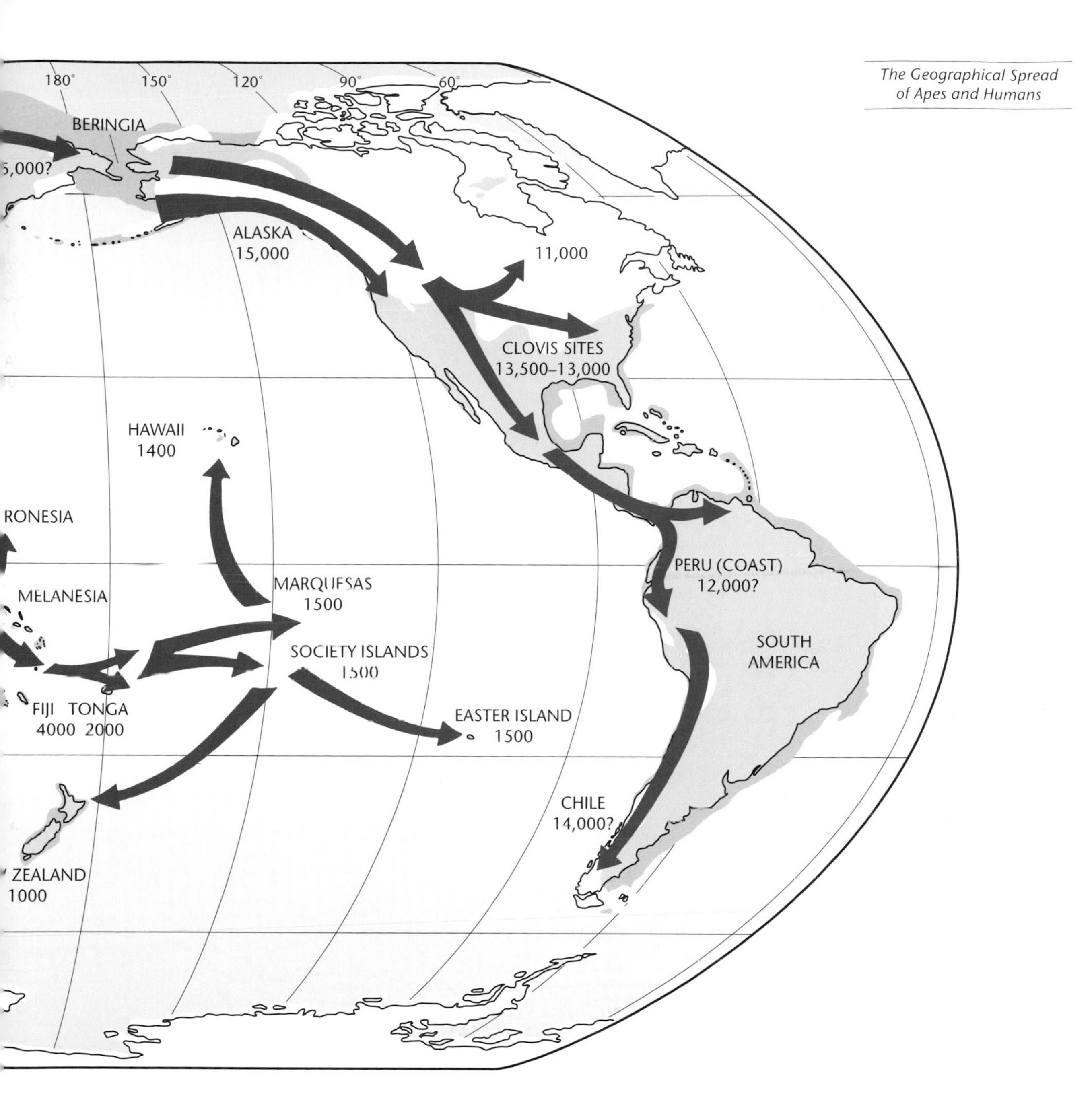

New Zealand from the rest of the world, and Australia and New Guinea from Southeast Asia. These regions were apparently never colonized by early humans; they remained uninhabited until the arrival of modern people.

The spread of modern humans from Africa, the so-called 'Out of Africa II', began at least 120,000 years ago, with an extension of early modern people from North Africa into the Middle East. It appears they subsequently spread eastwards through subtropical and tropical regions to eventually reach Australia by boat (see pp. 170–173) by 50,000 years ago. Europe was probably entered about 40,000 years ago, but there is only sparse data for the other regions of the world outside Africa. There is evidence of modern people in Sri Lanka about 30,000 years ago, and Japan some 20,000 years ago. Early modern people were certainly in China by about 30,000 years ago, and possibly arrived much earlier, but their time of arrival in the Americas is hotly disputed (see next section).

The spread of humans into the more extreme climates of the world, such as the polar deserts and high mountains, took place after the end of the last Ice Age, some 12,000 years ago, and it was only during this period that the more remote islands such as Madagascar, New Zealand and Polynesia were eventually reached by human colonizers.

The First Americans

(Above) The distinctive Clovis industry appeared in the archaeological record of the Americas about 13,000 years ago. The distinctive lanceolate points, first found near Clovis, New Mexico, were thinned using a hammer of bone or wood, and fluted at the base to fit a spear shaft. Some workers believe that the spread of Clovis artifacts records the first arrival of humans in North America, but evidence is growing that there was a short or long pre-Clovis phase. These points were found with a mammoth skeleton in Naco, Arizona.

(Above) The first Americans may have come by boat, hopping between islands south of the Bering Straits. Small, simple craft such as this Eskimo two-person kayak from Nunivak Island, Alaska, have been used in the area for many thousands of years. Compared with the treacherous land-based route, these early mariners could have rapidly travelled down the west coast of the Americas, perhaps as far as Monte Verde in southern Chile.

At an early stage in the European colonization of the Americas it was recognized that there were physical resemblances between Native Americans and the peoples of East Asia in features such as eye shape and hair form. These resemblances have been confirmed by anthropological studies, showing that they extend to details of tooth shape and DNA. But when and from where did people first arrive in the Americas, and how many different migrations were there? To reach the Americas, it is generally assumed that the first colonizers travelled across the Bering Straits between Siberia and Alaska at a time of low sea level. They may just have followed migrating herds of reindeer (caribou) on the exposed land bridge of Beringia, or they may have hopped by boat between islands immediately to the south. When they did arrive they entered a totally new world for humans, one that ran from the Arctic Circle virtually to the Antarctic, with almost every kind of climate and environment on Earth and its own unique plants and animals.

Clovis stone tools

It used to be believed that the first Americans, represented by the manufacturers of a distinctive stone tool industry called Clovis, only entered the continent some 13,000 or 12,000 years ago. From that date on, there were major extinctions of large mammals such as elephants, horses, camels and ground sloths throughout the Americas, and it has been generally assumed that the sudden impact of human hunting was responsible. However, some scientists believe that climatic and environmental changes or even disease were primarily responsible for these extinctions, rather than hunting. Moreover, there is growing evidence from sites such as Meadowcroft Rockshelter and Monte Verde that there was pre-Clovis occupation of the Americas.

The implications of Monte Verde

Sites in South America which have been dated as approximately contemporaneous with Clovis suggest that there were already culturally diverse peoples across the Americas by this time, and

(Right) Overhead view of the excavations at the enigmatic site of Meadowcroft Rockshelter, Pennsylvania. Radiocarbon dating of the earliest human occupation at this site controversially suggests that humans were living in the Americas well before the Clovis tradition flourished – perhaps by as much as several thousand years. However, these dates have been widely disputed, and Meadowcroft may, in fact, turn out to be a very early Clovis site.

therefore a much earlier time of entry must be postulated. Furthermore, an open site far to the south in Chile, called Monte Verde, appears to have evidence of human occupation from about 14,500 years ago. Archaeologists have argued that this requires an original entry into Alaska prior to the peak of the last Ice Age, about 20,000 years ago, otherwise there would have been insufficient time to travel the large distances involved. Indeed, some scientists believe this could have been as far back as 30,000 years. Others, however, believe that the earliest colonizers could have moved rapidly down the

(Left) Monte Verde in southern Chile was an open-air settlement on the banks of a small creek, dating back some 14,000 years. These timbers are the foundations of a long tent-like structure constructed with a frame of logs and planks and covered in animal hide. Pits, hearths, tools and food remains are associated with the structure, suggesting it was primarily for residential use.

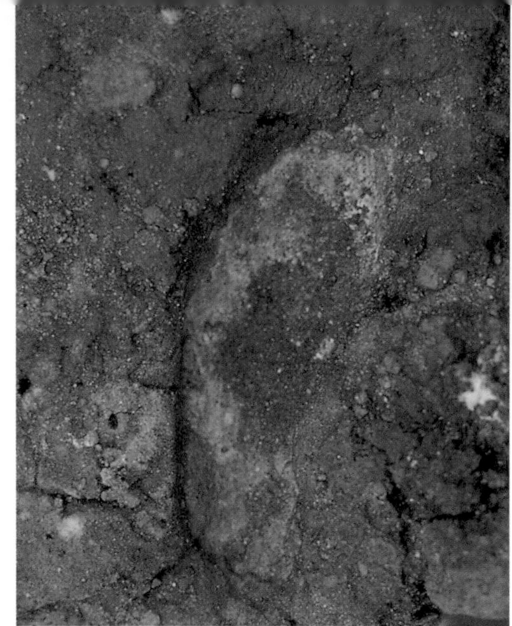

(Above) The 'toolkit' found at Monte Verde. These artifacts, made from bone, stone, and wood are very simple, and, indeed, some of them may not have been 'manufactured' at all. Human occupation at Monte Verde, though, is not in dispute – several human footprints such as this one (above right) have been excavated, as well as the wishbone-shaped foundation of a hut (right), littered with pieces of mastodon hide and meat. This was probably a non-residential area, set aside for the butchery of carcasses.

west coast of the Americas after about 18,000 years ago, only moving inland subsequently.

Physical diversity

There is also new evidence from human fossils at least 9000 years old in North, Central and South America that the early inhabitants of the Americas were physically diverse, and did not closely resemble present day Native Americans. They include, in the USA, the controversial skeleton of 'Kennewick Man' (Washington State) dated to about 9000 years and the subject of fierce legal dispute about ownership, and the 10,000 year old partly mummified Spirit Cave (Nevada) skeleton, found with preserved weaving and basketry. Farther south, a woman's skeleton from Mexico City (Peñon III) has been dated at more than 12,500 years, and another woman's skull found in a cave in the Brazilian state of Minas Gerais has been dated to about 10,500 years. The skulls are generally long-headed and low-faced, and this contrasts with many recent Americans. Recent research suggests that such distinctive populations may even have survived in isolated regions such as the Baja peninsula of Mexico within the last few hundred years. It is thus possible that the

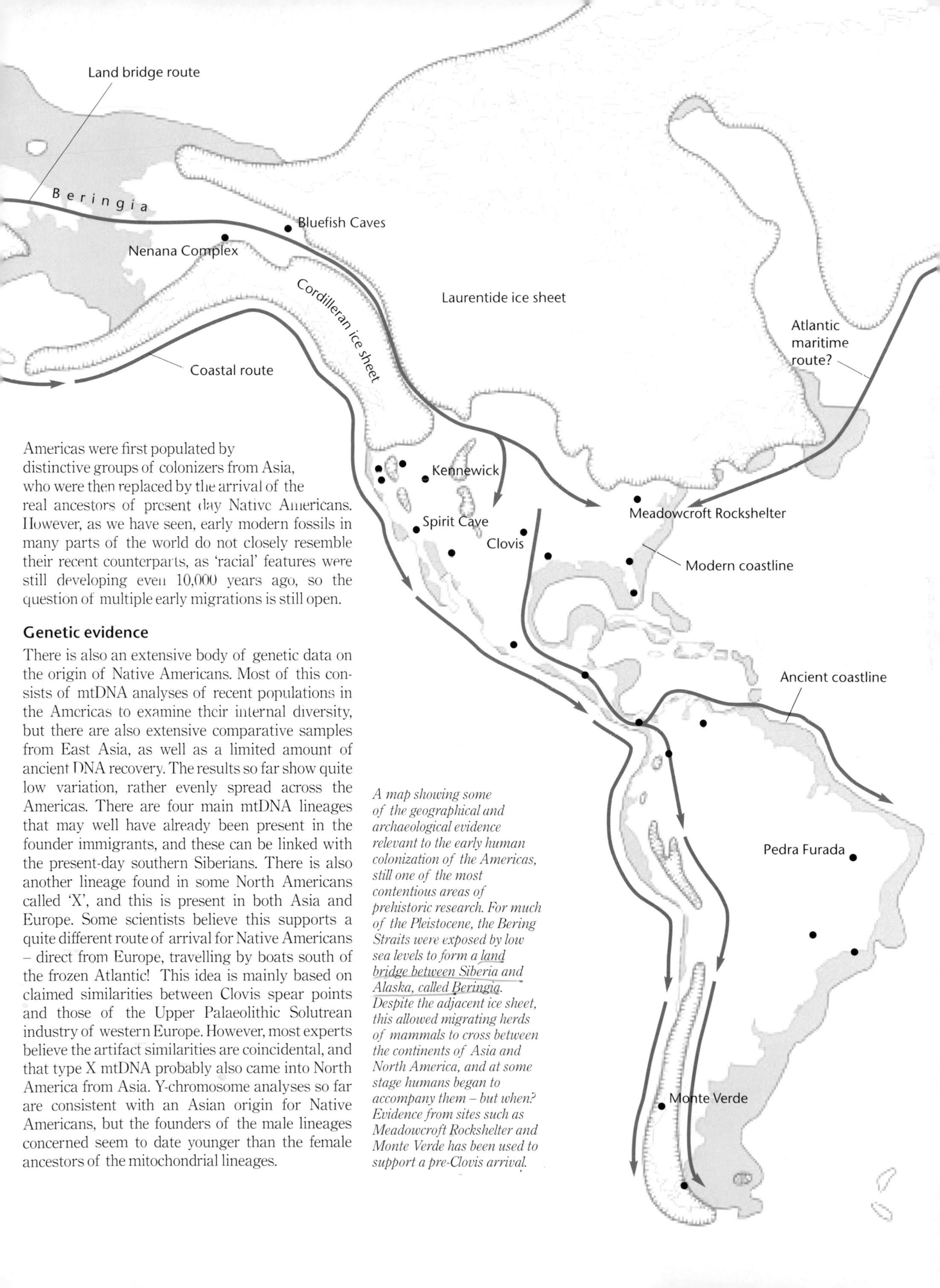

Americas were first populated by distinctive groups of colonizers from Asia, who were then replaced by the arrival of the real ancestors of present day Native Americans. However, as we have seen, early modern fossils in many parts of the world do not closely resemble their recent counterparts, as 'racial' features were still developing even 10,000 years ago, so the question of multiple early migrations is still open.

Genetic evidence

There is also an extensive body of genetic data on the origin of Native Americans. Most of this consists of mtDNA analyses of recent populations in the Americas to examine their internal diversity, but there are also extensive comparative samples from East Asia, as well as a limited amount of ancient DNA recovery. The results so far show quite low variation, rather evenly spread across the Americas. There are four main mtDNA lineages that may well have already been present in the founder immigrants, and these can be linked with the present-day southern Siberians. There is also another lineage found in some North Americans called 'X', and this is present in both Asia and Europe. Some scientists believe this supports a quite different route of arrival for Native Americans – direct from Europe, travelling by boats south of the frozen Atlantic! This idea is mainly based on claimed similarities between Clovis spear points and those of the Upper Palaeolithic Solutrean industry of western Europe. However, most experts believe the artifact similarities are coincidental, and that type X mtDNA probably also came into North America from Asia. Y-chromosome analyses so far are consistent with an Asian origin for Native Americans, but the founders of the male lineages concerned seem to date younger than the female ancestors of the mitochondrial lineages.

A map showing some of the geographical and archaeological evidence relevant to the early human colonization of the Americas, still one of the most contentious areas of prehistoric research. For much of the Pleistocene, the Bering Straits were exposed by low sea levels to form a land bridge between Siberia and Alaska, called Beringia. Despite the adjacent ice sheet, this allowed migrating herds of mammals to cross between the continents of Asia and North America, and at some stage humans began to accompany them – but when? Evidence from sites such as Meadowcroft Rockshelter and Monte Verde has been used to support a pre-Clovis arrival.

Evolution and Behaviour in Relation to the Environment

In the next few pages we shall try to connect up our knowledge of the evolution of locomotion, diet and behaviour in apes and humans into a coherent whole and relate the changes to what was going on in the environment at the same time. Primates are social animals, and generally speaking monkeys and apes display great complexity of social interactions. It is unusual, for example, to encounter a solitary primate, for nearly always there will be other members of the group close by. The type of group differs greatly between different species, but it nearly always serves the same functions, protection against predators and protection of food resources; having a home and keeping enemies away from it. Civilized societies have not changed that much from our primate ancestors.

Living apes

Living apes mainly eat fruit; they live in trees; and they have complex social systems. The first two features impose limitations on the distribution of ape species. If they depend mainly on fruit, there must be fruit available in the environment, and in many environments this simple requirement is not met. Most frugivores live in tropical forests for the simple reason that this is where fruit is readily available all year. So living apes are restricted to the tropics and almost entirely to tropical rainforest. Only the chimpanzee is able to live in more marginal environments where the climate is more seasonal and fruit is in short supply for large parts of the year. They can do this partly because their social system is flexible enough to allow small groups of more active individuals to explore new country and penetrate further into less favourable territory, and partly because they have the capacity to learn about their environment, so that they know, perhaps, that in times of food shortage, there is a possible food source miles away through apparently barren territory.

The second requirement for tree-living apes is the presence of trees. Most apes depend on trees both for protection from predators and for food, but only the gibbons habitually use trees as pathways for movement. The great apes, chimpanzees, gorillas and orang utans, generally come to the ground when they want to cover any distance, for they are too large and heavy to move from one tree to another. This is true even when they live in dense forest, and it is an absolute requirement when they live in more open types of forest. It should come as no surprise, therefore, to find some degree of ground-living behaviour in the course of ape evolution in any type of environment, but this would be especially expected if the apes are associated with more open forest environments.

Almost all living apes are restricted to living in tropical rainforest environments – accept the chimpanzee, which is able to survive in more marginal conditions such as this savannah area of East Africa.

The Social Organization of Living Apes and Humans

Bonobo

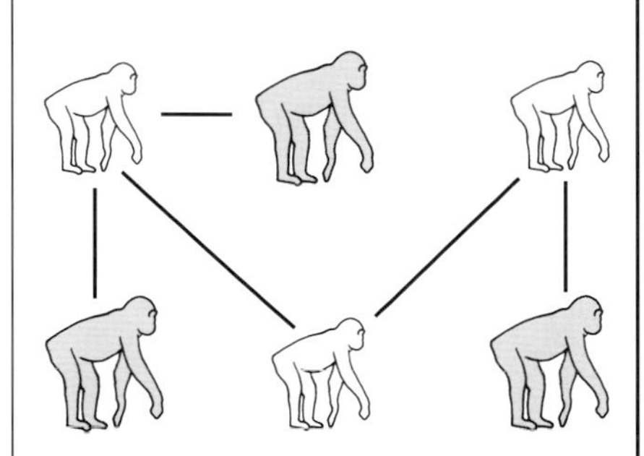

Bonobos live in egalitarian family groups with the strongest relationships between females, although females also bond with males. A male's rank is inherited from his mother, to whom he remains closely bonded. This female-dominated society is unique in primates and rare in mammals as a whole.

Chimpanzee

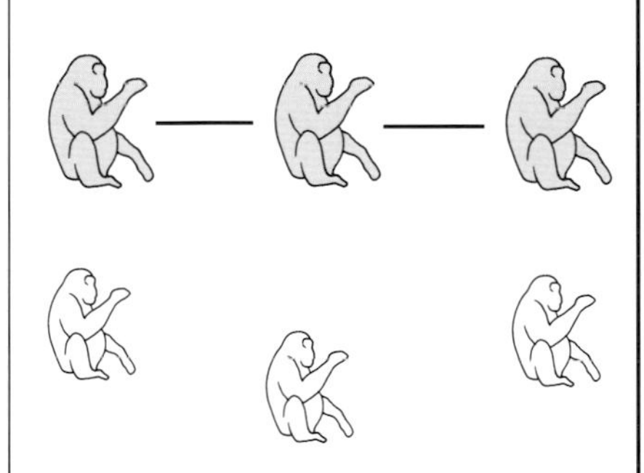

In contrast to bonobos, chimpanzee groups are dominated by the males. They defend a territory within which the females live, protected from outside intruders by the males. Males and females form short-term bonds, and females are not strongly bonded to other females.

Gibbon

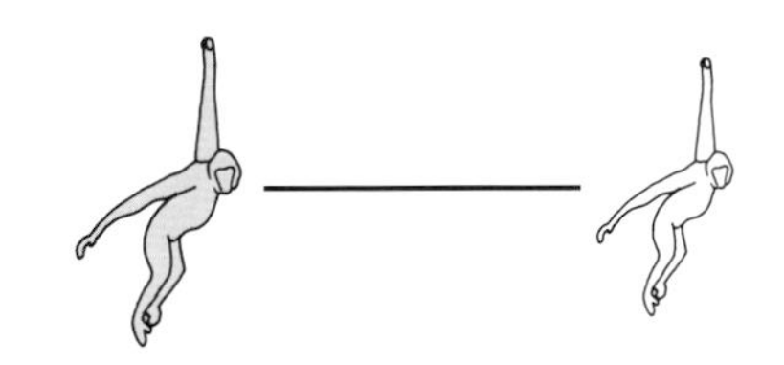

Gibbons live in monogamous family groups, with males and females sharing most activities, including territorial defence.

Human

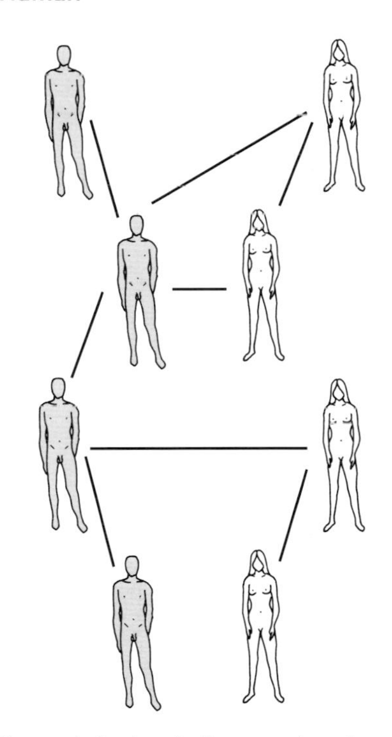

Human behaviour is diverse and overlaps with many of the apes' social systems: many human societies are based on monogamy, like gibbons, while others have polygamy as standard behaviour, like gorillas, or polygyny like chimpanzees.

Gorilla

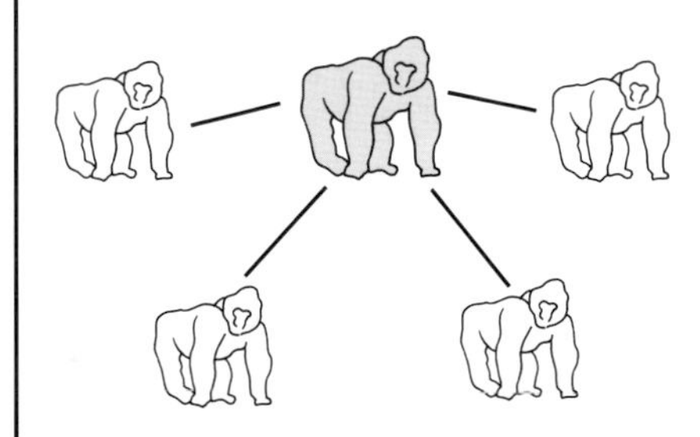

Gorillas practise polygamy, with groups led by a dominant male. Several females and their young are included in the group, and male offspring are tolerated until they mature sufficiently to challenge the dominant male.

Orang utan

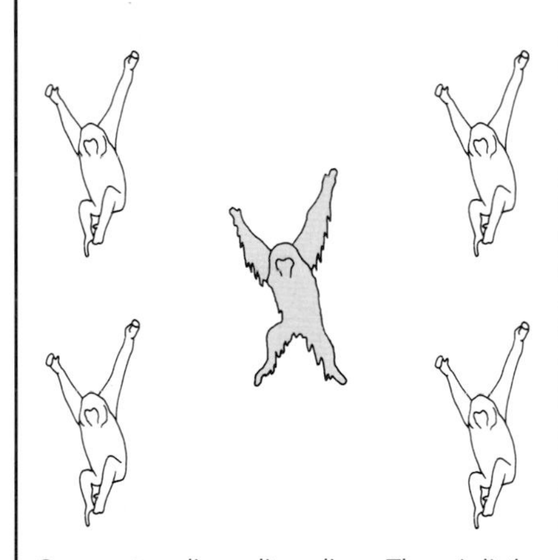

Orang utans live solitary lives. There is little contact between males and females, with each establishing independent home ranges, that of the males being larger and containing within it the ranges of one or more females.

The different social structures found in living populations of apes and humans.

Social systems among the apes

The apes have a variety of different social systems. Gibbons live in monogamous groups with a pair of adults and their offspring. This is actually quite an uncommon social system in mammals, although it is more common among birds. There is little difference in size between males and females because the males do not have to defend themselves against the attacks of other males. In polygamous groups with one male and several females this is not the case, for there are always extra males seeking their own harem and they are ready and waiting to exploit any weakness in group males. In these cases size is important and there is strong selection for

(Right) Gibbons live in monogamous family groups. Males and females are similarly sized, although in this particular species of gibbon, Hylobates concolor, *the female turns a lighter colour after adolescence.*

(Below) The dominant male in a small gorilla group is distinguished by its silver back. The 'silverback' is considerably larger than the other gorillas in the group – once other males reach maturity, and begin to turn silver themselves, they are driven away.

increased body size in the males to defend themselves and their group of females. Gorillas live in small polygamous groups, with one male, several females and their offspring. The male leader of a group is distinguished both by its large size, maybe twice that of the females, and by its silver back – the hairs on its back turn white and are very conspicuous. The younger black-backed male offspring are tolerated while they retain their black colour, but as soon as they start to become silver, they are driven away from the group.

Chimpanzees have the most complex social system of any living ape. The common chimpanzees live in loosely coherent groups in which individuals come together in subgroups or even move about by themselves for a time, depending on the availability of food or the danger from enemies such as predators or other chimpanzees. All the members of the group occupy the same home range and they all know each other, and at times they may aggregate together. The core of chimpanzee society are groups of males and most social co-operation occurs between bonded males, very often closely related individuals that have grown up together. Females transfer out of the groups in which they were born and often lack close bonding with other individuals, except with their own offspring and when they are sexually receptive. This is a very flexible social system and is one that would appear to offer greater ability to make use of less favourable environmental conditions, giving them greater flexibility, and they keep in touch with each other by a varied and rich repertoire of calls. Their diet of fruit, as in all primates that eat fruit, predisposes them to have good memories and the capacity to transfer knowledge of plentiful feeding places to

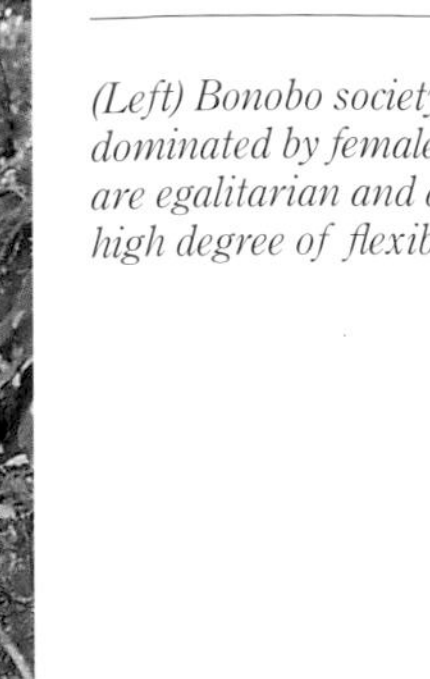

(Left) Bonobo society is dominated by females. Groups are egalitarian and display a high degree of flexibility.

their offspring, for fruits may be widely separated in their territory and different trees may produce fruit at different times of year. In this they face the same problems as all frugivores, but chimpanzees can live in more seasonal environments where these problems are greatly magnified.

Bonobos also seem to display greater flexibility in use of food sources, eating a greater variety of herbaceous food, and this may relate to the differences in their social system compared with chimpanzees. Females bond together in bonobo societies rather than the males, group sizes are

(Below) Chimpanzees have highly complex social systems, the males being dominant and forming the core of the society. Chimpanzees cope well in less favourable environments, and this is partly due to their flexible group structure. They are also able to supplement their diet of fruit with other sources of food, and they have learnt how to make and use tools for feeding – in this case (left), the chimpanzee is fishing for termites in a termite mound using a twig that it has prepared for this purpose.

Orang utans lead largely solitary lives, although there is much competition between males for access to females.

larger, both the permanent group and the temporary foraging groups, and the latter typically include both sexes and all age groups. This female-dominated society is unique in primates and rare in mammals as a whole. Human society appears to be derived not from a bonobo-like source but from a male-dominated chimp-like ancestor.

Orang utans also have loosely coherent social groups, but in this case they appear not to have the close bonding observed in chimpanzees. They live largely solitary lives, moving and feeding on their own, but they belong to a population of known individuals where males compete for the females. This has led to great size differences between males and females, greater even than in gorillas, with males more than twice the weight of females.

With living apes it is fairly straightforward to come to some understanding of the interactions between environment and behaviour. What now of fossil apes? To answer this, we are going to look at three stages in ape and human evolution where we have the best evidence. Even the best, however, falls far short of what is known of the living apes, and much of the uncertainty about evolution that still exists today is the result of the inadequate fossil record, which is both patchy and fragmentary.

Early Miocene apes

Proconsul lived in Africa during the early Miocene 20 to 18 million years ago. It ate soft fruits found in its tropical forest environment and was a generalized four-footed climbing ape living in trees. There were at least four species of *Proconsul*, however, and, using the same basic life form, they did different things. In addition there were closely related apes with rather different adaptations. The two species from Rusinga Island, for example, had a variable environment and are found associated with both closed forest and more open woodland environments. Their limbs were more robust and they would have been less agile in trees compared to the species from Songhor and Koru, two sites of similar age about 48 km (30 miles) from Rusinga where the environments were all tropical rainforest. The Songhor/Koru species of *Proconsul* lived in trees and ate similar diets of fruit, but differences in the bones of the hands and feet showed that the Songhor and Koru apes were more agile and used the environment in subtly different ways. There were also leaf-eating species of *Rangwapithecus* present in these much richer environments, and several species of smaller ape-like primates.

Little is known of the social interactions of *Proconsul*. None of the species exhibits big differences in size between males and females, and it is likely that they lived in groups with many males just as many monkeys do today. These groups of living monkeys can often be large, with 100 or more individuals, and sometimes the groups include two or more species. Differences between males and females in the forest-living species are similar to those seen in *Proconsul* species. No living ape has a social system of this type, but in most anatomical respects the species of *Proconsul* were more monkey-like than ape-like, and it may be that their behaviour was monkey-like as well. The conclusion about the early Miocene apes, and *Proconsul* in particular, is pretty much as one would expect in an ape ancestor. It lived in trees and ate fruit in tropical forest environments. Related species ate more leaves. The minor differences between the different species are no more than are found in groups of closely related living species.

Late Miocene apes

By the time of the late Miocene, 10 to 9 million years ago, the fossil apes were very different. There were at least two separate lineages of fossil ape adapted to very different life styles in different environments. These are represented by *Ankarapithecus* on the one hand and *Dryopithecus* on the other. A number of characters of the teeth and skull all indicate that *Ankarapithecus meteai* had powerful adaptations for chewing food that differ from *Proconsul*. These relate to the cheek teeth, the dimensions of the upper and lower jaws, and the development of the muscles used for chewing. Its teeth had thick enamel, the function of which was related to chewing hard food items, strengthening the molar teeth against the stresses imposed by breaking down hard resistant objects such as nuts and woody fruits and seeds. Humans have thick enamel, but the living apes do not. The teeth of *Ankarapithecus* were also very large associated with the thick enamel, which increases the processing area of the teeth. The same thing has been seen in other thick-enamelled fossil apes and human ancestors, for example *Sivapithecus* and the australopithecines, but it is not seen on any living ape.

Projecting interlocking canine teeth in the front of the jaws may restrict movement of the jaws during chewing, and this may in turn restrict the grinding capacity of species with this feature. *Ankarapithecus* has relatively small canines, once thought to be a key feature in human evolution but now recognized in several groups of fossil ape, for example in *Graecopithecus* (otherwise known as *Ouranopithecus*). Related to this are the flattened, robust cheek region and robust jaws in *Ankarapithecus*, and the combination of these characters in both fossil apes and australopithecines indicates a strong functional relation between them in contrast to the living apes, which have more lightly built jaws and teeth.

Ankarapithecus is taken here as being representative of a whole group of fossil apes which had similar adaptations. Where evidence is available, the members of this group all display some degree of ground adaptation. They were undoubtedly four-footed apes adapted for moving on the tops of supports, whether trees or the ground, and to this extent they retained the monkey-like adaptations of *Proconsul*. They would have been just as at home in trees as were the earlier apes, but their primary form of locomotion when covering distance was probably on the ground.

One other characteristic all these fossil apes had in common was that they lived in places where climates were either non-tropical or highly seasonal. Many of them are found only in Europe and Asia, for example *Graecopithecus* and *Sivapithecus* in Greece and Pakistan, both well outside the tropics today. In the past, however, these areas had subtropical climates, and there would have been subtropical forests and woodlands present in which the fossil apes may have lived, just as there are apes living in the subtropical monsoon forests of India, Burma and southwest China. Subtropical climates are generally highly seasonal with heavy rainfall falling during the summer months and a long dry season in between. These climates can be highly productive, for when rain falls during the summer there is plenty of water available during the growing season, but the indications are that the summer rainfalls in these environments were all relatively low, and forests and woodlands would have been

African woodlands have mostly been destroyed and replaced by agriculture, but this Acacia *woodland is an example of the kind of environment late Miocene apes and Pliocene hominins may have occupied.*

less dense than most subtropical and tropical forests today. It was this condition that probably necessitated the adaptations to life on the ground mentioned above, and the changes in the teeth and jaws enabled these fossil apes to survive on poor quality diets during the long dry seasons that would have occurred every year.

The other group of late Miocene apes belong to the genera *Oreopithecus* and *Dryopithecus*, with interesting differences from the first group. Their teeth and jaws had changed little from those of *Proconsul*, and their diet cannot be distinguished by present methods. Buttressing of their mandibular symphyses (chin region) was different, and their skulls were also more strongly buttressed. On the other hand, their limb bones were very different, being adapted for suspension beneath branches of trees just like the living apes, and they would have rarely – if ever – come to the ground. The adaptations of their legs and arms are, in fact, very similar to those of the orang utan, the most arboreal of the great apes. In most respects, *Dryopithecus* and *Oreopithecus* were similar, although dentally they were distinct, and it has also been suggested that *Oreopithecus* may have had some adaptations to life on the ground similar to those of humans by walking upright on two legs.

The environment occupied by *Dryopithecus* in southern Europe was subtropical, like that of the thick-enamelled apes, but the climate was wetter and probably encompassed evergreen subtropical forest. In this it was more like the subtropical monsoon forests of Burma and the wetter regions of eastern India where gibbons still survive today. The *Dryopithecus* species were fruit-eating and totally tree-living, and it is likely that *Dryopithecus* may have occupied a similar position to that of gibbons in its Miocene environment.

The earliest human ancestors

The two groups of late Miocene apes do not seem to have been closely related to one another, but together they set the scene for the emergence of human ancestors towards the end of the Miocene and into the Pliocene. Exactly when this event occurred is not known, but it is placed during the time interval 7 to 5 million years ago, right at the end of the Miocene. The earliest human ancestors are usually recognized by their adaptations for bipedalism, that is walking upright on two legs.

In recent years, a number of fossils have been described as being ancestral to later humans, but the evidence for some of these is ambiguous, mainly because of lack of evidence on the question of bipedalism. *Orrorin* from northern Kenya appears to be a thick-enamelled species with postcrania like that of late Miocene apes, but nothing is known of its environment or its behaviour, and the evidence for bipedalism is speculative. Evidence from the fauna and sediments associated with this fossil suggests a heavily wooded tropical woodland or forest. *Ardipithecus ramidus* from Ethiopia is said to be a human ancestor, and some evidence has been published showing that it may have been bipedal, but no leg bones have yet been published. A subspecies has been described making it the next oldest after *Orrorin*, and it appears from associated fossils that their environment was not one of open

(Right) A family group of Dryopithecus *hanging out in the trees of a subtropical swamp forest in southern Europe.*

(Far right) An Australopithecus afarensis *group. They are likely to have led at least partially arboreal lifestyles.*

savannah but was woodland. There are fossil remains from trees associated with *Ardipithecus*, and tree-dwelling animals like colobus monkeys are the most common constituent of the fauna. The fauna indicates that habitats became drier through time, and it is probably significant that hominins become less common as the habitat dries. *Kenyanthropus* from Kenya is similar in age to the earliest *Ardipithecus*, but it has thicker molar enamel. Like *Ardipithecus*, however, *Kenyanthropus* lived in a well-watered mosaic of habitats dominated by woodland and forest-edge habitats. Both, however, have relatively small teeth, like the *Dryopithecus* lineage rather than the *Ankarapithecus*/*Sivapithecus* lineage. *Australopithecus anamensis* from Kanapoi in Kenya has parts of limb bones that show very clearly that it was bipedal, but it is slightly later in time than *Ardipithecus*. It also has the characteristic thick-enamelled adaptations described above for the late Miocene apes: thick enamel on its back teeth, which are also enlarged, robust jaws and massive chewing muscles. *Ardipithecus* lacks all these characters and in fact it looks much more like an ape, but not like one of the late Miocene thick-enamelled apes – more like a chimpanzee.

The early australopithecines

The evidence from these very early human fossils is rather meagre, but what is available strongly suggests that the early australopithecines lived in woodland or forest habitats and all were at least partly arboreal in life style. More evidence is available for the slightly later form, *A. afarensis*, which may have had an extensive range from South to central and northeast Africa from deposits between 4 and 3 million years ago. In Laetoli (Tanzania) there are jaws and teeth that are similar to the Kanapoi specimens, and as we have seen (p. 187), there are fossilized footprints attributed to *A. afarensis* that show beyond any shadow of a doubt that this early human was bipedal. There is much controversy, however, over the exact form of these footprints, for while some scientists claim they are completely modern in their shape, others claim that the big toe is partly divergent as in the apes. The fossils from Hadar in Ethiopia are more complete and include the partial skeleton known as Lucy. From this and other fossils, it is apparent that as well as being fully bipedal, *A. afarensis* would also have been quite at home in trees, for its finger bones are long and curved so that it would have had a powerful grasp for use in climbing. Its foot could probably be used for gripping branches as well as the hand. Of course this is what makes monkeys and apes so agile in trees, because they can hold on to branches with both hands and feet. This evidence, if it is correct, means that these early humans could both live on the ground, where they walked bipedally, and in the trees, where they could suspend themselves like the suspensory apes. Their environment in the Afar Depression was one of mosaic wetlands with abundant woodland and forest, while at Laetoli the environment was probably drier, with no standing water and the dominant vegetation woodland and wooded grassland.

The later australopithecines

The later australopithecines from East and South Africa became adapted for more open country environments. The process can be seen in operation at Olduvai Gorge, where the environments reconstructed for Middle Bed 1 have been shown by several independent lines of evidence to have been closed woodland verging on forest, while by the time of Upper Bed 1 the environment had become much drier. An elaborate theory has been built up purporting to show that the evolution of bipedalism itself was linked with the move to savanna environments. By becoming bipedal, so the theory runs, humans removed most of their body from proximity with the hot ground. In fact, there is no evidence for such an environmental change during the early stages of human evolution. The late Miocene fossil apes and the Pliocene human ancestors lived in much the same kind of environment, tropical to subtropical woodlands and forest; they ate similar foods, consisting largely of fruit; and they moved about both in trees and on the ground to similar extents, except some species of ape became bipedal and others continued on all-fours; or did they? There are possible answers to these questions, but they are not concerned with the adaptation to environment but are more to do with social development and tool use. This topic is considered next.

Tools and Human Behaviour: the Earliest Evidence

The scientist Jacob Bronowski said that most animals leave traces of their bodies when they die, but only humans leave traces of what their minds have created. For over 2 million years humans have left such traces in the form of stone tools, and we can assume that these were only a part of what early people made and used. Many materials such as wood, bamboo and animal hides are not as durable as stone or bone, and are only preserved under special circumstances. So, much evidence of past human creativity has been lost, and it is quite possible that even before stone tools were first made, there was a phase of toolmaking we cannot recognize which involved the use of leaves, wood or bones. Raymond Dart believed that the australopithecines produced an 'osteodontokeratic' culture (from bones, teeth and horns), but the evidence he put forward from South African sites is not now accepted. However, our closest living relatives, the chimpanzees, have been observed making tools from grass stems for termite fishing, and more recently, they have even been observed in Guinea using rocks as tools to process food, further closing the behaviour gap between them and us. Recently, it has been suggested that some early hominins, such as the robust australopithecines, could have shown similar behaviour.

(Opposite below) The open site of Bilzingsleben in eastern Germany was occupied by humans (probably Homo heidelbergensis*) about 400,000 years ago, around an ancient spring and river. Bones of large mammals such as elephant and rhinoceros show signs of butchery, and it has been claimed that arrangements of stones were made by humans, possibly indicating hut structures.*

The first stone tools

The first known stone tools, often made of volcanic rock, have been excavated in East African sites dating from nearly 2.5 million years ago. These have been called Pebble Tool industries, or 'Oldowan', after Olduvai Gorge, where they were first recognized. The tools concerned were very simple – rounded river cobbles or pebbles were collected, and then a few flakes were struck off them by the use of another stone. Sometimes the cutting edge on the pebble was used, and in other cases, the flakes themselves were used as simple knives. There is generally no way of determining what the tools were being used for, but in some cases they have been found in association with animal bones which carry cut marks, so we can assume that they were involved in the butchery of carcasses. Some unmodified pebbles or cobbles seem to have been used to break open bones, presumably to extract the nutritious marrow. But in many other cases, these tools could have been used to process vegetable foods – for example, digging up and pounding tubers, slicing stems, and breaking open nuts and seeds. Perhaps containers of hide or wood were also made to carry food or water around, but we have no trace of these now.

It is usually assumed that early humans, of the species *Homo habilis* or *Homo rudolfensis*, were the manufacturers of the first known tools, even though fossil remains have not usually been found in direct association with the tools. However, it is certainly possible that australopithecines such as *A. garhi* were also tool-users or toolmakers, and this could include the robust species of South and East Africa, which co-existed with the earliest humans at sites like Olduvai Gorge and Koobi Fora. Pebble tools continue through the archaeological record of East Africa for nearly a million years, but different kinds of tools appear there about 1.5 million years ago.

The introduction of handaxes

These tools are called handaxes, or 'Acheulian', after the French site of St Acheul, where they have been found in large numbers. They seem to have been first made by the species *Homo ergaster* or *Homo erectus*, but later on were made by *Homo heidelbergensis*, at sites like Boxgrove, and by the ancestors of Neanderthals in Europe and *Homo sapiens* in Africa. The tools are usually almond or teardrop-shaped, but sometimes were broken across to make a chisel-like 'cleaver'. In Africa they

(Below) The oldest stone tools are known as choppers (Mode 1). These were made by detaching one or a few flakes from a cobble of rock or lava. They are often identified as tools rather than natural products because of evidence that they were transported over long-distances, as well as their context (for example, when found amongst cut and smashed animal bones).

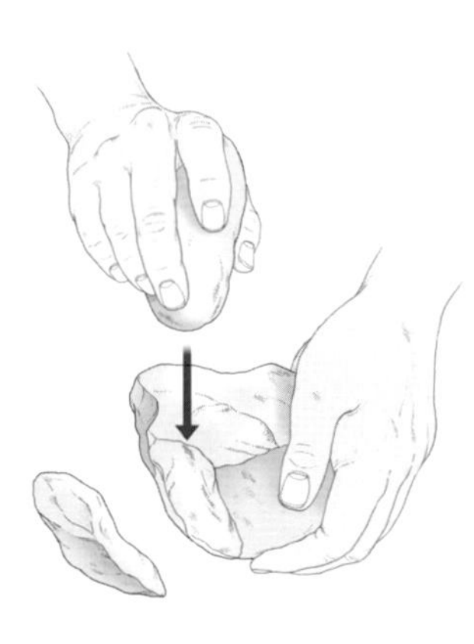

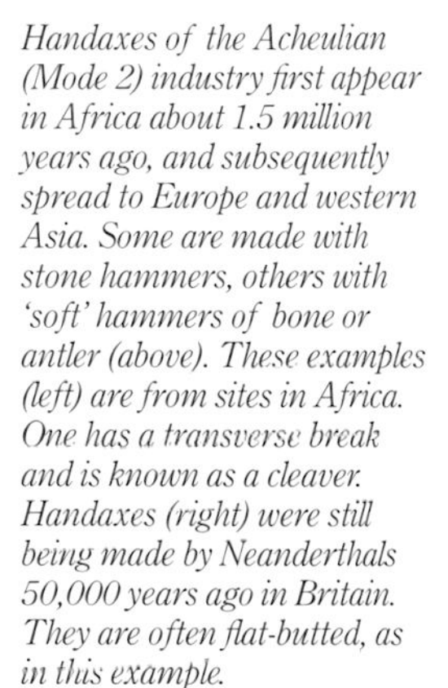

Handaxes of the Acheulian (Mode 2) industry first appear in Africa about 1.5 million years ago, and subsequently spread to Europe and western Asia. Some are made with stone hammers, others with 'soft' hammers of bone or antler (above). These examples (left) are from sites in Africa. One has a transverse break and is known as a cleaver. Handaxes (right) were still being made by Neanderthals 50,000 years ago in Britain. They are often flat-butted, as in this example.

were often made of volcanic rock, such as lava, while in other areas they were produced in local rocks such as chert or flint. Curiously, handaxes were hardly ever made by early people in the Far East, and there has been much speculation about the reason for this. Was it that the idea of handaxes never reached the region, were they a 'fashion' that never caught on, or did ancient Asians use other tools instead? Some scientists believe that tools made of perishable bamboo may have replaced the handaxe in eastern Asia.

But the handaxe was certainly important to *Homo erectus* and its successors in the western half of the inhabited world. Its form hardly changed in a million years, and its distinctive shape can be recognized from southern Africa to Israel, and from England to India. It was clearly a multipurpose tool, since it combined a point at one end, a blunt butt at the other, and scraping and cutting surfaces along the sides, and it also provided an additional block of raw material for the production of fresh and sharp flakes when these were needed. The people who made the handaxes clearly had a specific shape in mind, and often went far beyond a purely utilitarian form in the care with which they produced them.

(Below) There were distinct patterns in the distribution of handaxes and choppers about 500,000 years ago. The archaeologist Hallam Movius gave his name to a line marking the most easterly extent of handaxes. If this line is a real cultural marker, does it indicate the isolation of east Asian humans, or that there were different human adaptations in different environments?

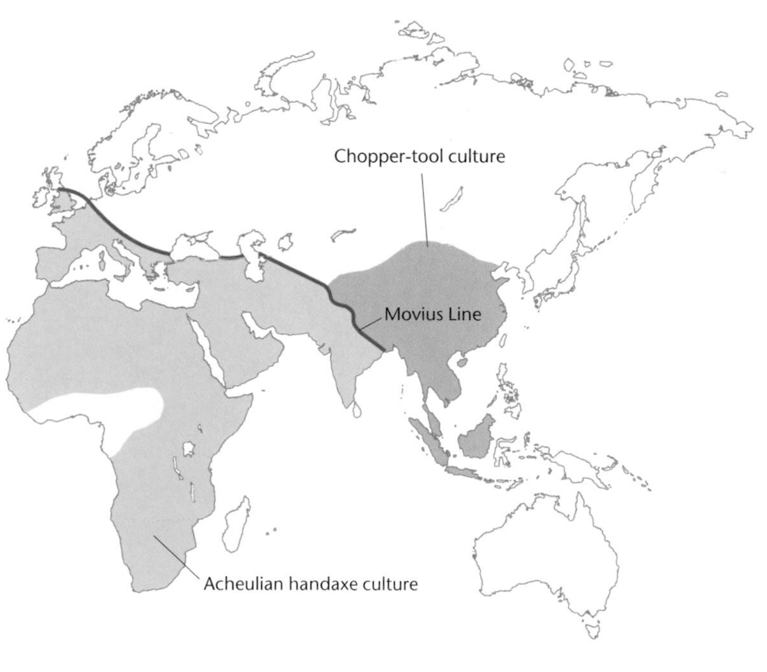

Tools and Human Behaviour: the Middle Palaeolithic

The Old Stone Age, or 'Palaeolithic' (*c.* 2.5 million to 12,000 years ago), is often divided into three stages, although these European-derived stages are of disputed value elsewhere. The Lower Palaeolithic (*c.* 2.5 million to 300,000 years ago) is the earliest period of stone tool production, from its pebble tool beginnings to the handaxe industries. The next stage, the Middle Palaeolithic (*c.* 300,000 to 40,000 years ago), covers the stone tool industries made by the Neanderthals in Europe and western Asia, and those made by the first modern humans in Africa and western Asia. The final stage, the Upper Palaeolithic (*c.* 40,000 to 10,000 years ago) is considered in the next section. A comparable system of classification, based on modes of production, distinguishes pebble tool industries as Mode 1, handaxe industries as Mode 2, Levallois production as Mode 3, Upper Palaeolithic technology as Mode 4 and microliths (very small tools, often mounted on hafts) as Mode 5.

(Below) A characteristic feature of many Middle Palaeolithic/Mousterian (Mode 3) industries is the use of the prepared-core or Levallois technique, where the knapper mapped out the shape of the target flake needed in initial preparations, and was then able to strike it off in its predetermined shape with a single blow. The characteristic humped shape of many prepared cores means that they are sometimes also known as tortoise-cores.

The Levallois technique

About 300,000 years ago, at the beginning of the Middle Palaeolithic, a method called the Levallois technique was invented (named after the French site where it was first recognized). This technique allowed the stone tool manufacturer to map out the final shape of the flake which could then be struck off from its core with a single blow, and it allowed much more control over tool production. The Levallois method was the most important innovation of the Middle Palaeolithic, although to begin with, it was often used to produce traditional-looking handaxes. Later it was used in a variety of local stone 'industries' in Europe, Asia and Africa.

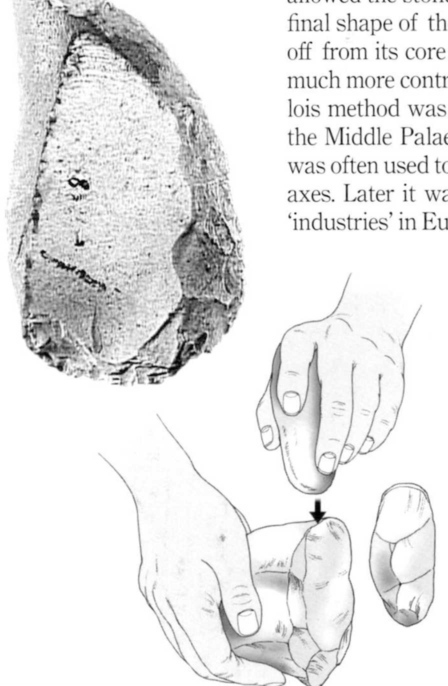

Neanderthal tools and behaviour

In Europe, the Middle Palaeolithic industries of the Neanderthals are also known as 'Mousterian', after the French cave of Le Moustier, one of the first sites where they were found. The Neanderthals made different kinds of flake tools, which we call by names such as 'scrapers', 'knives' and 'points', although we cannot be certain of how they were used. In rare cases, parts of wooden spears have been preserved at Neanderthal sites, and at Lehringen in Germany the end of a wooden spear was found in an elephant skeleton. The Neanderthals probably also mounted stone points on wooden handles to make short stabbing spears. We assume that they made other artifacts in wood, and probably dressed skins to make simple clothing. However, they seem to have made little use of bone, antler and ivory, even though these materials were everywhere around them. This was probably because these materials are much more difficult to work effectively than stone or wood.

The Neanderthals do not seem to have produced art, although natural pigments such as the iron oxide called red ochre or haematite are sometimes found at their sites, suggesting these may have been used to colour objects, or their bodies. There are also a few examples of humanly made scratches on bones and stones, but recent claims that a site in Slovenia contained part of a Neanderthal flute have been refuted by studies which suggest that the bone in question had been punctured by bears chewing on it.

However, despite claims to the contrary, it seems that the Neanderthals did bury their dead. At sites in Europe and western Asia, a number of Neanderthal skeletons have been excavated in circumstances which suggest intentional burial. But, as we have seen (p. 163), early modern people may have begun the practice of burying their dead

(Right) The Neanderthals made many different kinds of stone tools, and distinctive assemblages may represent either different activities being performed or else different traditions. The tools shown here are described as scrapers (top and left) and points (right), but there is little direct evidence of their actual function.

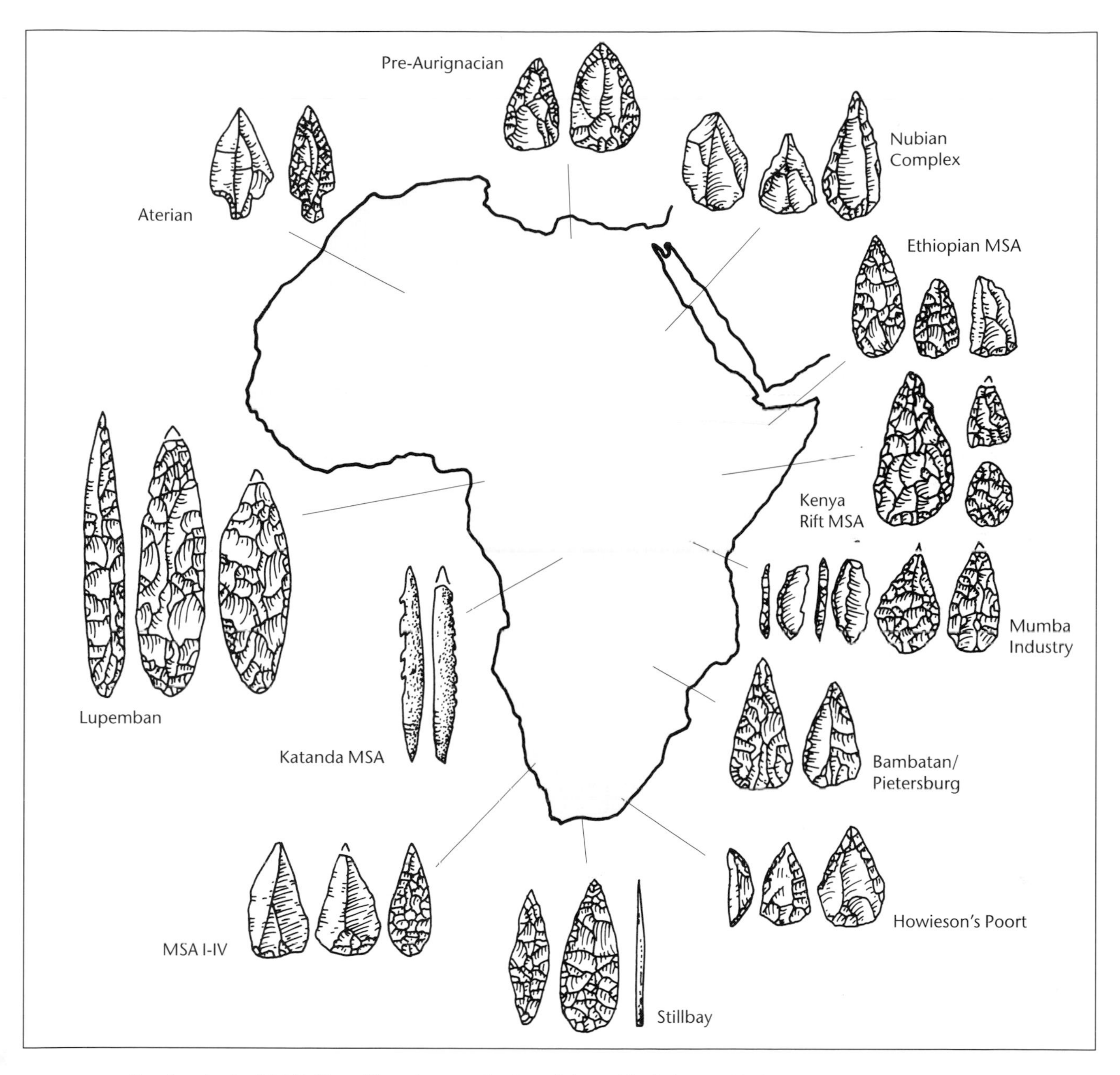

The Middle Stone Age industries of Africa show great variation – in fact much greater variation than equivalent industries in Europe or Asia. This map shows some of these variants from across the continent, ranging from the tanged points of the Aterian in the northwest through to the leaf points of the Stillbay in the south. Some Middle Stone Age industries retained handaxes, while others showed advanced features such as composite tools and bone working.

at an even earlier date, in the Middle East. There is no good evidence of human burials in the Far East until the arrival of modern humans in the region, thought to be after 70,000 years ago.

The African Middle Stone Age

In Africa, the Middle Palaeolithic is also known as the 'Middle Stone Age', abbreviated to MSA. In the north, stone tool industries were often similar to those made by the Neanderthals, but elsewhere there was more variety. The 'Sangoan' of central Africa was characterized by large pick-like tools, which it has been suggested could have been used for felling trees, while in the south, an industry called 'Howieson's Poort' was dominated by the production of long thin flakes, or blade tools, like those found in the much later Upper Palaeolithic of Europe. There is also evidence of a greater use of red ochre and bone working in southern African MSA sites, which some archaeologists interpret as a sign of increasing complexity in behaviour. However, the first early modern people known to have emerged from Africa – those found at the Middle Eastern sites of Skhul and Qafzeh, around 100,000 years ago, were similar to the Neanderthals in their technology. There are only hints of a greater behavioural complexity in the burial patterns – for example, a man at Skhul was buried with a large pig's jaw in his arms, and a child at Qafzeh was buried with an antlered deer's skull.

Tools and Human Behaviour: the Upper Palaeolithic

(Below) Blade technology has been compared to the famous Swiss Army Knife for its multiple uses and adaptability. Thin blades of stone were first struck off carefully prepared cores. These blades could then be modified to produce a range of specialized tools for cutting, piercing or engraving. Some of these, in turn, were then used to work materials such as wood, bone, antler or ivory to produce further specialized tools, or artwork.

About 40,000 years ago, there was a change in the dominant method of toolmaking in Africa and the Middle East, and this change soon spread to other areas, such as Europe. Whereas the usual procedure in the Lower and Middle Palaeolithic was to reduce a piece of rock down to only one or a few tools, the new method allowed many long thin flakes (or 'blades') to be systematically produced from a single original block of stone. The blades were often knocked off by the use of a pointed 'punch' made of bone or antler. They were then worked further to turn them into 'knives', 'scrapers', 'chisels', 'borers' etc. The industries concerned are called 'Upper Palaeolithic' in Europe and western Asia, and 'Later Stone Age' in Africa.

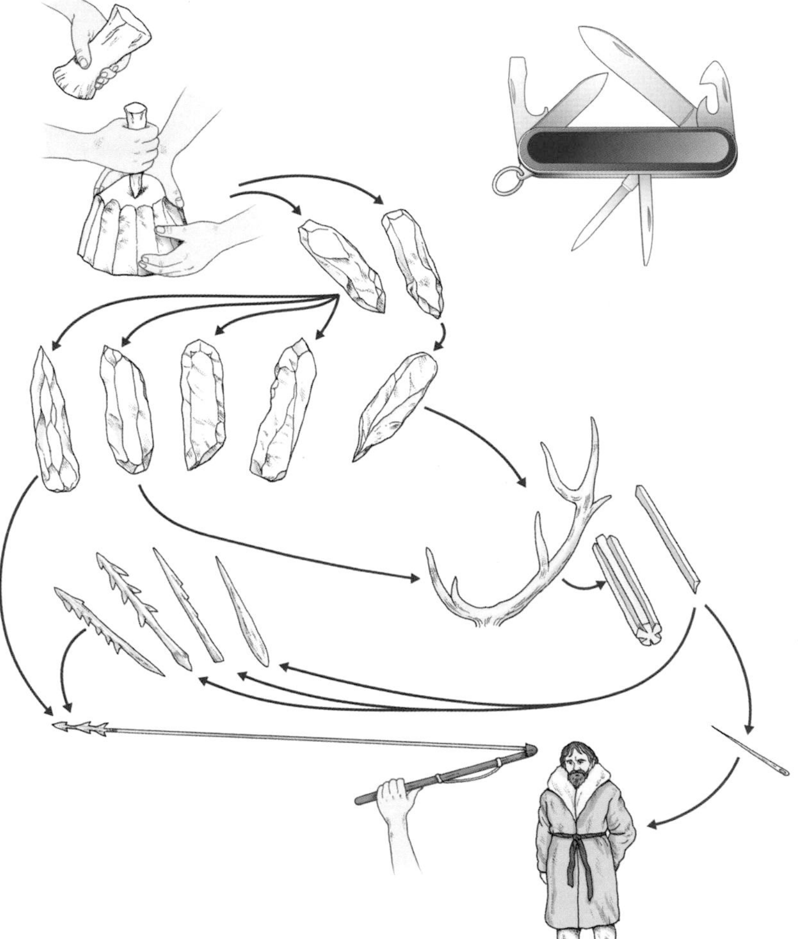

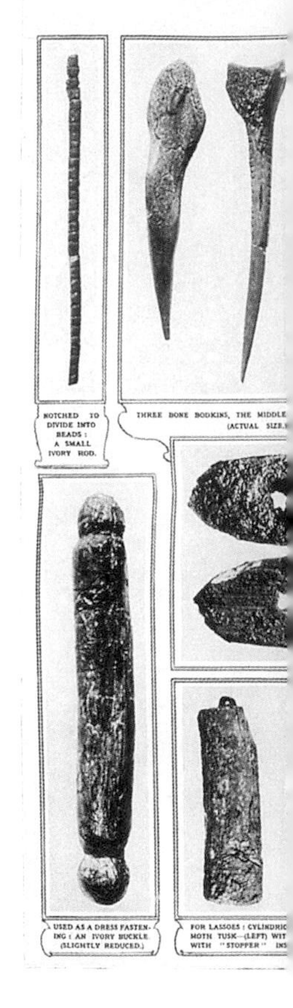

(Right) The site of Předmostí in the Czech Republic produced a large sample of human skeletons but also a remarkable range of stone, bone and ivory tools of the Gravettian industry. Most of this material was tragically destroyed during World War II, and survives only in photographs. These are some of the tools and weapons found there, some of which are of unknown function. The bone 'handle' (bottom right) has been reconstructed with a more recent stone axe head added to it.

(Below) This 'batôn' of reindeer antler was excavated from Gough's Cave in England. The spiral shaping of the hole supports the idea that 'batôns' could have in fact been used as pulleys.

New tools and art

Alongside the predominance of blades, there was also a great increase in the working of bone, antler and ivory, and evidence of clay working, ropes and even basketry. Composite tools made of several parts become more common, such as harpoons with detachable heads, and spearthrowers were used to increase the range of projectiles. Personal

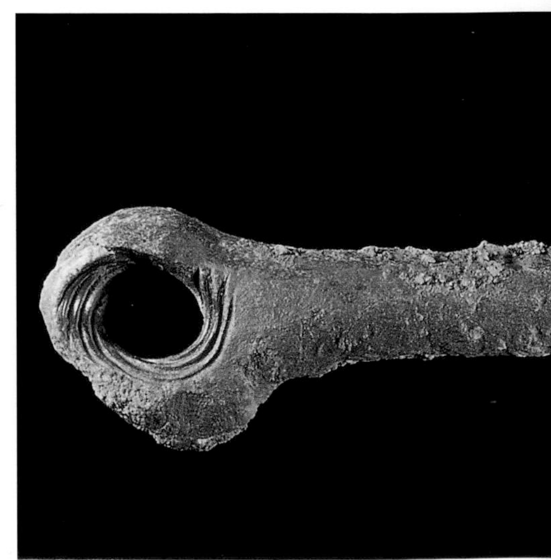

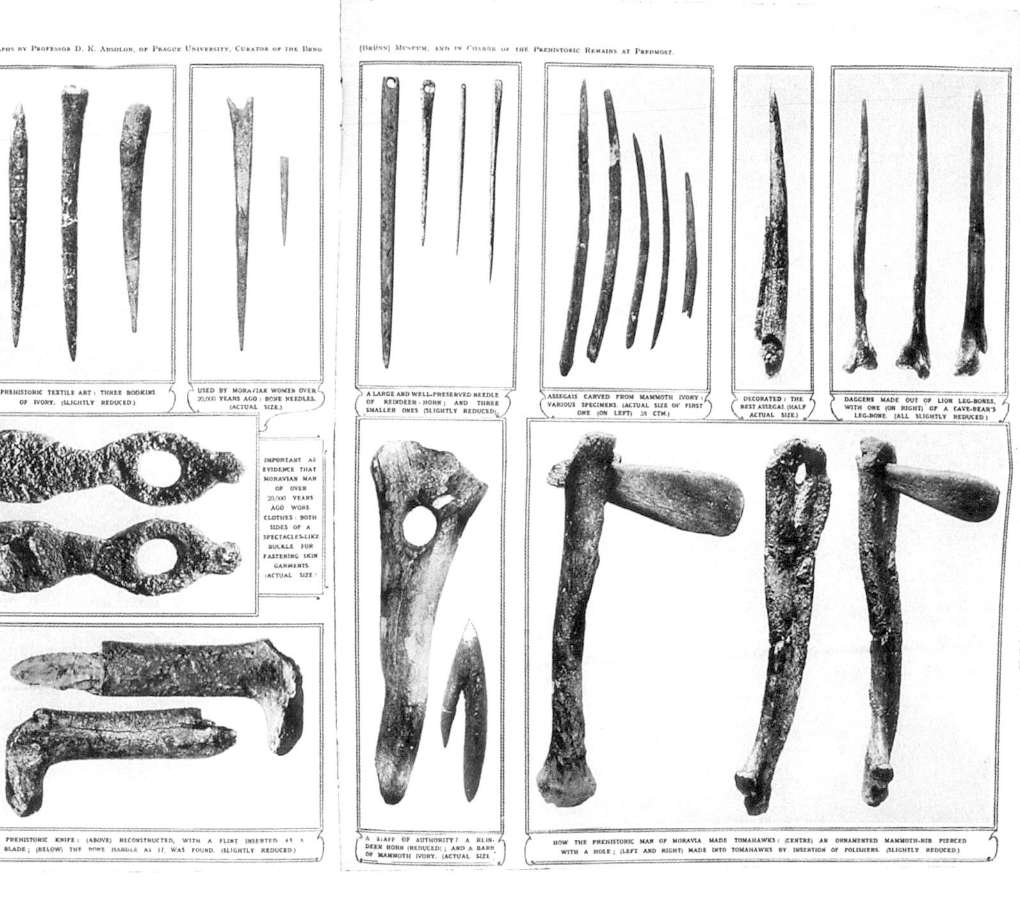
…OGRAPHS BY PROFESSOR D. K. ABSOLON, OF PRAGUE UNIVERSITY, CURATOR OF THE BRNO (BRÜNN) MUSEUM, AND IN CHARGE OF THE PREHISTORIC REMAINS AT PREDMOST.

PREHISTORIC TEXTILE ART: THREE BODKINS OF IVORY. (SLIGHTLY REDUCED.)

USED BY MORAVIAN WOMEN OVER 20,000 YEARS AGO: BONE NEEDLES. (ACTUAL SIZE.)

A LARGE AND WELL-PRESERVED NEEDLE OF REINDEER-HORN; AND THREE SMALLER ONES (SLIGHTLY REDUCED).

ASSEGAIS CARVED FROM MAMMOTH IVORY: VARIOUS SPECIMENS. (ACTUAL SIZE OF FIRST ONE (ON LEFT) 35 CTM.)

DECORATED: THE BEST ASSEGAI. (HALF ACTUAL SIZE.)

DAGGERS MADE OUT OF LION LEG-BONES, WITH ONE (ON RIGHT) OF A CAVE-BEAR'S LEG-BONE. (ALL SLIGHTLY REDUCED.)

IMPORTANT AS EVIDENCE THAT MORAVIAN MAN OF OVER 20,000 YEARS AGO WORE CLOTHES: BOTH SIDES OF A SPECTACLES-LIKE BUCKLE FOR FASTENING SKIN GARMENTS. (ACTUAL SIZE.)

A PREHISTORIC KNIFE: (ABOVE) RECONSTRUCTED, WITH A FLINT INSERTED AS A BLADE; (BELOW) THE BONE HANDLE AS IT WAS FOUND. (SLIGHTLY REDUCED.)

A STAFF OF AUTHORITY? A REINDEER HORN (REDUCED); AND A BARB OF MAMMOTH IVORY. (ACTUAL SIZE.)

HOW THE PREHISTORIC MAN OF MORAVIA MADE TOMAHAWKS: (CENTRE) AN ORNAMENTED MAMMOTH-RIB PIERCED WITH A HOLE; (LEFT AND RIGHT) MADE INTO TOMAHAWKS BY INSERTION OF POLISHERS. (SLIGHTLY REDUCED.)

(Below left) Modern humans are characterized by the use of symbolism and art, and this was certainly prevalent in the Upper Palaeolithic of Europe. But when and where did such behaviour begin? The site of Blombos Cave in South Africa has produced 75,000-year-old evidence of the use of iron oxide as a pigment, in the form of red ochre. It was fashioned into crayons, perhaps to decorate the body, and here a piece of red ochre has been engraved with markings, possible symbolic. When fresh, these patterns would have been very striking – the colour of blood.

(Below) Upper Palaeolithic and Later Stone Age harpoons were beautiful but also highly functional. The barbs meant that they remained deep in the wounded prey.

ornaments appear in the archaeological record, such as necklaces of shells in Australia, beads of ostrich eggshell in Africa, and pendants of ivory in Europe. There is also much greater evidence of the use of pigments, sometimes painted on objects, sometimes on cave walls, and sometimes on bodies at burial. This 'creative explosion' is seen by many archaeologists as marking the definite arrival of fully modern minds, although it must be remembered that not all parts of the world show the full suite of Upper Palaeolithic features at the same time – for example, although art, body adornment, bone working and complex burials are known from Australia at 30,000 years, blade tools of European and African type are missing. But there is now also evidence from sites such as Blombos Cave in South

Africa that artistic expression, in the form of engraved ochre, was already present over 75,000 years ago, during the Middle Stone Age.

Economic and social life

By the time of the Upper Palaeolithic, camps generally became larger and more permanent, and dwellings became more complex, including evidence of skin tents and even houses made of mammoth bones where wood was unavailable. Fire technology improved with the building of stone-lined hearths and ovens, and stone oil-lamps have been found in some caves. Techniques of food gathering also diversified, with the development of boats and fishing, and probably the production of nets, traps and pits. There is also evidence of the beginnings of social stratification, since some individuals were buried with many more grave goods than others. At a site called Sungir, in Russia, the skeletons of a man and two young children were discovered with thousands of ivory beads, which must once have decorated the clothes in which they were buried. These beads would have taken many hours to produce, suggesting that the children were the offspring of an important individual – perhaps a

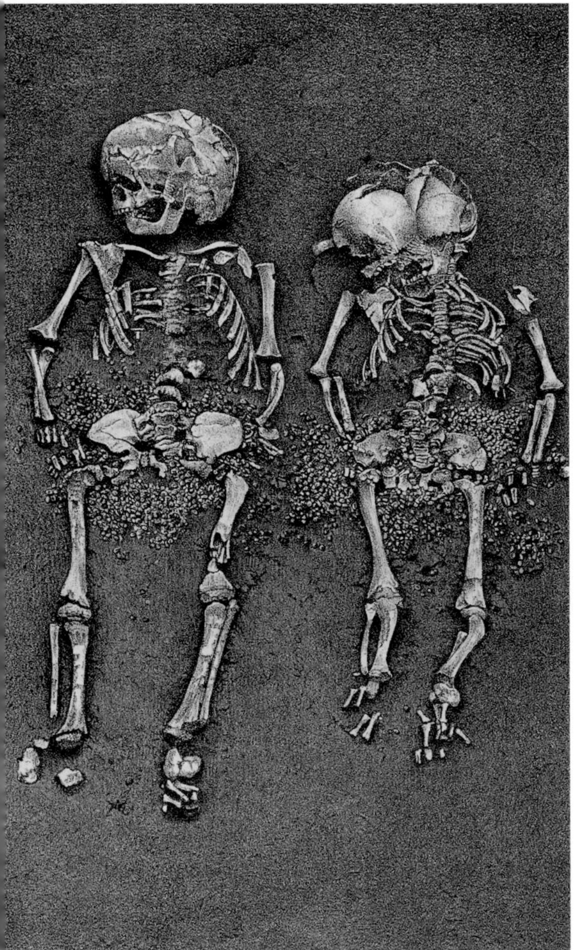

(Above) On the relatively cold and arid plains of ice age central and eastern Europe, wood was sometimes a scarce resource. Accordingly, peoples of the Middle and Upper Palaeolithic turned to large mammal bones for raw materials and even for fuel. The bones of large animals such as mammoth and woolly rhinoceros could have been collected from the landscape in large numbers and were used to construct large huts, as here at Mezhirich in the Ukraine.

(Left) This double burial of children, excavated in 1874, is from the Grotte des Enfants in Italy. The bodies are adorned with hundreds of sea shells and pierced animal teeth.

(Right) A burial complex at Sungir, Russia, contained the skeletons of two children in one grave, and a man's skeleton and a woman's skull in another. The man's skeleton, shown here, was associated with nearly 3000 ivory beads, which must have been on his clothing, and many ivory bracelets.

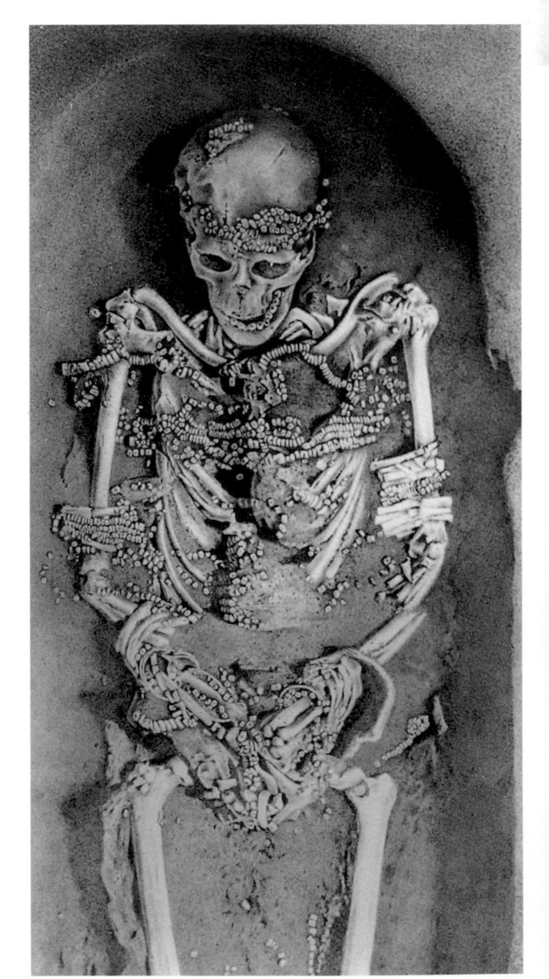

chief. A similar double burial of children is known from Grotte des Enfants in Italy. Burial patterns were also increasingly complex – for example, three teenagers were buried together at Dolní Věstonice. The burial pit was specially prepared and the bodies carefully arranged, with the addition of red ochre powder and objects such as wooden stakes.

European stone tool industries

In Europe, there was a succession of Upper Palaeolithic industries, most of which have been named after the French sites where they were first recognized. The earliest – the Aurignacian – occurred widely across the continent from about 35,000 years ago, and was associated with the first modern people (the Cro-Magnons) and the earliest representational art. In parts of Europe this was succeeded by the Gravettian, known from famous sites such as Předmostí and Dolní Věstonice, while industries such as the Solutrean and Magdalenian followed on (the famous cave of Lascaux was painted by early Magdalenian people). The Magdalenian continued until the end of the Ice Age, some 11,000 years ago, when the Upper Palaeolithic gave way to the Mesolithic, or Middle Stone Age (not to be confused with the much earlier African Middle Stone Age!).

(Above) The triple burial at Dolní Věstonice. A deformed and crippled body, perhaps a female, lies between two strongly built males, their skulls encircled with necklaces of pierced teeth and ivory.

(Below) Solutrean flint tools from Le Placard, France.

Lithic Industries

	Possible age range (years ago)
(Châtelperronian	38,000–33,000)
Aurignacian	35,000–29,000
Gravettian	29,000–22,000
Solutrean	22,000–17,000
Magdalenian	17,000–11,000

The First Artists

Some kind of artistic expression is universal in human societies today, whether it is expressed in music, dance, painting, carving, pottery, weaving, metalwork etc. Although in the so-called developed world art is often seen as a luxury, for most hunter-gatherer groups it is intricately bound up in their way of life, whether it forms part of their spiritual beliefs, is a mark of their territory or social identity, or whether it is created for trade with neighbouring groups. Because art, like language, is found everywhere humans live today, from Siberia to South Africa, from New Zealand to Greenland, it is thought to have been part of a general human

Map showing sites in Europe where Upper Palaeolithic art has been found.

inheritance which began to develop in Africa over 75,000 years ago.

Discovering Palaeolithic art

But when the first examples of European Palaeolithic art were found in the last century, archaeologists found it difficult to believe that people who seemed so technologically primitive had either the capability, the desire or the time to produce such sophisticated images. Therefore it was suggested that ancient materials had been engraved more recently – for example, by the Romans – or that the 'art' had actually been faked in recent times and planted in Palaeolithic sites. But gradually, examples of 'mobiliary art' – that is, art which was small enough to be carried around – came to be accepted as genuine; for example, the French site of La Madeleine contained part of a mammoth tusk which had actually been engraved with a picture of a mammoth. But acceptance of paintings on the cave walls (called 'parietal art') as genuine took longer, because some were remote from occupation sites, and many showed such breathtaking artistic skill that they were considered to be beyond the minds and abilities of Stone Age humans. However, comparisons of the styles in the mobiliary and parietal art showed they were consistent, so much so that some archaeologists believed

(Left) This small, carved ivory head from Pavlov, in the Czech Republic, is about 27,000 years old. The details in the face suggest that it might actually represent a portrait.

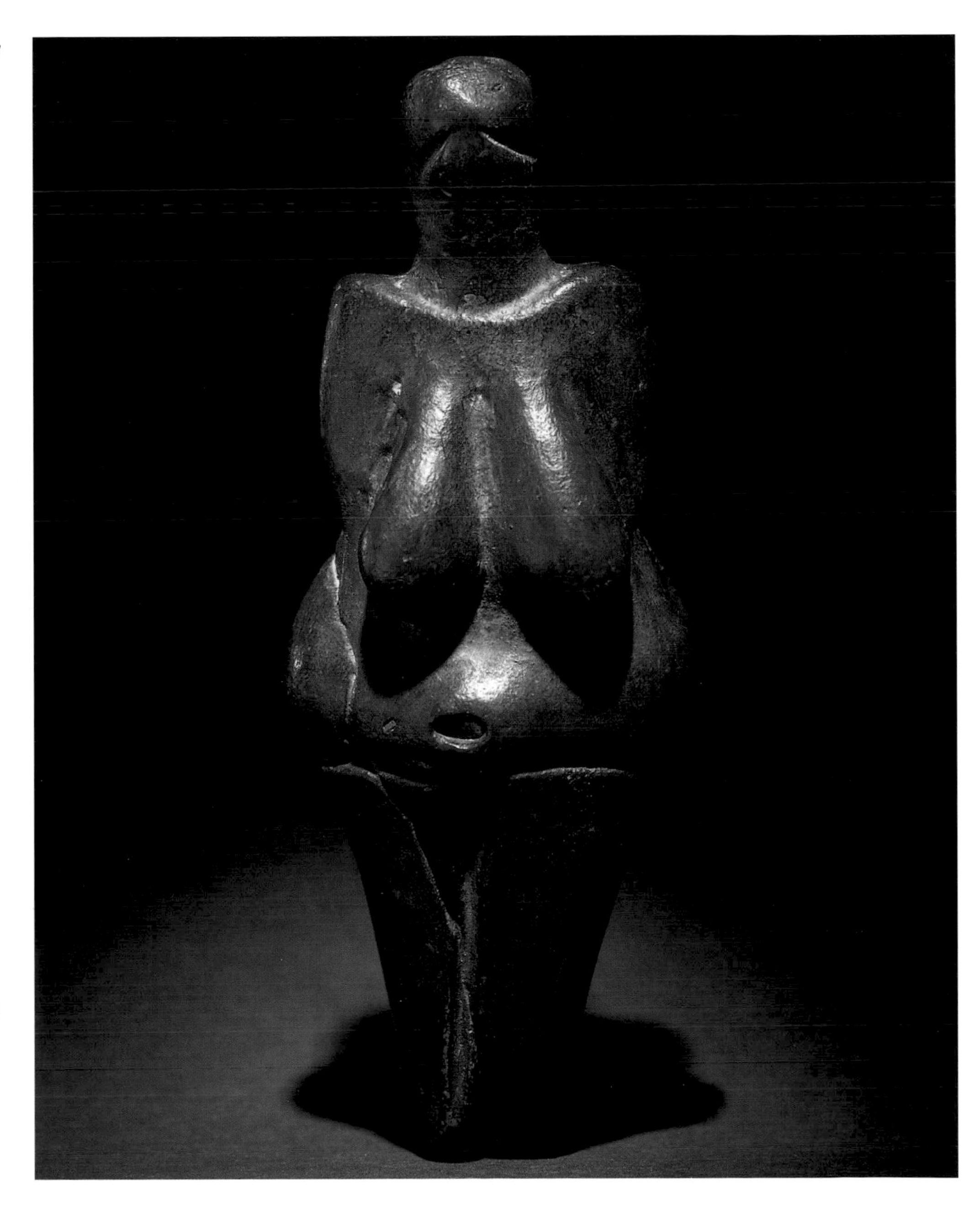

(Right) This Venus statuette from Dolní Věstonice is about 110 mm high and shows the well-proportioned female shape and lack of facial detail characteristic of these figurines. The statuette was moulded from clay and bone ash and then must have been fired at high temperature – one of the oldest known ceramic sculptures.

(Left) The Venus of Laussel was discovered in 1911, engraved in a rockshelter near Lascaux. There has been much speculation about the horn in her hand. Is it merely a stylistic representation of an ibex horn, or are its 13 marks significant?

(Right) The remarkable lion-headed statue from Hohlenstein-Stadel, Germany. The therianthrope (a composite animal-human) had the body of a man but the head of a lion, and must have been a sacred object for its Aurignacian makers.

they could even recognize the same artist at work, while in other cases, excavations uncovered paintings which had clearly been buried for many thousands of years.

What was the art for?

We now know that Palaeolithic art was produced by the Cro-Magnons in Europe for over 25,000 years. During that time period, there were probably many different reasons why the art was created, but archaeologists believe that some of it was used in ceremonies – perhaps religious ceremonies, or initiation rites. This is because the art was often painted on walls deep within cave systems, and was therefore not meant to be viewed easily. In addition, some of the painted chambers have been tested for their resonance, and they appear to have had special acoustic properties, so perhaps drumming and chanting was part of whatever took place in them.

(Right) This tiny mammoth-ivory head from Brassempouy in France is one of the most delicate of Upper Palaeolithic sculptures. It is not certain whether the 'hair' is styled, or whether this represents a hat.

(Above) The Axial Gallery in Lascaux, in the Dordogne, France.

(Right) One of the famous 'Chinese horses' from Lascaux. It is about 1.8 m (6 ft) across.

It has also been suggested that some of the art shows hallucinogenic characteristics, so it may have been produced under the effects of trances or drugs.

From animal paintings to human figurines

Most of the art shows animals which were familiar to the Cro-Magnons, especially deer and horses, but there are also representations of wild cattle, mammoth, rhinoceros, ibex and sometimes, rather schematic people. However, small human figurines called 'Venuses' have been found in Palaeolithic sites across Europe. These representations, which are usually of plump women, are known from engravings on stone and bone, from sculptures in materials such as ivory, and from statuettes moulded in clay, which was then baked. Male figures are less common, but one of the most remarkable is also one of the oldest. In 1939 around 200 worked ivory fragments were discovered in the cave of Hohlenstein-Stadel in southern Germany. However, it was not until 1969 that they were reconstructed into a 30-cm (12-in) high statue of a lion-headed man (illustrated on p. 218). The fragments came from an Aurignacian level in the cave dated to 32,000 years ago.

The famous cave art sites

The most famous painted caves are also among the youngest – the paintings of Altamira in Spain are about 15,000 years old, and those of Lascaux in France date from around 18,000 years. But three recent discoveries show just what still remains to be found, as well as how much we still have to learn about Palaeolithic art. The site of Cosquer was discovered on the Mediterranean coast of France far from any other painted caves, and can now only be reached by underwater diving. But when its art was

(Below) An antler carving of a bison from La Madeleine rockshelter, France, dating to about 13,000 years ago. The artist has cleverly shown the animal turning to lick its flank, whilst managing to maintain a sense of perspective.

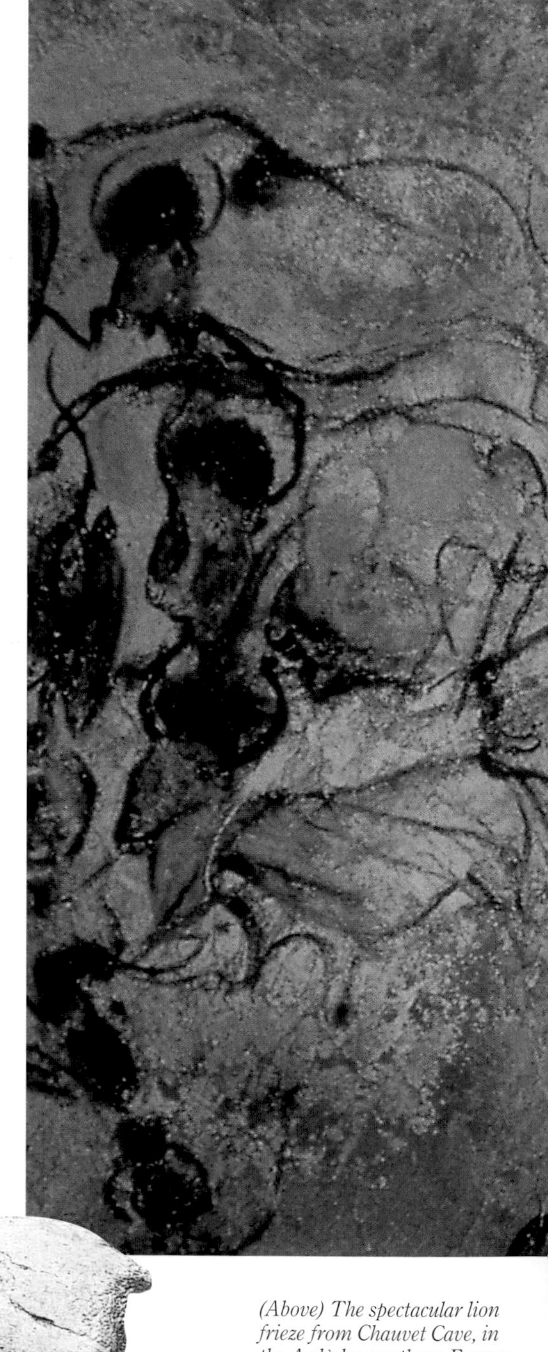

(Above) The spectacular lion frieze from Chauvet Cave, in the Ardèche, southern France, roughly 30,000 years old.

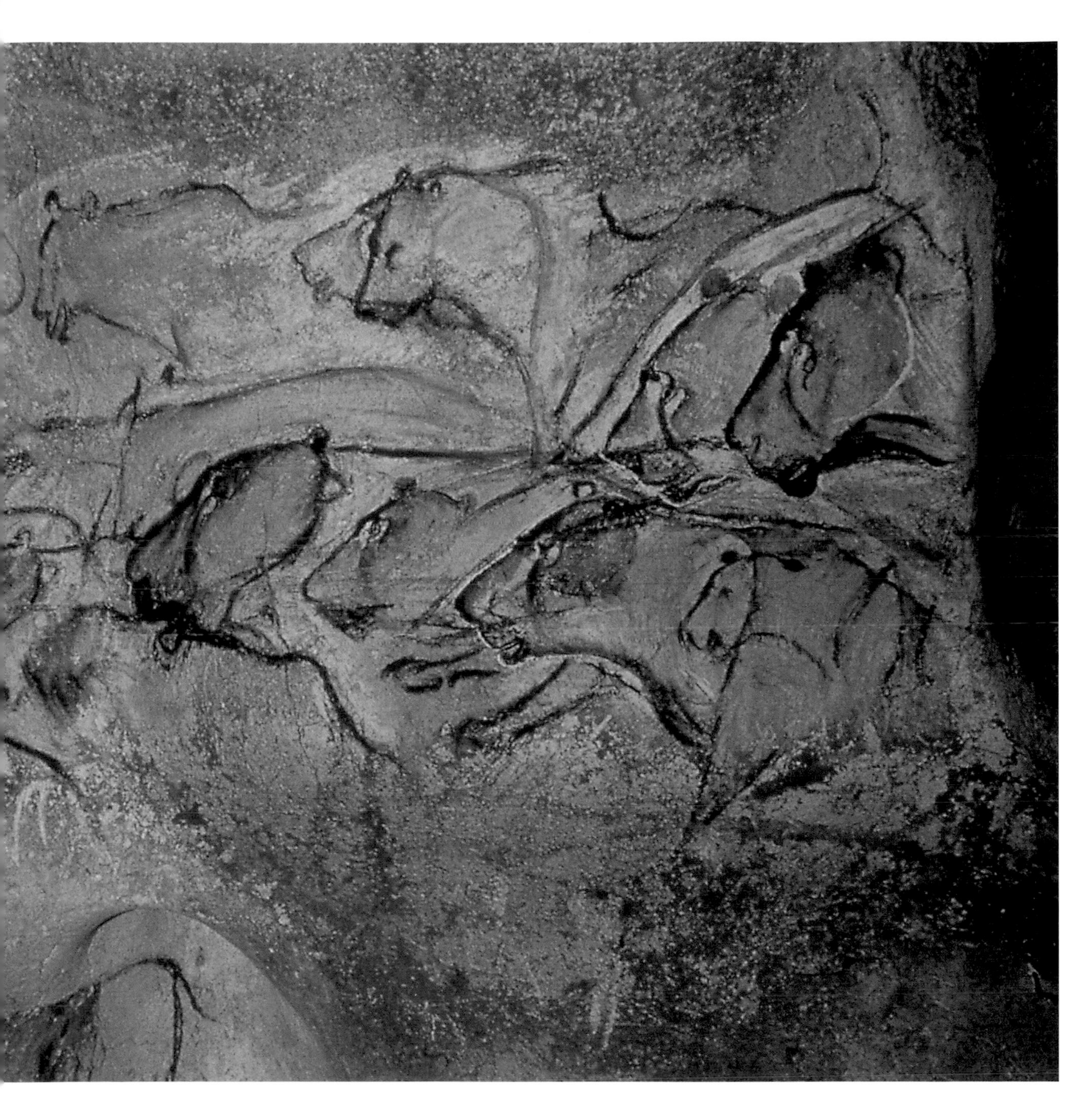

created between 27,000 and 18,000 years ago, near the climax of the last Ice Age, the sea level was probably at least 80 m (260 ft) lower than today, and all its chambers were dry. Its depictions include many stencilled hands, as well as pictures of the extinct penguin-like bird, the Great Auk. A second new site is Cussac Cave, near Lascaux, containing huge engravings on the walls, and several Cro-Magnon skeletons on the floors.

Even more remarkable is the art of Chauvet Cave, near Avignon in France. Discovered in 1994, this site is perhaps the most important painted cave yet found. Although the paintings are mostly single coloured, black or red, the range and skill of the depictions is astonishing, with rhinos charging each other, and the brooding heads of lions and bears. Chauvet provided the biggest surprise of all when charcoal pigments in the paintings were dated as about 31,000 years old, thus showing that some of the most sophisticated art was also some of the oldest. Clearly, the early Cro-Magnons who carved the lion-man of Hohlenstein-Stadel and painted Chauvet Cave were carrying on a tradition established long before, in Europe or elsewhere.

Reconstructing Ancient Human Behaviour 1

Archaeologists must try to reconstruct ancient behaviour from the fragmentary debris left behind long ago. This debris may be incomplete, badly disturbed, and open to different interpretations. For example, Mary Leakey excavated a semicircle of cobbles in Bed I at Olduvai Gorge, which she believed represented the base of a primitive hut or hunting hide. But other archaeologists cautioned that tree roots could sometimes collect cobbles during flash floods, and once the tree rotted away, a misleading arrangement of stones could result. Equally, the Leakeys considered that associations of animal bones and stone tools at Olduvai indicated that *Homo habilis* may have been a capable hunter, but other archaeologists interpreted the same evidence as indicating that *habilis* was only scavenging the kills of other carnivores.

The evidence from Boxgrove

A site like Boxgrove (see pp. 72–75) arguably gives clearer evidence of human hunting abilities. Here, half a million years ago, the butchery of large mammals was skilful, following a logical progression from skinning, through disarticulation to filleting and bone-smashing, and cut marks from stone tools always underlie any gnaw marks from other carnivores, who must therefore have scavenged the carcasses after the people had finished with them. Whether the animals were actually actively hunted by the humans, or whether the humans waited to drive off the real hunters, such as lions, is unclear, but current research favours the first alternative. The carcasses were apparently mainly from healthy mid-life adults, and in the case of powerful animals such as the rhinoceros, it

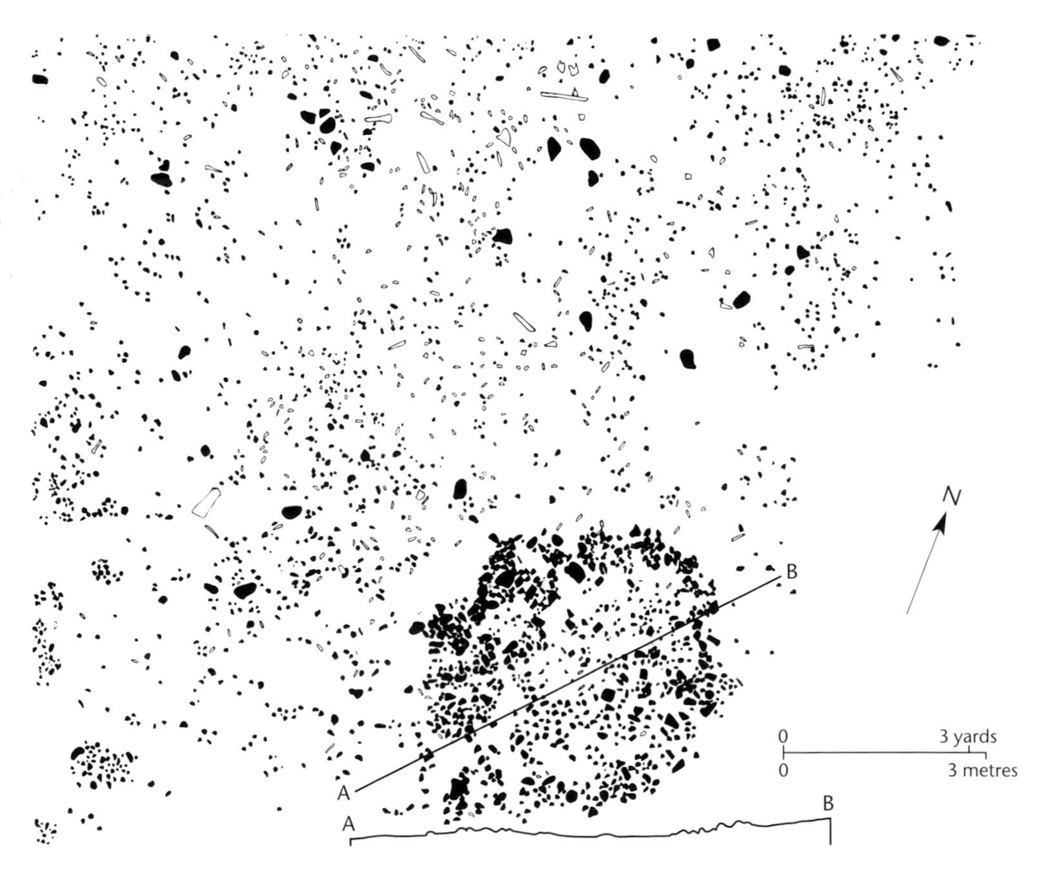

(Below right) One of the well-preserved hunting spears found at Schöeningen in Germany. It measures some 2 m (6.5 ft) long.

(Below) This circular concentration of basalt fragments at Olduvai Gorge was interpreted by Mary Leakey as the base of a structure, but could merely indicate the location of a tree.

(Opposite) This horse bone from Boxgrove shows extensive traces of butchery marks made by stone tools. Such evidence can be used to show the sequence of a skeleton's dismemberment. Careful examination may also show damage from the teeth of predators or scavengers, and the sequence of modification can be critical in determining whether humans or other animals had first access to the carcass.

(Right) Excavation of this fireplace at Vanguard Cave, Gibraltar, shows the Neanderthals cooked some meaty bones, but probably ate the marrow raw.

seems unlikely they had any natural predators other than the humans. Additionally, more direct evidence of hunting comes from an apparent spear-point hole on a horse shoulder blade.

Indirect support for the use of spears at Boxgrove has recently come from a somewhat later site in Germany called Schöeningen, where several beautifully made and preserved full-length wooden spears were excavated among a collection of horse skeletons. However, we can only reconstruct, and therefore glimpse, a small part of the lives of the Boxgrove people, as they obtained some of their food. We do not know whether they wore clothing, built shelters, or had the use of fire, as no evidence of these survive at Boxgrove.

Neanderthals and Cro-Magnons

For later peoples, such as the Neanderthals, we have more information, because we have been able to excavate areas where they actually lived, as well as hunted. An example of this comes for the Gibraltar caves (see pp. 76–79), where we can directly compare the way of life of the Neanderthals and the succeeding early modern people in the same sites. The manner in which they used the caves in Gibraltar does seem to contrast. Although their diets seemed similar, to judge from the ibex, deer, rabbit and bird bones in their respective occupation levels, the fireplaces preserved from the Middle Palaeolithic of both Gorham's and Vanguard caves seemed literally to be only that – a place where one or a few Neanderthals dug a hole in the sand and briefly lit a fire. After a short time, perhaps one night, they moved on. But in the Upper Palaeolithic levels at Gorham's, the early modern people seemed to light their fires in the same places over many years, and there was evidence of beach cobbles being used to line the hearths. Their settlements seemed longer-term and more intensive, and perhaps consisted of larger groups. This is consistent with the general pattern of both Neanderthal and Cro-Magnon occupation across the continent of Europe.

Archaeologists such as Lewis Binford have used such evidence to propose that Neanderthals and other premodern peoples had completely different social structures than living peoples. Instead of family groups with males and females in regular association, Binford has argued that Neanderthal men and women may have lived largely separate lives. The women and their children foraged locally, mainly on plant and small mammal resources, while groups of men hunted and scavenged widely across the landscape, only returning occasionally to the women with parts of carcasses. This view seems too extreme from the limited evidence we have, but it would certainly be unwise to assume that the social structures of today, which are often based on long-term pairing of males and females, had a long evolutionary history.

(Below) Early human skeletons may show evidence of injuries, such as wounds or fractures. Neanderthals in particular seem to have been prone to such injuries, evidence of their potentially hard and dangerous lives. Experts have compared the distribution of Neanderthal skeletal injuries with those of different athletes and sportspeople. The closest match was found with rodeo riders, who regularly get into close proximity to wild animals. This suggested that many Neanderthal injuries could have been caused by dangerous animals during hunting. However, the pattern is also consistent with elevated levels of interpersonal violence.

Related to mating

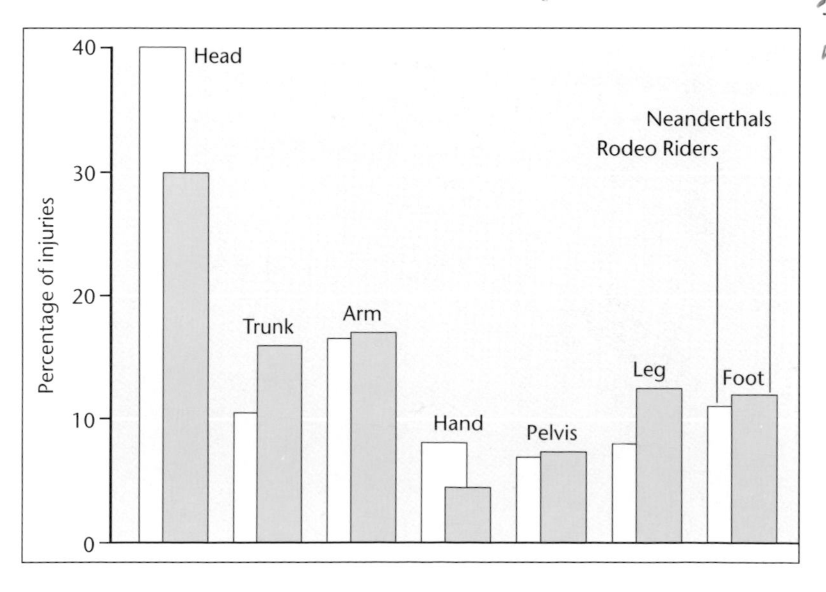

Reconstructing Ancient Human Behaviour 2

As we have seen, the direct physical evidence from which we can reconstruct ancient human behaviour is limited to the bones and stones left behind, and the sites in which they are preserved, and there are clearly major analytical problems in interpreting these selective and fragmentary patterns. However, as we have also seen, we now have a large and growing body of knowledge about the behaviour and capabilities of our closest relatives, the great apes, from both captive and free-living examples, and these can provide models for the reconstruction of the behaviour of ancient human populations. Some researchers have even extended comparisons to use other primates such as baboons, or non-primates such as social carnivores (e.g. wolves or hyaenas). There is also a large body of data about the social structures, behaviours and adaptations of present-day hunter-gatherers and foragers that can provide comparative frameworks for reconstructing the behaviour of our ancient predecessors, although as these are modern humans, increasing caution must be used in comparisons as we go further and further back into the past. While, on the face of it, the life styles and technology of recent hunter-gatherers may appear simple enough to draw analogies with Palaeolithic humans, we know that their social structures, language and religious systems can be every bit as complex as those in 'developed' societies, and it seems unlikely that pre-modern humans had anything comparable.

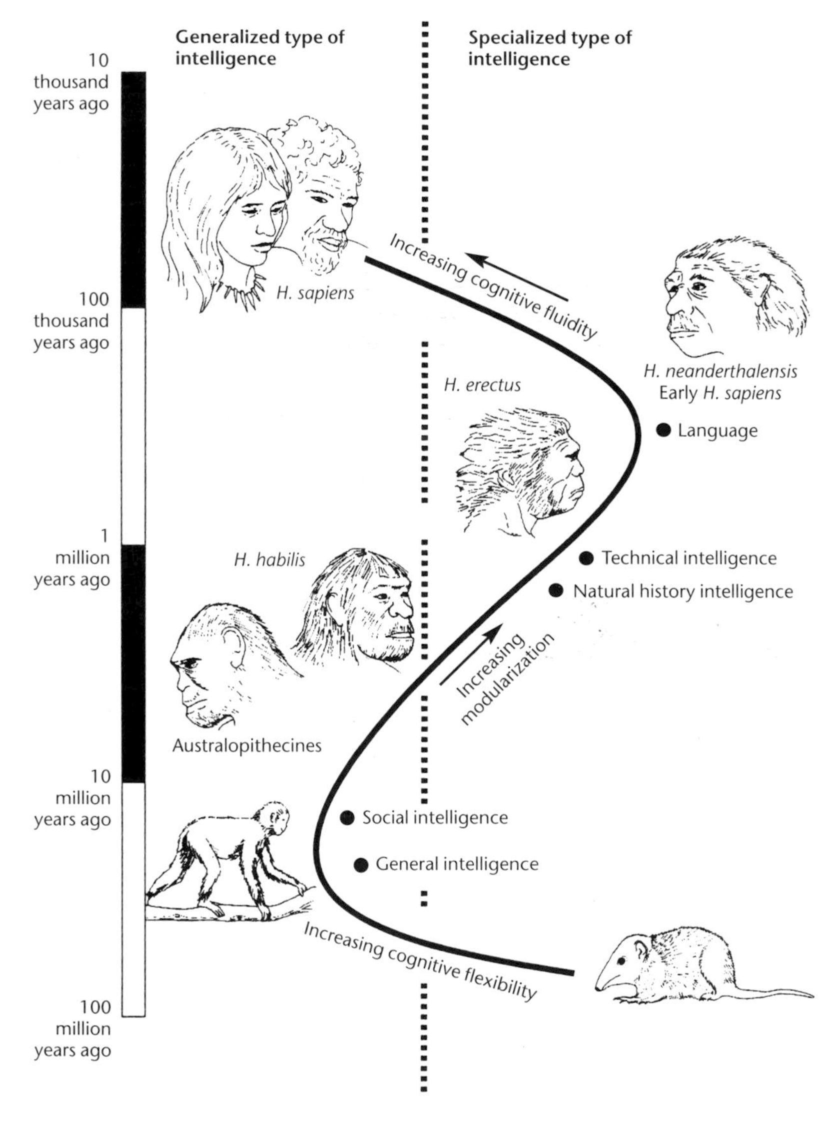

The archaeologist Steve Mithen's view of the evolution of intelligence. He believes that earlier humans were intelligent, but that different domains of intelligence were combined in a new way in Homo sapiens.

Evolutionary psychology

Recently, an entirely different approach to understanding how our behaviour has evolved has developed – the field of evolutionary psychology. Evolutionary psychology attempts to understand our behaviour by examining the selective forces that may have shaped it. Some researchers believe that we have evolved mental modules in our brains that respond to environmental or social cues. These cues may elicit specific behaviours in reacting to threats, social situations, sexual signals, infant demands etc. and it is argued that typical human responses have been selected for in our evolutionary past because they improved survival and reproductive success. Other researchers believe that there is much greater flexibility in our responses but nevertheless, we (often unconsciously) weigh up choices to optimize the benefits to ourselves and our kin, whether the benefits come in the form of resources, social advantage or reproductive success.

A simple example of how evolutionary psychology operates is how it looks at the evolution of behavioural differences between males and females. Fertile men can potentially have many offspring right through their adult lifespan, whereas women are much more restricted by birth intervals, their shorter reproductive life (given the menopause), and the need to nurture their offspring. Thus it is argued that past evolution has operated such that men are attracted by potential fertility in a prospective partner, while women are attracted by men who are likely to provide stability and resources after reproduction. There is certainly good evidence to back up such theoretical expectations from a range of human societies, yet it is clear that such imperatives are not the whole story, given variations such as partnerships that continue in the face of reproductive cessation or failure, and homosexual relationships.

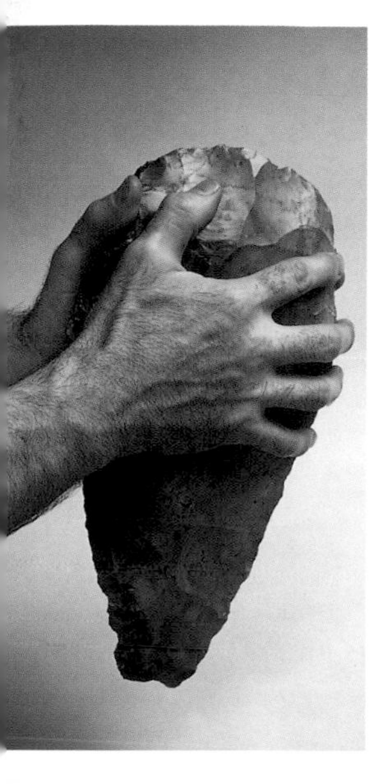

(Above) *Some handaxes are massive, such as this one from Furze Platt, near Maidenhead in England. It is not known whether these were functional, or perhaps for display.*

Some sites are particularly rich in their evidence of stone tools. It has been estimated, for example, that the Lower Palaeolithic site of Swanscombe in southern England has produced at least 100,000 handaxes. This photograph shows the site of Olorgesailie, in Kenya, where thousands of Early Stone Age artifacts have been uncovered. Such sites may accumulate these concentrations over long periods of time through natural processes such as water transport or the erosion of surrounding sediments by the action of wind or water.

An important concept for some researchers in evolutionary psychology is the Environment of Evolutionary Adaptedness (EEA) – that is, the past environment that produced the selective effect that is observed operating today. It is often assumed that the Pleistocene was the period that shaped most of our distinctive human behaviours today, and it clearly was a critical period in human prehistory, but some elements must have evolved before the Pleistocene, and others have perhaps developed in response to the major shifts that came, in some parts of the world, with the rise of agriculture. For some researchers, many of the maladies that affect people in the developed world such as mental illness, drug abuse, violent crime and child abuse are symptoms of mismatches between the environments in which we evolved and the environments in which we live now.

Some workers have applied evolutionary psychological concepts to help explain enigmas in past human behaviour, a good example being handaxes. The basic shape of these tools remained unchanged across a million years and many thousands of miles, and this is rather surprising, although we know from modern experiments that handaxes do make good butchery tools. Yet they are sometimes made so large or so beautifully that it is difficult to believe they were meant to be functional, and at some sites they occur in such profusion, and in such fresh states, that it does seem many of them were never used. Two researchers have independently suggested an explanation based on evolutionary psychology. Handaxe production was not just a practical matter but had social significance – in particular males used handaxe manufacture to signal attractiveness and fitness to potential mates. Thus handaxes acted as a kind of status symbol for males, and the ability to make them skilfully would have had selective value in terms of reproductive success.

The field of evolutionary psychology is still in its infancy and no doubt some of its proponents have an oversimplified view of human behaviour and overplay the importance of purely genetic and instinctive components compared with those that are flexible or environmentally or culturally influenced. But as the science develops it should throw new light on the evolution of critical human features such as language, symbolism and religious systems, for which we lack much hard data.

An Overview of Human Origins

In this book, we have surveyed some 30 million years of primate evolution, and over 5 million years of human evolution. What can we really say about our evolutionary past and our present success, what do we still have to learn, and what might lie ahead for the human species?

An undirected evolution

One of the main lessons of the human evolutionary story is how undirected it really was, and how insignificantly it began, and continued. If major changes had not impacted the Earth about 65 million years ago, the dominance of reptiles on land and sea would not have been disturbed, and the great radiation of the mammals, including our early primate ancestors, could not have begun. Thirty million years ago, our ancestors were small tree-dwelling monkey-like creatures, and 4 million years ago, our ancestors were probably still partly tree-living, but now in the form of apes who walked bipedally on the ground. The circumstances which led some of those apes down the road to humanity are still unknown, and chance events must have played an important part.

Even as recently (in geological terms) as 130,000 years ago, the proverbial observer from another planet would hardly have earmarked *Homo sapiens* as a species likely to dominate the planet one day. The species was restricted to one region – Africa – and was probably only present in low numbers, perhaps as low as 10,000 breeding individuals. The species was only capable of producing Middle Stone Age tools, lacked any means of independent food production and was largely at the mercy of an unstable planetary climate. Elsewhere on the planet

Humans have a natural curiosity and a tendency to disperse. This reached its zenith in expeditions into space, most spectacularly in the moon landings that began in 1969. The search for life on other worlds is one of the driving forces for such exploration. Is life unique to our planet – are we alone in the Universe? During the next few years it may be possible to say whether life exists elsewhere in our solar system – for example on Mars, or in the ice-capped oceans of Jupiter's moon, Europa.

there were other, equally successful (or depending on the viewpoint, unsuccessful) human species, such as the Neanderthals in Europe, and *Homo erectus* in Indonesia. That alien observer would have seen no reason to imagine that *Homo sapiens* would one day emerge from Africa, take over from the other species, progressively colonize every habitable region of the planet, and eventually even travel beyond it.

A complex process

Human evolution was certainly a complex process, and we are still only beginning to appreciate this complexity. Two million years ago, there were at least four human-like species living in Africa, and even 100,000 years ago, there were probably three species, one each in Africa, Europe and East Asia. Today we are alone, but this is a very unusual situation for a hominin species. From this perspective, it is easy to imagine that we were predestined to succeed, and that our qualities were those that were required for success. If a *Tyrannosaurus* could have thought like this, it might well have concluded that it, too, was the pinnacle of evolution with its great size, strength and ferocity. While we are justly proud of our large brains, it is as well to remember that the Neanderthals had brains about the same size as ours, but we are here now, and not them. However, if the events of the Ice Age had unfolded slightly differently, perhaps our species would not have emerged from Africa at all, and the Neanderthals might eventually have been the ones to colonize the rest of the world. Perhaps the first human foot on the Moon's surface might then have belonged to a Neanderthal!

An unpredictable future

The future of our species is as unpredictable as was our past. Mammalian species generally appear to have lifetimes measured in the hundreds of thousands or the low millions of years so, on that basis, we could expect a long, but finite future. Most species become extinct without issue, but some give rise to new species. Our chances of survival would

(Right) This satellite image shows the collapse of a huge section of the Larsen ice shelf (over 100 miles (160 km) across) in the Antarctic, caused by unusually warm summer temperatures. Pools of meltwater are clearly visible on the remaining shelf. Such collapses are becoming more frequent with global warming.

seem to rest partly in our own hands, and partly elsewhere. While we continue to overexploit the Earth because of our inflated numbers, and to threaten the planet with the consequences of nuclear war, the future looks gloomy. But just as challenging are the likely effects of future climatic change, since there may be greater climatic changes in the next 100 years than in the past 10,000.

In the last couple of centuries, we have been conducting an uncontrolled experiment with the

Our ancestors had to cope with the effects of rapid climate change many times in their evolution, but through our own actions, the immediate future is likely to bring more severe change on our planet than we have ever experienced before. This satellite image (left) shows the Brazilian state of Mato Grosso. Dark green areas are intact forest, cleared areas are brown, and plumes of smoke show extensive burning, indicating the continuing destruction of the rainforests, further contributing to greenhouse gases in the atmosphere. (Above) A typical scene of destruction in what was once a lush rainforest environment.

composition of the Earth's atmosphere, and global warming is a growing reality. We are probably entering a 'super-interglacial', warmer than humans have ever experienced before, and while that may sound appealing to those of us who live in the colder parts of the world, along with the higher temperatures would come major and unpredictable shifts in the patterns of the world's weather. One of the most significant lessons of the recent past is that the Earth's climate is very unstable, and such changes could include a sudden and severe chilling of the North Atlantic, even in the face of global warming, if the Gulf Stream is diverted southwards by increasing meltwaters from the Arctic. Let us hope that our descendants can cope with the chaos predicted by many of the world's expert climatologists. Perhaps by then, if our species is not too impoverished or preoccupied, it will have been able to establish viable colonies on other planets, and the human evolutionary story will start to unfold off the Earth, as well as continue on it.

(Below) The reconstructed figure in the centre is a Neanderthal woman, and she is surrounded by other species of ape and human, both past and present. With continuing destruction of their habitats, the apes, our closest living relatives, may soon share the fate of the Neanderthals (and all the other human species except our own) and become extinct.

Further Reading

Abbreviations

AHG *Annals of Human Genetics*
AJHG *American Journal of Human Genetics*
AJPA *American Journal of Physical Anthropology*
EA *Evolutionary Anthropology*
JHE *Journal of Human Evolution*
NG *National Geographic*
PNAS *Proceedings of the National Academy of Sciences*
SA *Scientific American*
YbPA *Yearbook of Physical Anthropology*

Websites

http://www.becominghuman.org/
http://encarta.msn.com/encnet/refpages/RefArticle.aspx?refid=761566394
http://www.pbs.org/wgbh/evolution/library/07/index.html
http://www.talkorigins.org/

General Reading

Aiello, L.C. & Dean, C., *Human Evolutionary Anatomy* (Academic Press, 1990)
Gore, R., 'The Dawn of Humans: The First Steps', *NG* 191: 72–99 (1997)
Hartwig, W.C. (ed.), *The Primate Fossil Record* (Cambridge University Press, 2002)
Johanson D.C. & Edgar B., *From Lucy to Language* (Simon and Schuster, 1996)
Jones, S., Martin, R. & Pilbeam, D., *Cambridge Encyclopaedia of Human Evolution* (Cambridge University Press, 1992)
Klein, R.G., *The Human Career: Human Biological and Cultural Origins* (University of Chicago Press, 1999)
Klein, R.G. & Edgar, B., *The Dawn of Human Culture* (John Wiley & Sons, 2002)
Lewin, R., *Human Evolution: An Illustrated Introduction* (Blackwell Scientific Publications, 1998)
New Look at Human Evolution (Special Issue *SA*, 2003)
Schick, K.D. & Toth, N., *Making Silent Stones Speak: Human Evolution and the Dawn of Technology* (Simon & Schuster, 1994)
Tattersall, I., *The Monkey in the Mirror: Essays on the Science of What Makes Us Human* (Oxford University Press, 2002)
Tattersall, I., Delson, E., Van Couvering, J. & Brooks, A.S. (eds.), *Encyclopedia of Human Evolution and Prehistory* (Garland Publishing, 1999)
Weaver, K.F., 'The Search for Our Ancestors', *NG* 168: 560–623 (1985)
Zihlman, A.L., *The Human Evolution Coloring Book* (Concepts Inc. Coloring, 2001)

I In Search of Our Ancestors

Living Apes and Their Environment

Fleagle, J., *Primate Adaptations and Evolution* (Academic Press, 1988)
Groves, C., *Primate Taxonomy* (Smithsonian Press, 2001)
Napier, J.H. & Napier, P.H., *A Handbook of Living Primates* (Academic Press, 1967)

Human Variations

Howells, W.W., 'Skull shapes and the map', *Papers of the Peabody Museum, Harvard* 79 (1989)
Lahr, M., *The Evolution of Modern Human Diversity: a Study of Cranial Variation* (Cambridge University Press, 1966)

Palaeoanthropology

Jones, S., Martin, R. & Pilbeam, D., *Cambridge Encyclopaedia of Human Evolution* (Cambridge University Press, 1992)
Lewin, R., *Human Evolution: An Illustrated Introduction* (Blackwell Scientific Publications, 1998)
Tattersall, I., Delson, E., Van Couvering, J. & Brooks, A.S. (eds.), *Encyclopedia of Human Evolution and Prehistory* (Garland Publishing, 1999)

The Geological Timescale

Lamb, S. & Sington, D., *Earth Story: The Shaping of Our World* (BBC Consumer Publishing, 1998)
Lewis C. & Knell, S. (eds.), *The Age of the Earth: from 4004 BC to AD 2002* (The Geological Society, 2002)

Dating the Past

Klein, R.G., *The Human Career: Human Biological and Cultural Origins* (University of Chicago Press, 1999)
Taylor, R. & Aitken, M. (eds.), *Chronometric Dating in Archaeology* (New York, 1997)

Studying Animal Function

Begun, D., Ward, C.V. & Rose, M. (eds.), *Function, Phylogeny and Fossils* (Plenum Press, 1997)
Aiello, L. & Dean, C., *Human Evolutionary Anatomy* (Academic Press, 1990)

Excavation and Analytical Techniques

Renfrew, C. & Bahn, P., *Archaeology: Theories, Methods and Practice* (Thames & Hudson, 2004)
Whybrow, P., *Travels with the Fossil Hunters* (Cambridge University Press, 2000)

New Techniques for Studying Fossils

Bocherens, H., Billiou, D., Mariotti, A., Toussaint, M., Patou-Mathis, M., Bonjean, D. & Otte, M., 'New isotopic evidence for dietary habits of Neanderthals from Belgium', *JHE* 40: 497–505 (2001)
Ponce de León, M. & Zollikofer, C., 'Neanderthal cranial ontogeny and its implications for late hominid diversity', *Nature* 412: 534–538 (2001)
Renfrew, C. & Bahn, P., *Archaeology: Theories, Methods and Practice* (Thames & Hudson, 2004)

Taphonomy: How Fossils are Preserved

Andrews, P., *Owls, Caves and Fossils* (University of Chicago Press, 1990)
Brain, C.K., *Hunters or the Hunted?* (University of Chicago Press, 1981)
Lyman, R., *Vertebrate Taphonomy* (Cambridge University Press, 1994)

What Fossils Tell us about Ancient Environments

Bromage, T. & Schrenk, F., *African Biogeography, Climate Change and Human Evolution* (Oxford University Press, 1999)
Vrba, E.S., Denton, G.H., Partridge, T.C. & Burckle, L.H., *Paleoclimate and Evolution with Emphasis on Human Origins* (Yale University Press, 1995)

Changing Climates

Bromage, T. & Schrenk, F., *African Biogeography, Climate Change and Human Evolution* (Oxford University Press, 1999)
Potts, R., 'Environmental hypotheses of hominin evolution', *YbPA* 41: 93–136 (1998)
Vrba, E.S., Denton, G.H., Partridge, T.C. & Burckle, L.H., *Paleoclimate and Evolution with Emphasis on Human Origins* (Yale University Press, 1995)

Site I: Rusinga Island

Andrews, P. & Van Couvering, J.H., 'Palaeoenvironments in the East African Miocene', in *Approaches to Primate Paleobiology* (Karger, 1975)
Walker, A. & Teaford, M., 'The hunt for *Proconsul*', *SA* 260: 76–82 (1988)

Site II: Paşalar

Andrews, P. (ed.), 'The Miocene Hominoid site at Paşalar, Turkey', *JHE* 19 (1990)
Andrews, P. & Alpagut, B. (eds.), 'Further Papers on the Miocene Site at Paşalar, Turkey', *JHE* 28 (1995)

Site III: Rudabánya

Begun, D. & Kordos, L., 'Revision of *Dryopithecus brancoi* based on the fossil material from Rudabánya', *JHE* 25: 271–285 (1993)
Kordos, L. & Begun, D., 'A late Miocene subtropical swamp deposit with evidence of the origin of the African apes and humans', *EA* 1: 45–57 (2002)

Site IV: Olduvai Gorge

Hay, R.L., *Geology of Olduvai Gorge* (University of California Press, 1996)
Klein, R.G., *The Human Career: Human Biological and Cultural Origins* (University of Chicago Press, 1999)

Site V: Boxgrove

Pitts, M. & Roberts, M., *Fairweather Eden: Life in Britain Half a Million Years Ago as Revealed by the Excavation at Boxgrove* (Arrow, 1998)
Roberts, M.B. & Parfitt, S.A., 'Boxgrove. A Middle Pleistocene Hominid site at Eartham Quarry, Boxgrove, West Sussex' (*English Heritage Archaeological Report* 17, 1999)

Site VI: Gibraltar

Stringer, C., 'Digging the Rock', in P. Whybrow (ed.), *Travels with the Fossil Hunters* (Cambridge University Press, 2000), pp. 42–59
Stringer, C., Barton, R.N. & Finlayson, C. (eds.), *Neanderthals on the edge: 150th anniversary conference of the Forbes' Quarry discovery, Gibraltar* (Oxbow Books, 2000)

II The Fossil Evidence

Origin of the Primates

Fleagle, J., *Primate Adaptations and Evolution* (Academic Press, 1988)
Groves, C., *Primate Taxonomy* (Smithsonian Press, 2001)
Szalay, F. & Delson, E., *Evolutionary History of the Primates* (Academic Press, 1979)

Early Anthropoids

Fleagle, J., *Primate Adaptations and Evolution* (Academic Press, 1988)
Fleagle, J. & Kay, R., *Anthropoid Origins* (Plenum Press, 1994)

What Makes an Ape?

Begun, D., Ward, C.V. & Rose, M. (eds.), *Function, Phylogeny and Fossils* (Plenum Press, 1997)
Conroy, G., *Primate Evolution* (W.W. Norton, 1990)

Ancestral Apes

Andrews, P., 'A Revision of the Miocene Hominoidea of East Africa', *Bulletin of the British Museum (Natural History)* 30: 85–225 (1978)
Hartwig, W.C. (ed.), *The Primate Fossil Record* (Cambridge University Press, 2002)

Proconsul and its Contemporaries

Andrews, P., 'A Revision of the Miocene Hominoidea of East Africa', *Bulletin of the British Museum (Natural History)* 30: 85–225 (1978)
Hartwig, W.C. (ed.), *The Primate Fossil Record* (Cambridge University Press, 2002)
Walker, A. & Teaford, M., 'The Kaswanga primate site: an early Miocene hominoid site on Rusinga Island, Kenya', *JHE* 17: 539–544 (1988)

Middle Miocene African Apes

Hartwig, W.C. (ed.), *The Primate Fossil Record* (Cambridge University Press, 2002)
Ward, S.C., Brown, B., Hill, A., Kelley, J. & Downs, W., '*Equatorius*, a new hominoid genus from the middle Miocene of Kenya', *Science* 285: 1382–1386 (1999)

The Exit from Africa

Andrews, P. (ed.), 'The Miocene Hominoid site at Paşalar, Turkey', *JHE* 19 (1990)
Hartwig, W.C. (ed.), *The Primate Fossil Record* (Cambridge University Press, 2002)

Ankarapithecus – a Fossil Enigma

Alpagut, B., Andrews, P., Fortelius, M., Kappelman, J., Temizsoy, I., Celebi, H. & Lindsay, W., 'A new specimen of *Ankarapithecus meteai* from the Sinap Formatoin of central Anatolia', *Nature* 382: 349–351 (1996)
Begun, D. & Gulec, E., 'Restoration of the type and palate of *Ankarapithecus meteai*: taxonomic and phylogenetic implications', *AJPA* 105: 279–314 (1998)
Hartwig, W.C. (ed.), *The Primate Fossil Record* (Cambridge University Press, 2002)

Orang utan Ancestors

Andrews, P. & Cronin, J., 'The relationships of *Sivapithecus* and *Ramapithecus* and the evolution of the orang utan', *Nature* 297: 541–546 (1982)
Pilbeam, D., 'Genetic and morphological records of the Hominoidea and hominid origins: a synthesis', *Molecular Phylogenetics and Evolution* 5: 155–168 (1996)

The Ancestry of the Living Apes

Hartwig, W.C. (ed.), *The Primate Fossil Record* (Cambridge University Press, 2002)
Kordos, L. & Begun, D., 'A late Miocene subtropical swamp deposit with evidence of the origin of the African apes and humans', *Evolutionary Anthropology* 11: 45–57 (2002)
Pilbeam, D., 'Genetic and morphological records of the Hominoidea and hominid origins: a synthesis', *Molecular Phylogenetics and Evolution* 5: 155–168 (1996)

Late Miocene Apes and Early Human Ancestors

Brunet, M. *et al.*, 'A new hominid from the Upper Miocene of Chad, Central Africa', *Nature* 418: 145–151 (2002)
Clarke, R.J., 'Newly revealed information on the Sterkfontein Member 2 *Australopithecus* skeleton', *South Afr. J. Sci.* 98: 523–526 (2002)
Haile-Selassi, Y., 'Late Miocene hominids from the Middle Awash, Ethiopia', *Nature* 412: 178–181 (2001)
Hartwig, W.C. (ed.), *The Primate Fossil Record* (Cambridge University Press, 2002)
Johanson, D., White T. & Coppens, Y., 'A new species of the genus *Australopithecus* (Primates: Hominidae) from the Pliocene of eastern Africa', *Kirtlandia* 28: 1–14 (1978)
Leakey, M.G., Feibel, C.S., McDougall, I. & Walker, A., 'New four million year hominid species from Kanapoi and Allia Bay, Kenya', *Nature* 376: 565–571 (1995)
Leakey, M.D. & Hay, R., 'Pliocene footprints in the Laetolil Beds at Laetoli, northern Tanzania', *Nature* 278: 317–323 (1979)
Leakey, M.D. & Harris, J.M., *Laetoli: A Pliocene Site in northern Tanzania* (Oxford University Press, 1987)
Senut, B., Pickford, M., Gommery, D., Mein, P., Cheboi, K. & Coppens, Y., 'First hominid from the Miocene (Lukeino Formation, Kenya', *C.R. Acad. Sci. Paris* 332. 137–144 (2001)
Ward, C.V., 'Interpreting the posture and locomotion of *Australopithecus afarensis*: where do we stand?', *Yrbk Phys. Anthrop.* 45: 185–225 (2002)
White, T., Suwa, G. & Asfaw, B., '*Australopithecus ramidus*, a new species of early hominid from Aramis, Ethiopia', *Nature* 371: 306–312 (1995)

Australopithecus africanus

Berger, L., 'The Dawn of Humans: Redrawing Our Family Tree?' *NG* 194: 90–99 (1998)
Falk, D., Redmond, J.C., Guyer, J., Conroy, G.C., Recheis, W., Weber, G.W., Seidler, H., 'Early hominid brain evolution: a new look at old endocasts', *JHE* 38: 695–717 (2000)
Sponheimer, M. & Lee-Thorp, J.A., 'Isotopic evidence for the diet of an early hominid, *Australopithecus africanus*', *Science* 283: 368–369 (1999)

Robust Australopithecines

Brain, C.K., 'Swartkrans, A Cave's Chronicle of Early Man', *Transvaal Museum Monograph*, No. 8 (1993)
Keyser, A., 'The Dawn of Humans: New Finds in South Africa', *NG*, 197: 76–83 (2000)
Suwa, G., Asfaw, B., Beyene, Y., White, T., Katoh, S., Nagaoka, S., Nakaya, H., Uzawa, K., Renne, P., & WoldeGabriel, G., 'The first skull of *Australopithecus boisei*', *Nature* 389: 489–492 (1997)

The Origins of Humans

Aiello, L.C. & Dunbar, R.I.M., 'Neocortex size, group size and the evolution of language', *Current Anthropology* 34: 184–194 (1993)
Aiello, L.C. & Wheeler, P., 'The expensive-tissue hypothesis: the brain and the digestive system in human and primate evolution', *Current Anthropology* 34: 184–193 (1995)
Elton, S., Bishop, L.C. & Wood, B.A., 'Comparative context of Plio-Pleistocene hominin brain evolution', *JHE* 41: 1–27 (2001)
Wood, B.A. & Richmond, B.G., 'Human evolution: taxonomy and palaeobiology', *Journal of Anatomy* 196: 19–60 (2000)

Early Homo

Asfaw, B., White, T., Lovejoy, O., Latimer, B., Simpson, S. & Suwa, G., '*Australopithecus garhi*: a new species of early hominid from Ethiopia', *Science* 284: 629–635 (1999)
Gore, R., 'New find (Dmanisi)' *NG* unpaginated news section (August 2002)
Heinzelin, J.D., Clark, J.D., White, T., Hart, W., Renne, P., WoldeGabriel, G., Beyene, Y. & Vrba, E., 'Environment and behavior of 2.5-million-year-old Bouri hominids', *Science* 284: 625–629 (1999)
Vekua, A., Lordkipanidze, D., Rightmire, G. Philip, Agusti, J., Ferring, R., Maisuradze, G., Mouskhelishvili, A., Nioradze, M., Ponce de Leon, M., Tappen, M., Tvalchrelidze, M. & Zollikofer C., 'A New Skull of Early *Homo* from Dmanisi, Georgia', *Science* 297: 85–89 (2002)
Wood, B. & Collard, M., 'The Human Genus', *Science* 284: 65–71 (1999)

Homo erectus

Asfaw, B., Gilbert, W.H., Beyene, Y., Hart, W.K., Renne, P.R., WoldeGabriel, G., Vrba, E. & White, T.D., 'Remains of *Homo erectus* from Bouri, Middle Awash, Ethiopia', *Nature* 416: 317–320 (2002)
Gore, R., 'Dawn of Humans: Expanding Worlds', *NG* 191: 84–109 (1997)
Ruff, C.B., Trinkaus, E. & Holliday, T.W., 'Body mass and encephalization in Pleistocene *Homo*', *Nature* 387: 173–176 (1997)

Models of Recent Human Evolution

Lewin, R., *The Origin of Modern Humans: A SA Library Vol* (W. H. Freeman, 2002)
Lieberman, D., McBratney, B. & Krovitz, G., 'The evolution and development of cranial form in *Homo sapiens*', *PNAS* 99: 1134–1139 (2002)
Stringer, C., 'Modern human origins: progress and prospects', *Philosophical Transactions of the Royal Society of London* 357B: 563–579 (2002)
Thorne, A. & Wolpoff, M., 'The Multiregional Evolution of Modern Humans', *SA* 266: 76–83 (1992)
Wolpoff, M. & Caspari, R., *Race and human evolution: a fatal attraction* (Simon & Schuster, 2002)

The Early Occupation of Europe: Gran Dolina

Atapuerca: nuestros antecesore (Junta de Castilla y León, 1999)
Manzi, G., Mallegni, F. & Ascenzi, A., 'A cranium for the earliest Europeans: phylogenetic position of the hominid from Ceprano, Italy', *PNAS USA* 98: 10011–10016 (2001)

Homo heildelbergensis

Gore, R., 'Dawn of Humans: The First Europeans', *NG* 192: 96–113 (1997)

Hublin, J.-J., 'Northwestern African Middle Pleistocene hominids and their bearing on the emergence of *Homo sapiens*', in L. Barham & K. Robson-Brown (eds.), *Human Roots: Africa and Asia in the Middle Pleistocene* (Western Academic and Specialist Press, 2001)

Rightmire, G.P., 'Comparison of Middle Pleistocene hominids from Africa and Asia', in L. Barham & K. Robson-Brown (eds.), *Human Roots: Africa and Asia in the Middle Pleistocene* (Western Academic and Specialist Press, 2001)

Atapuerca and the Origin of the Neanderthals

Arsuaga, J.L., *The Neanderthal's Necklace: In Search of the First Thinkers* (Wiley, 2002)

Arsuaga, J.L., Bermúdez de Castro, J.M. & Carbonell, E. (eds.), 'The Sima de los Huesos Hominid site', *JHE* 33: 105–421 (1997)

Atapuerca: nuestros antecesore (Junta de Castilla y León, 1999)

The Neanderthals

Gore, R., 'Neanderthals', *NG* 189: 2–35 (1996)

Shreeve J., *The Neandertal enigma: solving the mystery of human origins* (William Morrow, 1995)

Stringer, C. & Gamble, C., *In Search of the Neanderthals* (Thames & Hudson, 1993)

Trinkaus E. & Shipman P., *The Neandertals: changing the image of mankind* (Alfred E. Knopf, 1992)

Africa – Homeland of Homo sapiens?

Gore, R., 'Tracking the first of our kind', *NG* 192: 92–99 (1997)

Lieberman, D., McBratney, B.,M. & Krovitz, G., 'The evolution and development of cranial form in Homo sapiens', *PNAS* 99: 1134–1139 (2002)

Stringer, C., 'Human evolution: Out of Ethiopia', *Nature* 423: 692–695 (2003)

Stringer, C. & McKie, R., *African Exodus: The Origins of Modern Humanity* (Pimlico, 1998)

Asia – Corridor or Cul-de-sac?

Bar-Yosef, O. & Pilbeam, D., 'Geography of Neandertals and Modern Humans in Europe and the Greater Mediterranean' (*Peabody Museum Bulletins*, no. 8, 2000)

Brown, P., 'Chinese Middle Pleistocene hominids and modern human origins in East Asia', in L. Barham & K. Robson-Brown (eds.), *Human Roots: Africa and Asia in the Middle Pleistocene* (Western Academic and Specialist Press, 2001)

Gibbons, A., '*Homo erectus* in Java: a 250,000 year anachronism', *Science* 274, 1841–1842 (1996)

What Happened to the Neanderthals?

D'Errico, F., 'The invisible frontier. A multiple species model for the origin of behavioral modernity', *EA* 12: 188–202 (2003)

'Neanderthals meet modern humans', *Athena Review* 2: 13–64 (2001)

Shea, J., 'Neandertals, competition, and the origin of modern human behavior in the Levant', *EA* 12: 173–187 (2003)

Stringer, C. & Davies, W., 'Those elusive Neanderthals', *Nature* 413: 791–792 (2001)

The Cro-Magnons

Brown, P., 'The first modern East Asians?: another look at Upper Cave 101, Liujiang and Minatogawa 1', in K. Omoto (ed.), *Interdisciplinary Perspectives on the Origins of the Japanese* (International Research Center for Japanese Studies, 1999), pp. 105–131

Gore, R., 'The Dawn of Humans: People Like Us', *NG* 198: 90–117 (2000)

Holliday, T., 'Body proportions in Late Pleistocene Europe and modern human origins', *JHE* 32: 423–448 (1997)

The First Australians

Bowler, J.M., Johnston, H., Olley, J.M., Prescott, J.R., Roberts, R.G., Shawcross, W. & Spooner, N.A., 'New ages for human occupation and climatic change at Lake Mungo, Australia', *Nature* 421: 837–840 (2003)

Thorne, A., Grün, R., Mortimer, G., Spooner, N., Simpson, J., Mcculloch, M., Taylor, L. & Curnoe, D., 'Australia's oldest human remains: age of the Lake Mungo 3 skeleton', *JHE* 36: 591–612 (1999)

Homo floresiensis

Brown, P., *et al.*, 'A new small-bodied hominin from the Late Pleistocene of Flores, Indonesia', *Nature* 431: 1055–1061 (2004)

Lahr, M. & Foley, R., 'Human evolution writ small', *Nature* 431: 1043–1044 (2004)

Morwood, M. J., *et al.*, 'Archaeology and age of a new hominin from Flores in eastern Indonesia', *Nature* 431: 1087–1091 (2004)

Genetic Data on Human Evolution

Cavalli-Sforza, L.L., *Genes, Peoples and Languages* (North Point Press, 2000)

Cavalli-Sforza, L. L. & Feldman, M.W., 'The application of molecular genetic approaches to the study of human evolution', *Nature Genetics* 33 supplement: 266–275 (2003)

Jorde, L., Watkins, W., Bamshad, M., Dixon, M., Ricker, C., Seielstad, M. & Batzer, M., 'The distribution of human genetic diversity: a comparison of mitochondrial, autosomal, and Y-chromosome data', *AJHG* 66: 979–988 (2000)

Oppenheimer, S., *Out of Eden* (Constable & Robinson, 2003)

Underhill, P., Passarino, G., Lin, A., Shen, P., Lahr, M., Foley, R., Oefner, P. & Cavalli-Sforza, L., 'The phylogeography of Y chromosome binary haplotypes and the origins of modern human populations', *AHG* 65: 43–62 (2001)

Mitochondrial DNA

Cann, R., Stoneking, M. & Wilson, A., 'Mitochondrial DNA and human evolution', *Nature* 325: 31–36 (1987)

Ingman, M., Kaessmann, H., Pääbo, S. & Gyllensten, U., 'Mitochondrial genome variation and the origin of modern humans', *Nature* 408: 708 –713 (2000)

Neanderthal DNA

Hoss, M., 'Neanderthal population genetics', *Nature* 404: 453–454 (2000)

Richards, M. & Macaulay, V., 'Genetic data and the colonization of Europe: genealogies and founders', in C. Renfrew & K. Boyle (eds.), *Archaeogenetics* (McDonald Institute, 2000), pp. 139–151

Schmitz, R. W., Serre, D., Bonani, G., Faine, S., Hillgruber, F., Krainitzki, H., Paabo, S. & Smith, F.H., 'The Neanderthal type site revisited: Interdisciplinary investigations of skeletal remains from the Neander Valley, Germany', *PNAS* 99: 13342–13347 (2002)

III Interpreting The Evidence

Evolution of Locomotion in Apes and Humans

Fleagle, J., *Primate Adaptations and Evolution* (Academic Press, 1988)

Napier, J.H. & Napier, P.H., *A Handbook of Living Primates* (Academic Press, 1967)

Ward, C.V., 'Interpreting the posture and locomotion of *Australopithecus afarensis*: where do we stand?' *YbPA* 45: 185–225 (2002)

The Evolution of Feeding

Chivers, D., Wood, B. & Bilsborough, A. (eds.), *Food Acquisition and Processing in Primates* (Plenum Press, 1984)

Crowe, I., *The Quest for Food* (Tempus Publishing, 2000)

The Geographical Spread of Apes and Humans

Andrews, P., 'Fossil evidence on human origins and dispersal', in J.D. Watson (ed.), *Molecular Biology of Homo sapiens* (Cold Spring Harbor Symposia on Quantitative Biology, 1986), pp. 419–428

Bernor, R. Fahlbusch, V. & Mittmann, H-W., *The Evolution of Western Eurasian Neogene Mammal Faunas* (Columbia University Press, 1996)

Gamble, C., *Timewalkers: The Prehistory of Global Colonization* (Harvard University Press, 1996)

Kingdon, J., *Lowly Origins* (Princeton University Press, 2003)

Lahr, M. & Foley, R., 'Towards a theory of modern human origins: geography, demography, and diversity in recent human evolution', *YbPA* 41: 137–176 (1998)

Stringer, C. & Andrews, P., 'Genetic and fossil evidence for the origin of modern humans', *Science* 239: 1263–1268, and 241: 772–774 (1988)

The First Americans

Dillehay, T.D., 'Tracking the first Americans', *Nature* 425: 23–24 (2003)

Parfit, M., 'Who were the first Americans?', *NG* 198: 40–67 (2000)

Evolution and Behaviour in Relation to the Environment

Andrews, P., 'Palaeoecology and hominoid palaeoenvironments', *Biological Reviews* 71, 257–300 (1996)

Fleagle, J., Hanson, C. & Reed, K. (eds.), *Primate Communities* (Cambridge University Press, 1999)

Klein, R.G., *The Human Career* (Chicago University Press, 1999)

McGrew, W., Marchant, L. & Nishida, T. (eds.), *Great Ape Societies* (Cambridge University Press, 1996)

Napier, J.R. & Napier, P.H., *The Natural History of the Primates* (Cambridge University Press, 1985)

Vrba, E.S., Denton, G.H., Partridge, T.C. & Burckle, L.H., *Paleoclimate and Evolution with Emphasis on Human Origins* (Yale University Press, 1995)

Wrangham, R. & Petersen, D., *Demonic Males* (Bloomsbury Publishing, 1996)

Tools and Human Behaviour: The Earliest Evidence

Barham, L. & Robson-Brown, K., (eds.), *Human Roots: Africa and Asia in the Middle Pleistocene* (Western Academic and Specialist Press, 2001)
Dennell, R., 'The world's oldest spears', *Nature* 385: 767–768 (1997)
Klein, R.G., *The Human Career: Human Biological and Cultural Origins* (University of Chicago Press, 1999)
Kuman, K. & Clarke, R.J., 'Stratigraphy, artefact industries and hominid associations for Sterkfontein, Member 5', *JHE* 38: 827–847 (2000)
Semaw, S., Renne, P., Harris, J.W.K., Feibel, C.S., Bernor, R.L., Fesseha, N. & Mowbray, K., '2.5-million-year-old stone tools from Gona, Ethiopia', *Nature* 385: 333–336 (1997)

Tools and Human Behaviour: the Middle Palaeolithic

McBrearty, S. & Brooks A., 'The revolution that wasn't: a new interpretation of the origin of modern human behavior', *JHE* 39: 453–563 (2000)
Roebroeks, W. & Gamble, C. (eds.), *The Middle Palaeolithic Occupation of Europe* (University of Leiden, 1999)

Tools and Human Behaviour: the Upper Palaeolithic

Ambrose, S.H., 'Palaeolithic technology and human evolution', *Science* 291: 1748 (2001)
Klein, R.G., 'Archeology and the evolution of human behavior', *EA* 9: 17–36 (2000)
Klein, R.G. & Edgar, B., *The Dawn of Human Culture* (John Wiley & Sons, 2002)

The First Artists

Bahn, P. & Vertut, J., *Journey Through the Ice Age* (University of California Press, 2002)
Chauvet, J.M., Eliette Brunel Deschamps, E.B., Christian Hillaire, C. & Clottes, J., *Chauvet Cave: The Discovery of the World's Oldest Paintings* (Thames & Hudson, 1996)
Clottes, J. (ed.), *Return to Chauvet Cave: Excavating the Birthplace of Art: the First Full Report* (Thames & Hudson, 2003)
White, R., *Prehistoric Art: The Symbolic Journey of Humankind* (Harry N. Abrams, Inc., 2003)

Reconstruting Ancient Human Behaviour 1

Binford, L.R., *Bones: Modern Myths and Ancient Men* (Academic Press, 1981)
Renfrew, C. & Bahn, P., *Archaeology: Theories, Methods and Practice* (Thames & Hudson, 2005)

Reconstructing Ancient Human Behaviour 2

Buss, D.M., *Evolutionary Psychology. The New Science of Behavior* (Allyn & Bacon, 1999)
Mithen S., *The Prehistory of the Mind: The Cognitive Origins of Art, Religion, and Science* (Thames & Hudson, 1996)

An Overview of Human Origins

Bowler, J.M., Johnston, H., Olley, J.M., Prescott, J.R., Roberts, R.G., Shawcross, W. & Spooner, N.A., 'New ages for human occupation and climatic change at Lake Mungo, Australia', *Nature* 421: 837–840 (2003)
Jones, S., Martin, R. & Pilbeam, D., *Cambridge Encyclopaedia of Human Evolution* (Cambridge University Press, 1992)
Tattersall, I., *Becoming Human: evolution and human uniqueness* (Oxford University Press, 1999)
Thorne, A., Grün, R., Mortimer, G., Spooner, N., Simpson, J., Mcculloch, M., Taylor, L. & Curnoe, D., 'Australia's oldest human remains: age of the Lake Mungo 3 skeleton', *JHE* 36: 591–612 (1999)

Illustration Credits

Abbreviations:
a=above, b=below, c=centre, l=left, r=right.

J.M. Adovasio: 197a
After Airvaux and Pradel, Gravure d'une tête humaine, *Bull. Soc. Préhist. française* 81 (1984), p.212–215: 169a
Courtesy of Library Services, American Museum of Natural History, New York. Photo D.Finnin/C.Chesek, transparency no. 4936(7): 189b
Arizona State Museum, University of Arizona. Photo E.B. Sayles, neg. 3417-x-1: 196ar
Igor Astrologo after Krings *et al.*, Neanderthal DNA Sequences and the Origin of Modern Humans, *Cell* 90 (1997): 181br
Lewis Binford: 128
F. Blickle/Network: 228–229a
Annick Boothe: 32a (after Hedges and Gowlett in *Scientific American* 254(1) (1986), p. 84), 130r (after Lewin, *In the Age of Mankind* (1988), p. 181), 176ar
After Boule, M., L'homme fossile de la Chapelle-aux-Saints, Annales de Paléontologie (1911–13): 142–143b
Courtesy of The Boxgrove Project: 73bl, 73br, 74, 75br, 222ar
Photo The British Museum, London: 209ar
Peter Brown: 7c, 174l
Michel Brunet: 114–115b, 115a
Peter Bull Art Studio © Thames & Hudson Ltd, London: 17a, 18, 28 (adapted from Relethford, *The Human Species* (2002), pp. 110–111), 35a, 36 (after Fleagle, *Primate Adaptation and Evolution* (1988), p. 187), 46a, 49, 56l (after Foley, *Principles of Human Evolution* (2004), p64), 67a, 94 (after Boyd and Silk, *How Humans Evolved* (2000), p. 318), 95a, 105br, 167ar, 193, 208bl, 209ac, 210bl, 212bl
After Giovanni Caselli: 157a
University of Chicago Press, illustration from C.K. Brain, *The Hunters or the Hunted: an Introduction to African Cave Taphonomy* (1981), p. 268: 25ar
David Chivers: 51b
Margaret Collinson: 61al
Michael Day: 160ar
Professor H.J. Deacon: 159b
Christopher Dean: 45bl
Professor Vittorio Pesce Delfino, University of Bari: 180al
Photo courtesy Jacques Descloitres, MODIS Land Rapid Response Team at NASA (Goddard Space Flight Center): 228l
Tom Dillehay: 197b, 198
Simon S.S. Driver: 194–195
© Ric Ergenbright/Corbis: 70a
Eurelios: 43, 115br
Frank Lane Picture Agency: 202a (Terry Whittaker)
French Ministry of Culture and Communication, Regional Direction for Cultural Affairs – Rhône-Alpes, Regional Department of Archaeology: 10, 11a, 11b, 182–183, 220–221a
Colin Groves: 171a
Barbara Harrisson 107ar
Chris Henshilwood, African Heritage Research Institute, Cape Town, South Africa: 213bc
From *The Hornet*, 1871: 1
© David G. Houser/Corbis: 173b
Illustrated London News, 1921: 148ar, 148b, 169br, 212–213a
Courtesy C.M. Janis: 52 (Biological Reviews 57, p. 292, figure 6), 52–53 (Janis, Biological Reviews 63, p. 199)
Frank Kiernan/Language Research Center, Georgia State University, Atlanta: 131al
From Richard Klein, *The Human Career*, 1999 (2nd ed.): 138b, 165b, 188r, 222bl
Courtesy Dr. Ninel Korniets: 214al
George Koufos: 112al
Photo Laborie, Bergerac: 219a
Photo courtesy Landsat 7 Science Team & NASA (Goddard Space Flight Center): 227
After Leakey and Harris, *Laetoli: A Pliocene Site in Northern Tanzania* (1987): 187l
David Lordkipanidze: 6, 139a
Giorgio Manzi: 144
Lawrence Martin: 101br, 108ar, 191bl
After Martin, R.D., *Primate Origins and Evolution*, 1990: 186ar
Laura McLatchy: 97al

Robin Mckie: 164br
Courtesy Steven Mithen (from *Prehistory of the Mind* (1996): 224
ML Design © Thames & Hudson Ltd, London: 37r (after Conroy, *Primate Evolution* (1990), p. 35), 51a (after Coppens *et al.*, *Earliest Man and Environments in the Lake Rudolf Basin,* (1976), p. 427), 55b, 59a, 62l, 66al, 68b (after Johnson and Shreeve, *Lucy's Child* (1989), p. 10–11), 72b, 76a, 83a, 100 (after Agust *et al.*, *Hominoid evolution and climatic change in Eurasia* Vol. 1 (1999), pp. 14–16, figs 2.6, 2.4, 2.3), 114l, 126l, 131ar, 136al, 145a (after *National Geographic* (1997), vol. 192(1), p. 102), 146al (after Parés and Pérez-González, The Pleistocene Site of Gran Dolina, *Journal of Human Evolution* 37, p. 317), 156bl, 158, 170l, 199, 209br, 216b, 223b
© Warren Morgan/Corbis: 22al
Mike Morwood: 7a, 174r, 175b
Salvador Moya-Sola, Institut de Paleontologia Miquel Crusafont, Sabadell: 111al, 112ar, 186l
Musée d'Aquitaine, Bordeaux. Photo B. Biraben: 218al
NASA: 59b (Johnson Space Center – Earth Sciences and Image Analysis); 226 (Langley Research Center)
National Museum of Australia, Canberra. Photo H. Basedow: 172al
National Museums of Kenya, Nairobi: 96bl, 117, 120
The Natural History Museum, London: 4–5, 8–9, 20b, 21al, 22ar, 24br, 25br, 30, 38, 39a, 39b, 40a, 40b, 41, 45br, 47a, 47b, 48a, 48b, 50l, 50r, 54 (P. Snowball), 55a (P. Snowball), 58a, 58–59, 60a, 60b, 61b (Maurice Wilson), 62–63, 63a, 64al, 64cl, 64ar, 64b, 65, 65b, 66ar, 66b, 67b, 69b, 70b, 70–71, 71ar, 71cr, 75bl, 77al, 77ar, 78, 78–79, 79bl, 79br, 82cr, 82b, 83b, 84–85a, 86a (Maurice Wilson), 87a, 87b (P. Stafford), 89a, 89bl, 89br, 90r, 90br, 91a (Maurice Wilson), 91b, 92l, 92–93, 93ar, 93br, 95b, 96ar, 97ar, 97cr, 97b, 98a, 98b, 99c, 99b (Maurice Wilson), 101a, 102–103a (Tania King), 104ar, 104bl, 105al, 106 (Maurice Wilson), 107al, 108al, 109ar, 111ar, 111cr, 113b, 113c, 122c, 123b, 125ar (Maurice Wilson), 127ar, 129b (Maurice Wilson), 133a, 134b, 135ar, 135br, 137ar, 141b, 149l, 149cr, 149br, 155al, 157cl, 159al, 160l, 160–161b, 161ac, 161ar, 162ar, 163ar, 163cr, 163br, 165a, 167bl, 168a, 168bl, 168br, 169bl, 190r, 191ar, 192, 205, 208br, 209al, 210cr, 210br, 212–213b, 213br, 214bl, 216ar, 217, 223a, 225l, 229b
Natural History Photographic Agency: 34b (Nick Garbutt); 36–37 (Christophe Rattier); 86b (Kevin Schafer); 88 (Martin Harvey); 203bl (Steve Robinson)
Nature Picture Library: 19a (Bruce Davidson); 35b (Richard Du Toit); 203br (Karl Amman)
Novosti, London: 214br
Oxford Scientific Films/Photolibrary.com: 200, 204; 17bl (Brian Kenney); 17br (Daniel Cox); 19b, 184 (Stan Osolinski); 23 (David Cayless); 34a, 190l (Mike Birkhead); 69a (Tom Leach); 90l (Richard Packwood); 130l (Neil Bromhall); 202b (Andrew Plumptre); 203a (Richard Smithers)
Geoff Penna © Thames & Hudson Ltd, London: 12–13, 20a, 22b (after Relethford, *The Human Species* (2002), p. 172), 27, 31, 33a, 46b, 56r, 131b (after Relethford, *The Human Species* (2002), p. 329), 162br, 176–177b (after Mountain *et al.*, *Genes and Time* (1992), 177a, 178bl, 179ar (after Fagan, *People of the Earth* (2004) p. 105), 179br (after Oppenheimer, *Out of Eden* (2003), p. 62), 181ar
David Pilbeam: 107b, 108–109b,
Mike Pitts: 2 (background), 14–15, 73a, 75a
Ben Plumridge © Thames & Hudson Ltd, London: 16, 21ar, 21br, 24bl (after Gould, *The Book of Life* (2001) p. 226), 33b, 53a (after Kappelman, *Journal of Human Evolution* 20, p. 105), 89c (after Aiello and Dean, *An Introduction to Human Evolutionary Anatomy* (1990), 112br, 122–123a, 178cl, 185 (after Fagan, *People of the Earth* (2004), p. 38), 186ac, 188l, 189a, 201 (after de Waal, Bonobo Sex and Society, *Scientific American* (1995), p. 85), 211
Yoel Rak: 154al
Photo RNM: 167c, 167br, 210cl; 215b (J.G. Berizz); 220bl (R.G. Ojeda)
Richard Schlecht/National Geographic Image Collection: 61ar
Courtesy of the artist Peter Schouten and the National Geographic Society: 175a
Reprinted with permission from *Science*, vol. 169, p. 522, 28th July 1995, © AAAS: 121
Science Photo Library: 29 (NASA); 57b (British Antarctic Survey); 80–81, 187r (John Reader); 176l (Alfred Pasieka); 178–179a (CNRI); 180–181b (Volker Steger)
© John Sibbick: 2 (figure), 68a, 72a, 76–77, 102–103b, 110, 118–119, 122l, 132–133, 136–137, 150b, 155r, 166r, 170–171b, 206bl, 206–207b
E.L. Simons: 84bl, 84–85b, 85br
Smithsonian Institution, Washington D.C., Neg. No. SI34358B: 196l
Photo Thomas Stephan, Ulmer Musuem: 218r
Chris Stringer: 26, 32b, 45a, 124al, 124–125b, 125al, 125ac, 125br, 126–127a, 126br, 127ac, 127b, 134ar, 136ar, 138a, 140ar, 140b, 141a, 143a, 150ar, 151a, 151b, 152, 154–155b, 156–157b, 163bl, 164bl, 172b, 173a, 209bl
Chris Stringer/Musée de l'Homme, Paris: 156al, 156ac, 156ar, 166bl
Chris Stringer/Wu Xinzhi: 162bl
W. Suschitzky: 82l
A.J. Sutcliffe: 129a
Jiri Svoboda, Institute of Archaeology, Brno: 215a
Dr. Hartmut Thieme: 222br
Javier Trueba, Madrid Scientific Films: 145b, 146–147a, 146–147b, 153l, 153r
Alan Walker/Bob Campbell National Musuems of Kenya, Nairobi: 139r
Dr. Steven Ward: 99a
Bradford Washburn/Boston Museum of Science: 57a
Professor Gerhard Weber, Department of Anthropology, University of Vienna: 44
Tim White: 116
Klaus Will, University of Heidelberg: 142a
Pamela Willoughby: 225r
Christopher Zollikover & Marci Ponce de Leon, University of Zürich: 42

Acknowledgments

We would like to acknowledge the many friends and colleagues whose work we discuss or illustrate, the editorial, design and production staff of Thames & Hudson, the publishing, picture library and photographic staff of the Natural History Museum, London, and Philip de Ste. Croix for his original work on the manuscript.

Index

Page numbers in *italics* refer to illustrations